John A. Richards · Xiuping Jia

Remote Sensing Digital Image Analysis

An Introduction

Third Edition with 181 Figures

Springer

John A. Richards
Research School of Information Sciences
and Engineering
The Australian National University
Canberra ACT 0200
Australia
e-mail: John.Richards@anu.edu.au

Xiuping Jia
School of Electrical Engineering
University College
The University of New South Wales
Australian Defence Force Academy
Campbell ACT 2600
Australia
e-mail: x-jia@adfa.edu.au

ISBN 3-540-64860-7 3rd ed. Springer-Verlag Berlin Heidelberg New York
ISBN 3-540-58219-3 2nd ed. Springer-Verlag Berlin Heidelberg New York

Library Congress Cataloging-in-Publication Data applied for
Die Deutsche Bibliothek - CIP-Einheitsaufnahme
Richards, John A.:
Remote sensing digital image analysis : an introduction / John A. Richards ; Xiuping Jia. -
Berlin; Heidelberg; New York; Barcelona; Hong Kong; London; Milan; Paris; Singapore;
Tokyo: Springer, 1999
    ISBN 3-540-64860-7

Typesetting: Fotosatz-Service Köhler GmbH, Würzburg
Cover design: Struve & Partner, Heidelberg
SPIN:10647414   61/3020-5 4 3 2 1 0 - Printed on acid-free paper

Preface to the Third Edition

In the time since the second edition of this text was produced two significant trends have been apparent. First, access to image processing technology has continued to improve significantly with most students and practitioners now having readily available inexpensive workstations and powerful software for analysing and manipulating image data.

The second change has been the dramatic increase in the numbers of satellite, aircraft and sensor programs. Perhaps most significant is the widespread availability of hyperspectral data and the special challenges presented by that data type for information extraction.

Accordingly, this third edition has been written to reflect those trends while, at the same time, preserving the important elements of image processing and analysis algorithms of significance in remote sensing applications.

The major changes between the previous edition and this are an update of Chapter 1, and the introduction of a new Chapter 13 dealing with methods for analysing hyperspectral data sets.

Chapter 12 has also been significantly altered to provide a focus on the interpretation of data sets that are mixed and could contain, for example, different types of imagery along with other spatial data types found in a geographic information system. The previous knowledge-based material has been retained but material originally covered in Chapter 8 dealing with multi-source data analysis has been combined with knowledge-based methods to create this chapter on data fusion.

The authors wish to express their appreciation to their colleagues for the assistance they have received in preparing this new edition. In particular David Langrebe of Purdue University remains a great supporter and both authors have had the benefit of working with Dave over the years, including periods spent at Purdue.

The authors are also enormously grateful to their families for their understanding and support in making the completion of this third edition possible.

Canberra, Australia, August 1998
John A. Richards
Xiuping Jia

Preface to the Second Edition

Possibly the greatest change confronting the practitioner and student of remote sensing in the period since the first edition of this text appeared in 1986 has been the enormous improvement in accessibility to image processing technology. Falling hardware and software costs, combined with an increase in functionality through the development of extremely versatile user interfaces, has meant that even the user unskilled in computing now has immediate and ready access to powerful and flexible means for digital image analysis and enhancement. An understanding, at algorithmic level, of the various methods for image processing has become therefore even more important in the past few years to ensure the full capability of digital image processing is utilised.

This period has also been a busy one in relation to digital data supply. Several nations have become satellite data gatherers and providers, using both optical and microwave technology. Practitioners and researchers are now faced, therefore, with the need to be able to process imagery from several sensors, together with other forms of spatial data. This has been driven, to an extent, by developments in Geographic Information Systems (GIS) which, in turn, have led to the appearance of newer image processing procedures as adjuncts to more traditional approaches.

The additional material incorporated in this edition addresses these changes. First, Chapter 1 has been significantly revised to reflect developments in satellite and sensor programs. Removal of information on older systems has been resisted since their data is part of an important archive which still finds value, particularly for historical applications.

Chapter 8, dealing with supervised classification methods, has been substantially increased to allow context classification to be addressed, along with techniques, such as evidential processing and neural networks, which show promise as viable image interpretation tools. While the inclusion of these topics has caused the chapter to become particularly large in comparison to the others, it was felt important not to separate the material from the more traditional methods. The chapter is now presented therefore in two parts – the first covers the standard supervised classification procedures and the second the new topics.

A departure from the application of classical digital image processing to remote sensing over the last five years has been the adoption of knowledge-based methods, where qualitative rather than quantitative reasoning is used to perform interpretations. This is the subject of a new chapter which seeks to introduce reasoning based on knowledge as a means for single image interpretation, and as an approach that can deal successfully with the mixed spatial data types of a GIS.

Besides these changes, the opportunity has been taken to correct typographical and related errors in the first edition and to bring other material up to date.

As with the first edition, the author wishes to record his appreciation to others for their support and assistance. Ashwin Srinivasan, now with the Turing Institute in Glasgow, as a very gifted graduate student, developed much of the material on which Chapter 12 is based and, during his time as a student, helped the author understand the mechanisms of knowledge processing. He also kindly read and offered comments on that chapter. The author's colleague Don Fraser similarly read and provided comments on the material on neural networks. Philip Swain and David Landgrebe of Purdue University continue to support the author in many ways: through their feedback on the first edition, their inter-action on image processing, and Dave's provision of MultiSpec, the Macintosh computer version of the LARSYS software package. Finally, the author again expresses gratitude to his family for their constant support, without which the energy and enthusiasm needed to complete this edition might not have been found.

Canberra, Australia, March 1993   John A. Richards

Preface to the First Edition

With the widespread availability of satellite and aircraft remote sensing image data in digital form, and the ready access most remote sensing practitioners have to computing systems for image interpretation, there is a need to draw together the range of digital image processing procedures and methodologies commonly used in this field into a single treatment. It is the intention of this book to provide such a function, at a level meaningful to the non-specialist digital image analyst, but in sufficient detail that algorithm limitations, alternative procedures and current trends can be appreciated. Often the applications specialist in remote sensing wishing to make use of digital processing procedures has had to depend upon either the mathematically detailed treatments of image processing found in the electrical engineering and computer science literature, or the sometimes necessarily superficial treatments given in general texts on remote sensing. This book seeks to redress that situation.

Both image enhancement and classification techniques are covered making the material relevant in those applications in which photointerpretation is used for information extraction and in those wherein information is obtained by classification. It grew out of a graduate course on digital image processing and analysis techniques for remote sensing data given annually since 1980 at the University of New South Wales. If used as a graduate textbook its contents with the exception of Chap. 7 can be covered substantially in a single semester. Its function as a text is supported by the provision of exercises at the end of each chapter. Most do not require access to a computer for solution. Rather they are capable of hand manipulation and are included to highlight important issues. In many cases some new material is introduced by means of these exercises.

Each chapter concludes with a short critical bibliography that points to more detailed treatments of specific topics and provides, where appropriate, comment on techniques of marginal interest to the mainstream of the book's theme.

Chapter 1 is essentially a compendium of data sources commonly encountered in digital form in remote sensing. It is provided as supporting material for the chapters that follow, drawing out the particular properties of each data source of importance. The second chapter deals with radiometric and geometric errors in image data and with means for correction. This also contains material on registration of images to maps and images to each other. Here, as in all techniques chapters, real and modelled image data examples are given. Chapter 3 establishes the role of computer processing both for photointerpretation by a human analyst and for machine analysis. This may be skipped by the remote sensing professional but is an important position chapter if the book is to be used in teaching.

Chapters 4 and 5 respectively cover the range of radiometric and geometric enhancement techniques commonly adopted in practice, while Chap. 6 is addressed to multispectral transformations of data. This includes the principal components transformation and image arithmetic. Chapter 7 is given over to Fourier transformations. This material is becoming more important in remote sensing with falling hardware costs and the ready availability of peripheral array processors. Here the properties of discrete Fourier analysis are given along with means by which the fast Fourier transform algorithm can be used on image data.

Chapters 8, 9 and 10 provide a treatment of the tools used in image classification, commencing with supervised classification methods, moving through commonly used clustering algorithms for unsupervised classification and concluding with means for separability analysis. These are drawn together into classification methodologies in Chap. 11 which also provides a set of case studies.

Even though the treatment provided is intended for the non-specialist image analyst, it is still necessary that it be cast in the context of some vector and matrix algebra. Otherwise it would be impracticable. Consequently, an appendix is provided on essential results on vectors and matrices, and all important points in the text are illustrated by simple worked examples. These demonstrate how vector operations are evaluated. Beyond this material it is assumed the reader has a passing knowledge of basic probability and statistics including an appreciation of the multivariate normal distribution.

Several other appendices are provided to supplement the main presentation. One deals with developments in image processing hardware and particularly the architecture (in block form) of interactive image display sub-systems. This material highlights trends towards hardware implementation of image processing and illustrates how many of the algorithms presented in the book can be executed in near real time.

Owing to common practice, some decisions have had to be taken in relation to definitions even though they could offend the purist. For example the term "pixel" strictly refers to a unit of digital image data and not to an area on the ground. The latter is more properly called an effective ground resolution element. However because the practice of referring to ground resolution elements as pixels, dimensioned in metres, is so widespread, the current treatment seeks not to be pedantic but rather follows common practice for simplicity. A difficulty also arises with respect to the numbering chosen for the wavebands in the Landsat multispectral scanner. Historically these have been referred to as bands 4 to 7 for Landsats 1 to 3. From Landsat 4 onwards they have been renumbered as bands 1 to 4. The convention adopted herein is mixed. When a particular satellite is evident in the discussion, the respective convention is adopted and is clear from the context of that discussion. In other cases the convention for Landsat 4 has been used as much as possible.

Finally, it is a pleasure to acknowledge the contributions made by others to the production of this book. The manuscript was typed by Mrs Moo Song and Mrs Ailsa Moen, both of whom undertook the task tirelessly and with great patience and forbearance. Assistance with computing was given by Leanne Bischof, at all times cheerfully and accurately. The author's colleagues and students also played their part, both through direct discussion and by that process of gradual learning that occurs over many years of association. Particular thanks are expressed to two people. The author counts himself fortunate to be a friend and colleague of Professor Philip Swain of Purdue University, who in his own way, has had quite an impact on the author's thinking about digital data analysis, particularly in remote sensing. Also, the author has had the good fortune to work with Tong Lee, a graduate student with extraordinary insight and ability, who also has contributed to the material through his many discussions with the author on the theoretical foundations of digital image processing.

The support and encouragement the author has received from his family during the preparation of this work has been immeasurable. It is fitting therefore to conclude in gratitude to Glenda, Matthew and Jennifer, for their understanding and enthusiasm.

Kensington, Australia, May 1986   John A. Richards

Contents

Chapter 3 – The Interpretation of Digital Image Data   75

Chapter 4 – Radiometric Enhancement Techniques   89

Chapter 5 – Geometric Enhancement Using Image Domain Techniques . . 113

Chapter 6 – Multispectral Transformations of Image Data 133

Chapter 8 – Supervised Classification Techniques 181

Chapter 13 – Interpretation of Hyperspectral Image Data 313

Appendix A – Satellite Altitudes and Periods   339

Appendix B – Binary Representation of Decimal Numbers 341

Appendix C – Essential Results from Vector and Matrix Algebra   343

Chapter 1

Sources and Characteristics of Remote Sensing Image Data

1.1 Introduction to Data Sources

1.1.1 Characteristics of Digital Image Data

Remote sensing image data of the earth's surface acquired from either aircraft or spacecraft platforms is readily available in digital format; spatially the data is composed of discrete picture elements, or *pixels*, and radiometrically it is quantised into discrete brightness levels. Even data that is not recorded in digital form initially can be converted into discrete data by use of digitising equipment.

The great advantage of having data available digitally is that it can be processed by computer either for machine assisted information extraction or for enhancement before an image product is formed. The latter is used to assist the role of photointerpretation.

A major characteristic of an image in remote sensing is the wavelength band it represents. Some images are measurements of the spatial distribution of reflected solar radiation in the ultraviolet, visible and near-to-middle infrared range of wavelengths. Others are measurements of the spatial distribution of energy emitted by the earth itself (dominant in the so-called *thermal* infrared wavelength range); yet others, particularly in the microwave band of wavelengths, measure the relative return from the earth's surface of energy actually transmitted from the vehicle itself. Systems of this last type are referred to as *active* since the energy source is provided by the remote sensing platform; by comparison remote sensing measurements that depend upon an external energy source, such as the sun, are called *passive*.

From a data handling and analysis point of view the properties of image data of significance are the number and location of the spectral measurements (or spectral bands or channels) provided by a particular sensor, the spatial resolution as described by the pixel size, in equivalent ground metres, and the radiometric resolution. The last describes the range and discernable number of discrete brightness values and is sometimes referred to alternatively as dynamic range or signal to noise ratio. Frequently the radiometric resolution is expressed in terms of the number of binary digits, or bits, necessary to represent the range of available brightness values. Thus data with 8 bit radiometric resolution has 256 levels of brightness (see Appendix B).

Together with the frame size of an image, in equivalent ground kilometres (which is determined usually by the size of the recorded image swath), the number of spectral bands, radiometric resolution and spatial resolution determine the data volume provided by a particular sensor and thus establish the amount of data to be processed, at least in principle. As an illustration consider the Landsat Thematic Mapper instrument. It has 7 wavelength bands with 8 bit radiometric resolution, six of which have 30 m spatial resolution and one of which has a spatial resolution of 120 m (the thermal band, for which the wavelength is so long that a larger aperture is necessary to collect sufficient signal energy to maintain the radiometric resolution). An image frame of 185 km × 185 km therefore represents 2.37 million pixels in the thermal band and 38 million pixels in the other six bands. At 8 bits per pixel a complete 7 band image is composed of 1.848×10^9 bits or 1.848 Gbit; alternatively and more commonly the data value would be expressed as 231 Mbytes, where 1 byte is equivalent to 8 bits.

The later sections of this chapter provide an overview of common remote sensing missions and their sensors in terms of the data-related properties just described. This is necessary to indicate orders of magnitude and other parameters when determining time requirements and other figures of merit used in assessing image analysis procedures; it will also place the analytical material in context with the data gathering phase of remote sensing. Before proceeding to such an overview of typical data sources it is of value to examine the spectral dimension in some detail since the choice of the spectral bands for a particular sensor significantly determines the information that can be extracted from the data for a particular application.

1.1.2 Spectral Ranges Commonly Used in Remote Sensing

In principle, remote sensing systems could measure energy emanating from the earth's surface in any sensible range of wavelengths. However technological considerations, the selective opacity of the earth's atmosphere, scattering from atmospheric particulates and the significance of the data provided exclude certain wavelengths. The major ranges utilized for earth resources sensing are between about 0.4 and 12 μm (referred to below as the visible/infrared range) and between about 30 to 300 mm (referred to below as the microwave range). At microwave wavelengths it is often more common to use frequency rather than wavelength to describe ranges of importance. Thus the microwave range of 30 to 300 mm corresponds to frequencies between 1 GHz and 10 GHz. For atmospheric remote sensing, frequencies in the range 20 GHz to 60 GHz are encountered.

The significance of these different ranges lies in the interaction mechanism between the electromagnetic radiation and the materials being examined. In the visible/ infrared range the reflected energy measured by a sensor depends upon properties such as the pigmentation, moisture content and cellular structure of vegetation, the mineral and moisture contents of soils and the level of sedimentation of water. At the thermal end of the infrared range it is heat capacity and other thermal properties of the surface and near subsurface that control the strength of radiation detected. In the microwave range, using active imaging systems based upon radar techniques, the roughness of the cover type being detected and its electrical properties, expressed in terms of complex permittivity (which in turn is strongly influenced by moisture content) determine the magnitude of the reflected signal. In the range 20 to 60 GHz, atmospheric oxygen and water vapour have a strong effect on transmission and thus can be inferred by measurements in that range. Thus each range of wavelength has its own strengths in terms of the information it can contribute to the remote sensing process. Consequently we find systems available that are optimised for and operate in particular spectral ranges, and provide data that complements that from other sensors.

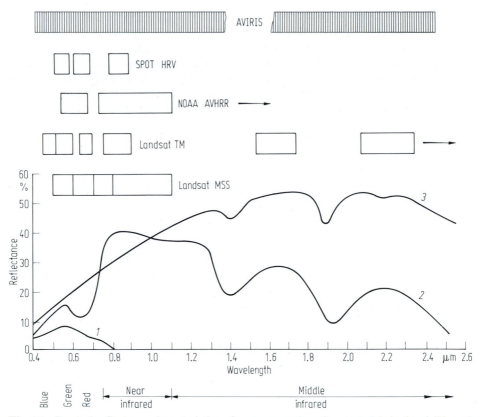

Fig. 1.1. Spectral reflectance characteristics of common earth surface materials in the visible and near-to-mid infrared range. *1* Water, *2* vegetation, *3* soil. The positions of spectral bands for common remote sensing instruments are indicated. These are discussed in the following sections.

Figure 1.1 depicts how the three dominant earth surface materials of soil, vegetation and water reflect the sun's energy in the visible/reflected infrared range of wavelengths. It is seen that water reflects about 10% or less in the blue-green range, a smaller percentage in the red and certainly no energy in the infrared range. Should the water contain suspended sediments or should a clear water body be shallow enough to allow reflection from the bottom then an increase in apparent water reflection will occur, including a small but significant amount of energy in the near infrared range. This is a result of reflection from the suspension or bottom material.

As seen in the figure soils have a reflection that increases approximately monotonically with wavelength, however with dips centred at about 1.4 µm, 1.9 µm and 2.7 µm owing to moisture content. These water absorption bands are almost unnoticeable in very dry soils and sands. In addition to these bands clay soils also have hydroxyl absorption bands at 1.4 µm and 2.2 µm.

The vegetation curve is considerably more complex than the other two. In the middle infrared range it is dominated by the water absorption bands at 1.4 µm, 1.9 µm and 2.7 µm. The plateau between about 0.7 µm and 1.3 µm is dominated by plant cell structure while in the visible range of wavelengths it is plant pigmentation that is the major determinant. The curve sketched in Fig. 1.1 is for healthy green vegetation. This has chlorophyll absorption bands in the blue and red regions leaving only green reflection of any significance. This is why we see chlorophyll pigmented plants as green.

An excellent review and discussion of the spectral reflectance characteristics of vegetation, soils, water, snow and clouds can be found in Hoffer (1978). This includes a consideration of the physical and biological factors that influence the shapes of the curves, and an indication of the appearances of various cover types in images record- ed in different wavelength ranges.

In wavelength ranges between about 3 and 14 μm the level of solar energy actually irradiating the earth's surface is small owing to both the small amount of energy leaving the sun in this range by comparison to the higher levels in the visible and near infrared range (see Fig. 1.2), and the presence of strong atmospheric absorption bands between 2.6 and 3.0 μm, 4.2 and 4.4 μm, and 5 and 8 μm (Chahine, 1983). Consequently much remote sensing in these bands is of energy being emitted from the earth's surface or objects on the ground rather than of reflected solar radiation.

Figure 1.2 shows the relative amount of energy radiated from perfect black bodies of different temperatures. As seen, the sun at 6000 K radiates maximally in the visible and near infrared regime but by comparison generates little radiation in the range around 10 μm. Incidentally, the figure shown does not take any account of how the level of solar radiation is dispersed through the inverse square law process in its travel from the sun to the earth. Consequently if it is desired to compare that curve to others corresponding to black bodies on the earth's surface then it should be appropriately reduced.

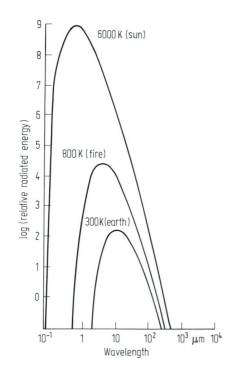

Fig. 1.2. Energy from perfect radiators (black bodies) as a function of wavelength

The earth, at a temperature of about 300 K has its maximum emission around 10 to 12 μm. Thus a sensor with sensitivity in this range will measure the amount of heat being radiated from the earth itself. Hot bodies on the earth's surface, such as bushfires, at around 800 K, have a maximum emission in the range of about 3 to 5 μm. Consequently to map fires, a sensor operating in that range would be used.

Real objects do not behave as perfect black body radiators but rather emit energy at a lower level than that shown in Fig. 1.2. The degree to which an object radiates by comparison to a black body is referred to as its emittance. Thermal remote sensing is sensitive therefore to a combination of an object's temperature and emittance, the last being wavelength dependent.

Microwave remote sensing image data is gathered by measuring the strength of energy scattered back to the satellite or aircraft in response to energy transmitted. The degree of reflection is characterized by the scattering coefficient for the surface material being imaged. This is a function of the electrical complex permittivity of the material and the roughness of the surface in comparison to a wavelength of the radiation used (Ulaby, Moore & Fung, 1982).

Smooth surfaces act as so-called specular reflectors (i. e. mirror-like) in that the direction of scattering is predominantly away from the incident direction as shown in Fig. 1.3. Consequently they appear dark to black in image data. Rough surfaces act as diffuse reflectors; they scatter the incident energy in all directions as depicted in Fig. 1.3, including back towards the remote sensing platform. As a result they appear light in image data. A third type of surface scattering mechanism is often encountered in microwave image data, particularly associated with manufactured features such as buildings. This is a corner reflector effect, as seen in Fig. 1.3, resulting from the right angle formed between a vertical structure such as a fence, building or ship and a horizontal plane such as the surface of the earth or sea. This gives a very bright response.

Media, such as vegetation canopies and sea ice, exhibit so-called volume scattering behaviour, in that backscattered energy emerges from many, hard to define sites within

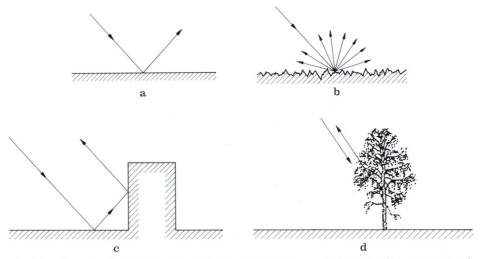

Fig. 1.3. a Specular, **b** diffuse, **c** corner reflector and **d** volume scattering behaviour, encountered in the formation of microwave image data

the volume, as depicted in Fig. 1.3. This leads to a light tonal appearance in radar imagery.

In interpreting image data acquired in the microwave region of the electromagnetic spectrum it is important to recognise that the four reflection mechanisms of Fig. 1.3 are present and modify substantially the tonal differences resulting from surface complex permittivity variations. By comparison, imaging in the visible/infrared range in which the sun is the energy source, results almost always from diffuse reflection, allowing the interpreter to concentrate on tonal variations resulting from factors such as those described in association with Fig. 1.1.

A comprehensive treatment of the essential principles of microwave remote sensing will be found in the three volume series by Ulaby, Moore and Fung (1981, 1982, 1985).

1.1.3 Concluding Remarks

The purpose of acquiring remote sensing image data is to be able to identify and assess, by some means, either surface materials or their spatial properties. Inspection of Fig. 1.1 reveals that cover type identification should be possible if the sensor gathers data at several wavelengths. For example, if for each pixel, measurements of reflection at 0.65 μm and 1.0 μm were available (i.e. we had a two band imaging system) then it should be a relatively easy matter to discriminate between the three fundamental cover types based on the relative values in the two bands. For example, vegetation would be bright at 1.0 μm and very dark at 0.65 μm whereas soil would be bright in both ranges. Water on the other hand would be black at 1.0 μm and dull at 0.65 μm. Clearly if more than two measurement wavelengths were used more precise discrimination should be possible, even with cover types spectrally similar to each other. Consequently remote sensing imaging systems are designed with wavebands that take several samples of the spectral reflectance curves of Fig. 1.1. For each pixel the set of samples can be analysed, either by photointerpretation, or by the automated techniques to be found in Chap. 8 and 9, to provide a label that associates the pixel with a particular earth surface material.

A similar situation applies when using microwave image data; viz. several different transmission wavelengths can be used to assist in identification of cover types by reason of their different scattering behaviours with wavelength. However a further data dimension is available with microwave imaging owing to the coherent nature of the radiation used. That relates to the polarizations of the transmitted and scattered radiation. The polarization of an electromagnetic wave refers to the orientation of the electric field during propagation. For radar systems this is chosen to be parallel to the earth's surface on transmission (a situation referred to as *horizontal polarization*) or at right angles to both the earth's surface and the direction of propagation (somewhat inappropriately called *vertical polarisation*). On scattering, some polarization changes can occur and energy can be received as horizontally polarized and/or vertically polarized. The degree of polarization rotation that occurs can be a useful indicator of surface material.

Another consequence of using coherent radiation in radar remote sensing systems, of significance to the interpretation process, is that images exhibit a degree of "speckle". This is a result of constructive and destructive interference of the reflections from surfaces that have random spatial variations of the order of one half a wavelength, or so. Noting that the wavelengths commonly employed in radar remote sensing are between about 30 mm and 300 mm it is usual to find images of most common cover types showing a considerably speckled appearance. Within a homogeneous region for example, such as a crop field, this causes adjacent radar image pixels to have large differences in brightness, a factor which complicates machine-assisted interpretation.

1.2 Weather Satellite Sensors

Weather satellites and those used for earth resources sensing operate in much the same bands of wavelength. Perhaps the major distinction in the image data they provide lies in the spatial resolutions available. Whereas data acquired for earth resources purposes generally has a pixel size less than 100 m × 100 m, that used for meteorological applications usually has a much coarser pixel – often of the order of 1 km × 1 km. This is the distinction used herein in order to separate the two types of sensor. Having made that distinction however it is important to note that because of the similarity in wavebands, meteorological satellite data such as that from the NOAA Advanced Very High Resolution Radiometer (AVHRR) does find application in remote sensing when large synoptic views are required.

1.2.1 Polar Orbiting and Geosynchronous Satellites

Two broad types of weather satellite are in common use. One is of the polar orbiting, or more generally low earth orbit, variety whereas the other is at geosynchronous altitudes. The former typically have orbits at altitudes of about 700 to 1500 km whereas the geostationary altitude is approximately 36,000 km (see Appendix A). Typical of the low orbit satellites are the current NOAA series (also referred to as Advanced TIROS-N, ATN), and their forerunners the TIROS, TOS and ITOS satellites. The principal sensor of interest from this book's viewpoint is the NOAA AVHRR mentioned above. This is described in Sect. 1.2.2 following.

The Nimbus satellites, while strictly test bed vehicles for a range of meteorological and remote sensing sensors, also orbit at altitudes of around 1000 km. Nimbus sensors of interest include the Coastal Zone Colour Scanner (CZCS) and the Scanning Multichannel Microwave Radiometer (SMMR). Only the former is treated below.

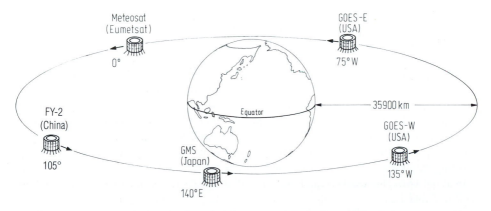

Fig. 1.4. Positions of the five geosynchronous meteorological satellites that provide global weather watch capabilities

Geostationary meteorological satellites have been launched by the United States, Russia, India, China, ESA and Japan. These are placed in equatorial geosynchronous orbits. Fig. 1.4 shows the current five operational satellites.

1.2.2 The NOAA AVHRR
(Advanced Very High Resolution Radiometer)

The AVHRR has been designed to provide information for hydrologic, oceanographic and meteorologic studies, although data provided by the sensor does find application also to solid earth monitoring. An earlier version of the AVHRR contained four wavelength bands. Table 1.1 however lists the bands available on the current generation of instrument (NOAA 14).

Table 1.1. NOAA advanced very high resolution radimeter

Spatial resolution	1.1 km at nadir
Dynamic range	10 bit
Swath width	2394 km
Spectral bands:	
band 1	$0.58 - 0.68$ μm
band 2	$0.725 - 1.0$ μm
band 3	$3.55 - 3.93$ μm
band 4	$10.3 - 11.3$ μm
band 5	$11.4 - 12.4$ μm

1.2.3 The Nimbus CZCS (Coastal Zone Colour Scanner)

The CZCS was a mirror scanning system, carried on Nimbus 7, designed to measure chlorophyll concentration, sediment distribution and general ocean dynamics including sea surface temperature. Its characteristics are summarised in Table 1.2.

1.2.4 GMS VISSR (Visible and Infrared Spin Scan Radiometer)
and GOES Imager

Geostationary meteorological satellites such as GMS (Japan) and the earlier GOES (USA) are spin stabilized with their spin axis oriented almost north-south. The primary sensor on these, the VISSR, scans the earth's surface by making use of the satellite spin to acquire one line of image data (as compared with an oscillating mirror in the case of AVHRR, CZCS, MSS and TM sensors), and by utilizing a stepping motor to adjust the angle of view on each spin to acquire successive line of data (on orbiting satellites it is the motion of the vehicle relative to the earth that displaces the sensor between successive scan lines). The characteristics of the VISSR are summarised in Table 1.3.

The most recent GOES environmental satellites are 3 axis stabilised and carry a GOES Imager with characteristics as shown in Table 1.3.

Table 1.2. Nimbus coastal zone colour scanner

Spatial resolution	825 m at nadir
Dynamic range	8 bit
Swath width	1566 km
Spectral bands:	
channel 1	0.433– 0.453 µm
channel 2	0.510– 0.530 µm
channel 3	0.540– 0.560 µm
channel 4	0.660– 0.680 µm
channel 5	0.700– 0.800 µm
channel 6	10.5 –12.5 µm

Table 1.3. VISSR and GOES Imager characteristics

	Band (µm)	Spatial resolution at nadir (km)	Dynamic range (bits)
VISSR			
	0.55 – 0.75 (visible)	1.25	6
	10.5 – 12.5 (thermal infrared)	5	8
GOES Imager			
	0.55 – 0.75	1	10
	3.80 – 4.00	4	10
	6.50 – 7.00	8	10
	10.20 – 11.20	4	10
	11.50 – 12.50	4	10

1.3 Earth Resource Satellite Sensors in the Visible and Infrared Regions

1.3.1 The Landsat System

The Landsat earth resources satellite system was the first designed to provide near global coverage of the earth's surface on a regular and predictable basis.

The first three Landsats had identical orbit characteristics, as summarised in Table 1.4. The orbits were near polar and sun synchronous – i.e., the orbital plane precessed about the earth at the same rate that the sun appears to move across the face of the earth. In this manner data was acquired at about the same local time on every pass.

Table 1.4. Landsat 1, 2, 3 orbit characteristics

Orbit:	Sun synchronous, near polar; nominal 9:30 am descending equatorial crossing; inclined at about 99° to the equator
Altitude:	920 km (570 mi)
Period:	103 min
Repeat Cycle:	14 orbits per day over 18 days (251 revolutions)

All satellites acquired image data nominally at 9:30 a.m. local time on a descending (north to south) path; in addition Landsat 3 obtained thermal data on a night-time ascending orbit for the few months that its thermal sensor was operational. Fourteen complete orbits were covered each day, and the fifteenth, at the start of the next day, was 159 km advanced from orbit 1, thus giving a second day coverage contiguous with that of the first day. This advance is daily coverage continued for 18 days and then repeated. Consequently complete coverage of the earth's surface was given, with 251 revolutions in 18 days.

The orbital characteristics of the second generation Landsats, commencing with Landsats 4 and 5, are different from those of their predecessors. Again image data is acquired nominally at 9:30 a.m. local time in a near polar, sun synchronous orbit; however the spacecraft are at the lower altitude of 705 km. This lower orbit gives a repeat cycle of 16 days at 14.56 orbits per day. This corresponds to a total of 233 revolutions every cycle. Table 1.5 summarises the Landsat 4, 5 orbit characteristics. Unlike the orbital pattern for the first generation Landsats, the day 2 ground pattern for Landsats 4 and 5 is not adjacent and immediately to the west of the day 1 orbital pattern. Rather it is displaced the equivalent of 7 swath centres to the west. Over 16 days this leads to the repeat cycle.

Landsat 6, launched in 1993, was not successfully placed in orbit and was lost over the Atlantic Ocean. Landsat 7, a similar satellite in all respects, is a planned replacement.

Table 1.5. Orbit parameters for Landsats 4, 5, 7

Orbit:	Near polar, sun synchronous; nominal 9:30 am descending equatorial crossing (10:00 am for Landsat 7)
Altitude:	705 km
Period:	98.9 min
Repeat Cycle:	14.56 orbits per day over 16 days (total of 233 revolutions)

Whereas Landsats 1, 2 and 3 contained on-board tape recorders for temporary storage of image data when the satellites were out-of-view of earth stations, Landsats 4 and 5 do not, and depend on transmission either to earth stations directly or via the geosynchronous communication satellite TDRS (Tracking and Data Relay Satellite). TDRS is a high capacity communication satellite that is used to relay data from a number of missions, including the Space Shuttle. Its ground receiving station is in White Sands, New Mexico from which data is relayed via domestic communication satellites.

1.3.2   The Landsat Instrument Complement

Three imaging instruments have been used with the Landsat satellites to date. These are the Return Beam Vidicon (RBV), the Multispectral Scanner (MSS) and the Thematic Mapper (TM). Table 1.6 shows the actual imaging payload for each satellite along with historical data on launch and out-of-service dates. Two different RBV's were used: a multispectral RBV package was incorporated on the first two satellites, whilst a panchromatic instrument with a higher spatial resolution was used on Landsat 3. The MSS on Landsat 3 also contained a thermal band; however this operated only for a few months.

Table 1.6. Landsat payloads, launch and out of service dates

Satellite	Imaging Instruments			Launched	Out-of-service
Landsat 1	RBV^m	MSS		23 Jul 1972	6 Jan 1978
Landsat 2	RBV^m	MSS		22 Jan 1975	27 Jul 1983
Landsat 3	RBV^p	MSS^t		5 Mar 1978	7 Sept 1983
Landsat 4		MSS	TM	16 Jul 1982	–
Landsat 5		MSS	TM	1 Mar 1984	–
Landsat 6			ETM	lost on launch	–
Landsat 7			ETM+	1999	–

m – multispectral RBV
p – panchromatic RBV
t – MSS with thermal band

The MSS will not be used after Landsat 5. With the launch of Landsat 7 an Enhanced Thematic Mapper + (ETM+) will be carried.

The following sections provide an overview of the three Landsat instruments, especially from a data characteristic point-of-view.

1.3.3 The Return Beam Vidicon (RBV)

As the name suggests the RBV's were essentially television camera-like instruments that took "snapshot" images of the earth's surface along the ground track of the satellite. Image frames of 185 km × 185 km were acquired with each shot, repeated at 25 s intervals to give contiguous frames in the along track direction at the equivalent ground speed of the satellite.

Three RBV cameras were used on Landsats 1 and 2, distinguished by different transmission filters that allowed three spectral bands of data to be recorded as shown in Table 1.7. On Landsat 3 two RBV cameras were used; however both operated panchromatically and were focussed to record data swaths of 98 km, overlapped to give a total swath of about 185 km. By so doing a higher spatial resolution of 40 m was possible, by comparison to 80 m for the earlier RBV system.

Historically the spectral ranges recorded by the RBV's on Landsats 1 and 2 were referred to as bands 1, 2 and 3. The MSS bands (see following) in the first generation of Landsats were numbered to follow on in this sequence.

1.3.4 The Multispectral Scanner (MSS)

The Multispectral Scanner was the principal sensor on Landsats 1, 2 and 3 and was the same on each spacecraft with the exception of an additional band on Landsat 3. The MSS is a mechanical scanning device that acquires data by scanning the earth's surface in strips normal to the satellite motion. Six lines are swept simultaneously by an oscillating mirror and the reflected solar radiation so monitored is detected in four wavelength bands for Landsats 1 and 2, and five bands for Landsat 3, as shown in Table 1.7. A schematic illustration of the six line scanning pattern used by the MSS is shown in Fig. 1.5. It is seen that the sweep pattern gives rise to an MSS swath width of 185 km thereby corresponding to the image width of the RBV. The width of each scan line corresponds to 79 m on the earth's surface so that the six lines simultaneously

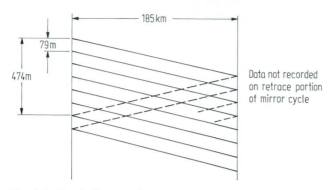

Fig. 1.5. The six line scanning pattern used by the Landsat multispectral scanner. Dimensions are in equivalent measurements on the ground. This scanning pattern is the same in each of bands 4 to 7. The same six line pattern is used on Landsats 4 and 5 except that the strip width is 81.5 m and 82.5 m respectively

correspond to 474 m. Approximately 390 complete six-line scans are collected to provide an effective image that is also 185 km in the along track direction. For Landsats 1 and 2, 24 signal detectors were required to provide four spectral bands from each of the six scan lines. A further two were added for the thermal band data of Landsat 3. Those detectors are illuminated by radiation reflected from the oscillating scanning mirror in the MSS, and produce a continuously varying electrical signal corresponding to the energy received along the 79 m wide associated scan line. The optical aperture of the MSS and its detectors for bands 4 to 7 is such that at any instant of time each detector sees a pixel that is 79 m in size also along the scan line. Consequently the effective pixel size (or instantaneous field of view IFOV) of the detectors is 79 m × 79 m. At a given instant the output from a detector is the integrated response from all cover types present in a 79 m × 79 m region of the earth's surface. Without any further processing the signal from the detector would appear to be varying continuously with time. However it is sampled in time to produce discrete measurements across a scan line. The sampling rate corresponds to pixel centres of 56 m giving a 23 m overlap of the 79 m × 79 m pixels, as depicted in Fig. 1.6. The thermal infrared band on Landsat 3, band 8, has an IFOV of 239 m × 239 m. As a result there are only two band 8 scan lines corresponding to the six for bands 4 to 7, as indicated above.

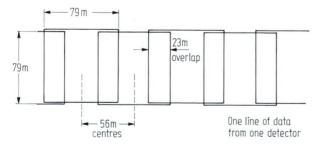

Fig. 1.6. The relationship between instantaneous field of view and pixel overlap for Landsat MSS pixels

The IFOV's of the multispectral scanners on Landsats 4 and 5 have been modified to 81.5 m and 82.5 m respectively although the pixel centre spacing of 56 m has been retained. In addition the bands have been renamed as bands 1, 2, 3 and 4, corresponding to bands 4, 5, 6 and 7 from the earlier missions.

After being spatially sampled, the data from the detectors is digitised in amplitude into 6 bit words. Before so-encoding, the data for bands 4, 5 and 6 is compressed allowing decompression into effective 7 bit words upon reception at a ground station.

1.3.5 The Thematic Mapper (TM) and Enhanced Thematic Mapper + (ETM+)

The Thematic Mapper is a mechanical scanning device as for the MSS, but has improved spectral, spatial and radiometric characteristics. Seven wavelength bands are used, with coverage as shown in Table 1.7. Note that band 7 is out of place in the progression of wavelengths, it having been added, after the initial planning phase, at the request of the geological community. The Enhanced Thematic Mapper + to be carried on Landsat 7 will also include a panchromatic band and improved spatial resolution on the thermal band.

Table 1.7. Characteristics of the Landsat imaging devices

Instrument	Spectral bands (μm)			IFOV (m)	Dynamic range (bits)
RBV^m	1.	0.475– 0.575	(blue)	79 × 79	
	2.	0.580– 0.680	(red)	79 × 79	
	3.	0.689– 0.830	(near IR)	79 × 79	
RBV^p		0.505– 0.750	(panchromatic)	40 × 40	
MSS	4.^{a}	0.5 – 0.6	(green)	79 × 79	7
	5.	0.6 – 0.7	(red)	79 × 79	7
	6.	0.7 – 0.8	(near IR)	79 × 79	7
	7.	0.8 – 1.1	(near IR)	79 × 79	6
	8.^{b}	10.4 –12.6	(thermal)	237 × 237	
TM	1.	0.45 – 0.52	(blue)	30 × 30	8
	2.	0.52 – 0.60	(green)	30 × 30	8
	3.	0.63 – 0.69	(red)	30 × 30	8
	4.	0.76 – 0.90	(near IR)	30 × 30	8
	5.	1.55 – 1.75	(mid IR)	30 × 30	8
	7.^{c}	2.08 – 2.35	(mid IR)	30 × 30	8
	6.	10.4 –12.5	(thermal)	120 × 120	8
ETM+					
largely as for TM but with 60 × 60 IFOV on band 6					
plus		0.52 – 0.90	(panchromatic)	15 × 15	8

^{a} MSS bands 4 to 7 have been renumbered MSS bands 1 to 4 from Landsat 4 onwards. IFOV = 81.5, 82.5 m for Landsats 4, 5.
^{b} MSS band 8 was used only on Landsat 3.
^{c} TM band 7 is out of sequence since it was added last in the design after the previous six bands had been firmly established. It was incorporated at the request of the geological community owing to the importance of the 2 μm region in assessing hydrothermal alteration.

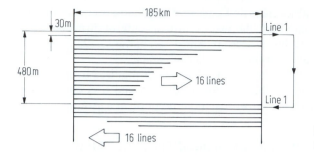

Fig. 1.7. Scanning characteristics of the Landsat Thematic Mapper

Whereas the MSS of all Landsats scans and obtains data in one direction only, the TM acquires data in both scan directions again with a swath width of 185 km. Sixteen scan lines are swept simultaneously giving a 480 m strip across the satellite path, as illustrated in Fig. 1.7. This permits a lower mirror scan rate compared with the MSS and thus gives a higher effective dwell time for a given spot on the ground, making possible the higher spatial resolution and improved dynamic range.

1.3.6 The SPOT HRV, HRVIR, HRG and Vegetation Instruments

The early French SPOT satellites (Systeme Pour d'Observation de la Terre), carried two imaging devices referred to as HRV's. These instruments utilize a different technology for image acquisition from that employed in the Landsat MSS and TM devices. Rather than using oscillating mirrors to provide cross-track scanning during the forward motion of the space platform, the SPOT HRV instruments consist of a linear array of charge coupled device (CCD) detectors. These form what is commonly referred to as a "push broom" scanner. Each detector in the array scans a strip in the along track direction. By having several thousand such detectors a wide swath can be imaged without the need for mechanical scanning. Moreover, owing to the long effective dwell time this allows for each pixel, a higher spatial resolution is possible. A trade-off however is that charge coupled device technology was not available for wavelengths into the middle infrared range at the time of early SPOT development. Consequently the spectral bands provided by the HRV are not unlike those of the Landsat MSS.

The HRV covers a ground swath width of 60 km; two instruments are mounted side by side in the spacecraft to give a total swath width of 117 km, there being a 3 km overlap of the individual swaths.

Two imaging modes are possible. One is a multispectral mode and the other panchromatic. The imaging characteristics of these are summarised, along with the satellite orbital properties, in Table 1.8.

An interesting property of the HRV is that it incorporates a steerable mirror to allow imaging to either side of nadir. This allows daily coverage for a short period along with a stereoscopic viewing capability.

SPOT 4, launched in March 1998, carries two instruments – the HRVIR (High Resolution Visible and Infrared) and the Vegetation instrument. Characteristics of both are summarised in Table 1.8.

Table 1.8. Spot satellite and sensor characteristics

Orbit:	Near polar, sun synchronous; nominal 10.30 am descending equatorial crossing
Altitude:	832 km
Period:	101 min
Repeat Cycle:	26 days

Satellite	Imaging Instrument	Launched	Out-of-Service
SPOT 1	HRV (×2)	22 Feb 1986	5 Jan 1991
		reactivated 9 Jan 1997	
SPOT 2	HRV (×2)	22 Jan 1990	
SPOT 3	HRV (×2)	26 Sep 1993	14 Nov 1997
SPOT 4	HRVIR (×2), Vegetation	24 Mar 1998	
SPOT 5	HRG (×2)	planned for late 2002	

Instrument	Spectral bands (μm)	IFOV (m)	Swath (km)	Dynamic range (bits)
HRV ^{m}	0.50 – 0.59 (green)	20 × 20	60	8
	0.61 – 0.68 (red)	20 × 20	60	8
	0.79 – 0.89 (near IR)	20 × 20	60	8
HRV ^{p}	0.51 – 0.73	10 × 10	60	8
HRVIR ^{m}	0.50 – 0.59 (green)	20 × 20	60	8
	0.61 – 0.68 (red)	20 × 20	60	8
	0.78 – 0.89 (near IR)	20 × 20	60	8
	1.58 – 1.75 (mid IR)	20 × 20	60	8
HRVIR ^{p}	0.61 – 0.68	10 × 10	60	8
HRG ^{m}	0.50 – 0.59 (green)	10 × 10	60	8
	0.61 – 0,68 (red)	10 × 10	60	8
	0.78 – 0.89 (near IR)	10 × 10	60	8
	1.58 – 1.75 (mid IR)	10 × 10	60	8
HRG ^{p}	0.51 – 0.73	5 × 5 and 3 × 3	60	8
Vegetation	0.43 – 0.47 (blue)	1000 × 1000	2250	8
	0.61 – 0.68 (red)	1000 × 1000	2250	8
	0.78 – 0.89 (near IR)	1000 × 1000	2250	8
	1.58 – 1.75 (mid IR)	1000 × 1000	2250	8

m – multispectral mode
p – panchromatic mode

SPOT 5, planned for early next decade, will carry an instrument to be known as HRG (High Resolution Geometry) that uses the same wavebands as the HRVIR but with a higher spatial resolution and a return to the 0.51 – 0.73 μm panchromatic band. Its characteristics are also summarised in Table 1.8.

1.3.7 ADEOS (Advanced Earth Observing Satellite)

ADEOS was launched by the Japanese space agency NASDA in August 1996. It carries a number of imaging instruments and non-imaging sensors, including OCTS (Ocean Colour and Temperature Sensor), AVNIR (Advanced Visible and Near Infra-red Radiometer), NSCAT (NASA Spectrometer), TOMS (Total Ozone Mapping Spectrometer), POLDER (Polarization and Directionality of the Earth's Reflectance),

Table 1.9. ADEOS satellite and sensor characteristics

Orbit:	near polar, sun synchronous; nominal 10.30 am equatorial crossing
Altitude:	797 km
Period:	101 min
Repeat Cycle:	41 days (ADEOS 2 will have a 4 day repeat cycle)

Instrument	Spectral bands (μm)	IFOV (m)	Swath (km)	Dynamic range (bits)
AVNIR[m]	0.42 − 0.50	16 × 16	80	8
	0.52 − 0.60	16 × 16	80	8
	0.61 − 0.69	16 × 16	80	8
	0.76 − 0.89	16 × 16	80	8
AVNIR[p]	0.52 − 0.69	8 × 8	80	7
OCTS	0.412 ± 0.01	700 × 700 *	1400	10
	0.443 ± 0.01	700 × 700	1400	10
	0.490 ± 0.01	700 × 700	1400	10
	0.520 ± 0.01	700 × 700	1400	10
	0.565 ± 0.01	700 × 700	1400	10
	0.670 ± 0.01	700 × 700	1400	10
	0.765 ± 0.02	700 × 700	1400	10
	0.865 ± 0.02	700 × 700	1400	10
	3.55 − 3.88	700 × 700	1400	10
	8.25 − 8.80	700 × 700	1400	10
	10.3 − 11.4	700 × 700	1400	10
	11.4 − 12.7	700 × 700	1400	10

m – multispectral mode
p – panchromatic mode
* – at nadir

IMG (Interferometric Monitor for Greenhouse Gases), ILAS (Improved Limb Atmospheric Sensor) and RIS (Retroreflector in Space).

Characteristics of the OCTS and AVNIR are given in Table 1.9, along with spacecraft orbital details.

1.3.8 Sea-Viewing Wide Field of View Sensor (SeaWiFS)

In August 1997 the OrbView-2 satellite was launched, carrying the SeaWiFS sensor with characteristics as shown in Table 1.10. Its wavebands have been chosen with ocean-related applications in mind.

1.3.9 Marine Observation Satellite (MOS)

The Marine Observation Satellites MOS-1 and MOS-1b were launched by Japan in February 1987 and February 1990 respectively and were taken out of service in March 1995 and April 1996 respectively. While intended largely for oceanographic studies, the data from the satellites' two optical imaging sensors – the MESSR (Multispectrum Electronic Self Scanning Radiometer) and the VTIR (Visible and Thermal

Table 1.10. OrbView-2 satellite and SeaWiFS sensor characteristics

Orbit:	near polar, sun synchronous
Altitude:	705 km
Period:	98.9 min
Repeat Cycle:	1 day

Instrument	Spectral bands (µm)	IFOV (m)	Swath (km)	Dynamic range (bits)
SeaWiFS	0.402 – 0.422	1100 × 1100	2800	10
	0.433 – 0.453	1100 × 1100	2800	10
	0.480 – 0.500	1100 × 1100	2800	10
	0.500 – 0.520	1100 × 1100	2800	10
	0.545 – 0.565	1100 × 1100	2800	10
	0.660 – 0.680	1100 × 1100	2800	10
	0.745 – 0.785	1100 × 1100	2800	10
	0.845 – 0.885	1100 × 1100	2800	10

Infrared Radiometer) are of value to land based remote sensing as well. The satellites also carried a Microwave Scanning Radiometer (MSR) intended for water vapour, snow and ice studies. Two MESSRs were used to provide side by side observations. Each has a 100 km swath width; with an overlap in coverage of 15 km the total available swath is 185 km.

Orbital details of the MOS satellites and characteristics of their optical sensors are given in Table 1.11.

Table 1.11. MOS orbit and sensor characteristics

MOS:		Altitude	908 km
		Orbit	sun synchronous, 99.1° inclination 10–11 am equatorial crossing
		Repeat Cycle	17 days
MESSR:	Bands	0.51–0.59 µm	
		0.61–0.69 µm	
		0.73–0.80 µm	
		0.80–1.10 µm	
	IFOV	50 m × 50 m	
	Dynamic range	8 bit	
	Swath per		
	MESSR	100 km	
VTIR:	Bands	0.5– 0.7 µm	
		6.0– 7.0 µm	
		10.5–11.5 µm	
		11.5–12.5 µm	
	IFOV	900 m × 900 m for visible channel 2700 m × 2700 m for the others	
	Dynamic range	8 bit	
	Swath Width	1500 km	

1.3.10   Indian Remote Sensing Satellite (IRS)

A series of remote sensing satellites has been launched by India since March 1988.
Known as IRS (1A, 1B, P2, 1C, 1D), they carry imaging systems known as the LISS
(Linear Imaging Self Scanner), the WIFS (Wide Field Sensor) and a panchromatic
sensor as shown in Table 1.12. Orbital details of the satellites and characteristics of
the sensors are summarised in Table 1.12.

Table 1.12. IRS satellite and sensor characteristics

Orbit:	Near polar, sun synchronous; nominal 10.35 am equatorial crossing
Altitude:	904 km (1A, 1B), 817 km (1C), 736/825 km (1D)
Period:	101 min
Repeat Cycle:	22 days (1A, 1B), 24 days (1C, P2)

Satellite	Imaging Instrument	Launched	Out-of-Service
IRS-1A	LISSI	Mar 1988	
IRS-1B	LISSII	Aug 1991	
IRS-P2	LISSII	16 Oct 1994	
IRS-1C	LISSIII, Pan, WIFS		
IRS-1D	LISSIII, Pan, WIFS	29 Sep 1997	

Instrument	Spectral bands (μm)	IFOV (m)	Swath (km)	Dynamic range (bits)
LISSI, II	0.45 – 0.52	73 × 73 (LISSI)	146	7
	0.52 – 0.59	36 × 36 (LISSII)	146	7
	0.62 – 0.68		146	7
	0.77 – 0.86		146	7
LISSIII	0.52 – 0.59	23 × 23	142 – 146	7
	0.62 – 0.68	23 × 23	142 – 146	7
	0.77 – 0.86	23 × 23	142 – 146	7
	1.55 – 1.70	70 × 70	142 – 146	7
Pan	0.5 – 0.57	10 × 10	70	7
WIFS	0.62 – 0.68	188 × 188	774	7
	0.77 – 0.86	188 × 188	774	7

1.3.11   RESURS – O1

Russia has orbited a series of remote sensing satellites since 1985 under the name
RESURS – O1. Table 1.13 gives platform and sensor characteristics for the third in
the series. The principal sensor, from which commercially available imagery is produced, is the MSU-SK, which is a conically scanning instrument.

Table 1.13. RESURS – O1 – 3 satellite and sensor characteristics

Orbit:	Sun synchronous, circular
Altitude:	678 km
Period:	98 min
Repeat Cycle:	21 days

Instrument	Spectral bands (μm)	IFOV (m)	Swath (km)
MSU-SK	0.5 – 0.6	160	600
	0.6 – 0.7	160	600
	0.7 – 0.8	160	600
	0.8 – 1.1	160	600
	10.4 – 12.6	600	600

1.4 Aircraft Scanners in the Visible and Infrared Regions

1.4.1 General Considerations

Multispectral line scanners, similar in principle to the Landsat MSS and TM instruments, have been available for use in civil aircraft since the late 1960's and early 1970's. As with satellite image acquisition it is the forward motion of the aircraft that provides along track scanning whereas a rotating mirror or a linear detector array provides sensing in the across track direction.

There are several operational features that distinguish the data provided by aircraft scanners from that produced by satellite-borne devices. These are of significance to the image processing task. First, the data volume can be substantially higher. This is a result of having (i) a large number of spectral bands or channels available and (ii) a large number of pixels produced per mission, owing to the high spatial resolution available. Frequently up to 1000 pixels may be recorded across the swath, with many thousands of scan lines making up a flight line; each pixel is normally encoded to at least 8 bits.

A second feature of importance relates to field of view (FOV) – that is the scan angle either side of nadir over which data is recorded. This is depicted in Fig. 1.8. In the case of aircraft scanning the FOV, 2γ, is typically about 70 to 90°. Such a large angle is necessary to acquire an acceptable swath of data from aircraft altitudes. By comparison the FOV for the Landsats 1 to 3 is 11.56° while that for the Landsats 4 and 5 is slightly larger at about 15°. The consequence of the larger FOV with aircraft scanning is that significant distortions in image geometry can occur at the edges of the scan. Often these have to be corrected by digital processing.

Finally, the attitude stability of an aircraft as a remote sensing platform is much poorer than the stability of a satellite in orbit, particularly the Landsat 4 generation for which the pointing accuracy is 0.01° with a stability of 10^{-6} degrees per second. Because of atmospheric turbulence, variations in aircraft attitude described by pitch, roll and yaw can lead to excessive image distortion. Sometimes the aircraft scanner is mounted on a three axis stabilized platform to minimise these variations. It is more common however to have the scanner fixed with respect to the aircraft body and utilize a variable sampling window on the data stream to compensate for aircraft roll.

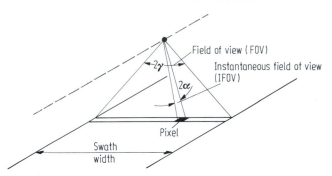

Fig. 1.8. The concept of field of view (FOV) and instantaneous field of view (IFOV)

Use of airborne multispectral scanners offers a number of benefits. Often the user can select the wavebands of interest in a particular application, and small bandwidths can be used. Also, the mission can be flown to specific user requirements concerning time of day, bearing angle and spatial resolution, the last being established by the aircraft height above ground level. As against these however, data acquisition from aircraft platforms is expensive by comparison with satellite recording since aircraft missions are generally flown for a single user and do not benefit from the volume market and synoptic view available to satellite data.

In the following sections the characteristics of a number of different aircraft scanners are reviewed. Those chosen are simply illustrative of the range of device available.

1.4.2 The Daedalus AADS 1240/1260 Multispectral Line Scanner

As an illustration of the properties of aircraft multispectral scanners and their data, the Daedalus AADS 1240/1260 is discussed here in a little detail. This is an example of an instrument with interchangeable detectors to permit data acquisition in a variety of waveband configurations. Image scanning is by means of a rotating mirror from which reflected light is passed via a series of mirrors and a dichroic lens to two sensor ports as shown in the diagram in Fig. 1.9. These ports can be fitted with two detectors in the

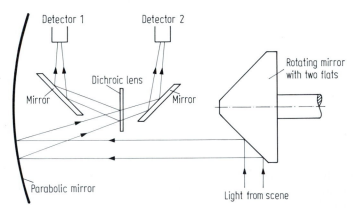

Fig. 1.9. Schematic diagram of the Daedalus infrared/visible aircraft multispectral scanner

Table 1.14. Spectral channels available with Daedalus visible and infrared multispectral line scanners

Ultraviolet	0.	0.32– 0.38 μm
Visible/Reflective IR	1.	0.38– 0.42 μm
	2.	0.42– 0.45 μm
	3.	0.45– 0.50 μm
	4.	0.50– 0.55 μm
	5.	0.55– 0.60 μm
	6.	0.60– 0.65 μm
	7.	0.65– 0.69 μm
	8.	0.70– 0.79 μm
	9.	0.80– 0.89 μm
	10.	0.92– 1.10 μm
Thermal	11.	3.0 – 5.0 μm
	12.	8.0 –14.0 μm

FOV = 86°
IFOV = 2.5 mrad
Dynamic range = 8 bit

thermal infrared range (ADDS 1240) or one infrared detector and a 10 channel spectrometer (AADS 1260). Alternatively a detector operating in the ultraviolet regime can be fitted to one of the ports.

The wavebands, or channels, possible with this arrangement are summarised in Table 1.14. All data are recorded as 8 bit words; spatial resolution and swath width depend upon aircraft altitude. As an illustration, at 1000 m above ground level, the sub-nadir resolution on the ground is 2.5 m if the angular IFOV of the detector is 2.5 mrad. At this altitude a typical swath width, with an angular FOV of 86°, is 1.87 km.

1.4.3 The Airborne Thematic Mapper (ATM)

Because of the versatility possible with aircraft scanner band positioning it is a relatively straightforward matter to choose channels that correspond to spaceborne sensors. For example inspection of Table 1.14 shows that the spectral bands of the Landsat MSS can be approximately simulated by recording the sum of channels 4 and 5 (for MSS band 4), the sum of channels 6 and 7 (MSS band 5), channel 8 (MSS band 6) and the sum of channels 9 and 10 (for MSS band 7).

Because of the intense interest in thematic mapper simulation studies prior to the launch of Landsat 4, an aircraft scanner with bands that approximately matched those of the TM was designed. Referred to as an Airborne Thematic Mapper (ATM), this has the bands shown in Table 1.15. Again the spatial resolution achievable is a function of the altitude at which the instrument is flown. At an altitude of 12.5 km with an IFOV of 2.5 mrad this produces the same equivalent ground pixel as the Landsat TM of 30 m.

1.4.4. The Thermal Infrared Multispectral Scanner (TIMS)

The thermal infrared range of wavelength, typically between 8 and 12 μm, contains spectral emission features that can be used to diagnose silicate rocks. Owing to stretching vibrations of the silicon-oxygen bond in the silicate crystal lattice, a broad

Table 1.15. Airborne thematic mapper spectral characteristics

Channel	Landsat TM equivalent band	
1. 0.42 – 0.45 µm		
2. 0.45 – 0.52 µm	1	FOV = 86°
3. 0.52 – 0.60 µm	2	IFOV = 2.5 mrad
4. 0.605– 0.625 µm		Dynamic range = 8 bit
5. 0.63 – 0.69 µm	3	
6. 0.695– 0.75 µm		
7. 0.76 – 0.90 µm	4	
8. 0.91 – 1.05 µm		
9. 1.55 – 1.75 µm	5	
10. 2.08 – 2.35 µm	7	
11. 8.5 –13.0[a] µm	6	

[a] Thermal band is broader to permit same aircraft IFOV as preceding channels.

minimum in emissivity occurs between 8 and 11 µm. The depth and position of this band is related to the crystal structure of the constituent minerals, varying with changing quartz content. Clay silicates additionally have aluminium-oxygen-hydrogen bending modes, while carbonates exhibit no distinguishing spectral variability in the thermal infrared range (Kahle and Goetz 1983).

To exploit these vibrational features, as means for identifying rock and soil types using multispectral scanner data, it is necessary to have finer spectral resolution in the thermal regime than is possible with scanners such as those discussed previously. By using a dispersive grating, a multichannel thermal scanner has been developed by Daedalus Enterprises Inc. for use by NASA/Jet Propulsion Laboratory for remote sensing of non-renewable resources. A summary of the characteristics of this instrument referred to as TIMS (Thermal Infrared Multispectral Scanner) is given in Table 1.16.

Table 1.16. Thermal infrared multispectral scanner characteristics

Channels	
1. 8.2– 8.6 µm	
2. 8.6– 9.0 µm	FOV = 76°
3. 9.0– 9.4 µm	IFOV = 2.5 mrad
4. 9.5–10.2 µm	Dynamic range = 8 bit
5. 10.2–11.2 µm	
6. 11.2–12.2 µm	

1.4.5 Imaging Spectrometers

Since the mid 1980's the availability of new detector technologies has made possible the development of aircraft scanners capable of recording image data in a large number, typically hundreds, of spectral channels. For a given pixel, enough samples of its reflectance properties may be obtained by these instruments to allow very accurate characterisation of the pixel's spectral reflectance curve over the visible and reflected infrared region. Because of the large number of channels the data sets are often referred to as hyperspectral.

Table 1.17. Imaging spectrometers

Instrument	Spectral range μm	Spectral resolution nm	Dynamic range bits	IFOV mrad	Pixels per line
GERIS (Geophysical) Environmental Research Corp.)	0.40 – 1.08 (24 channels) 1.0 – 2.0 (7 channels) (2.0 – 2.5) (32 channels)	25.4 120 16.5	16	2.5, 3.3, 4.6 (selectable)	512 or 1024
CASI (Itres Research)	0.4 – 0.9 (288 channels)	1.8	12	1.02 to 1.53	512
AVIRIS (Airborne Visible and Infrared Imaging Spectrometer – JPL)	0.4 – 0.72 (31 channels) 0.69 – 1.30 (63 channels) 1.25 – 1.87 (63 channels) 1.84 – 2.45 (63 channels)	9.7 9.6 8.8 11.6	12	1	550
MIVIS (Daedalus Enterprises Inc.)	0.433 – 0.833 (20 channels) 1.15 – 1.55 (8 channels) 2.00 – 2.50 (64 channels) 8.20 – 12.70 (10 channels)	20 50 ≤ 500	12	2	765
MAIS (Academia Sinica)	0.45 – 1.1 (32 channels) (1.40 – 2.50) (32 channels) 8.2 – 12.2 (7 channels)	20 30 400 to 800	8 8	3.0 4.5 3	
HYDICE (Hyperspectral Digital Image Collection Experiment. US Naval Research Labs)	0.4 – 2.5 (206 channels)	7.6 – 14.9	12	0.5	320
HYMAP (Integrated Spectronics Pty Ltd)	0.44 – 0.88 0.881 – 1.335 1.4 – 1.81 1.95 – 2.94 (128 bands total)	16 13 12 16	12	2.5 × 2.0	512

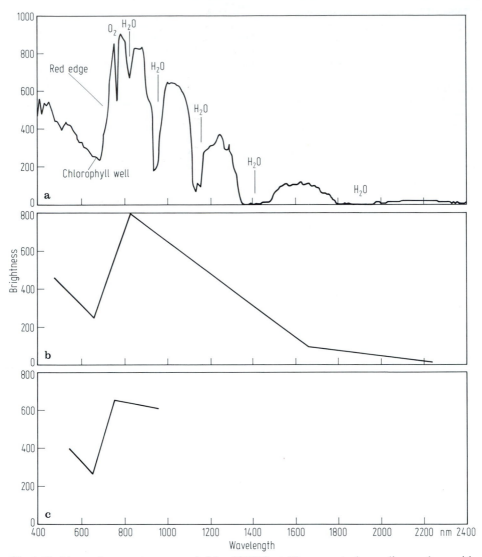

Fig. 1.10. Vegetation spectrum recorded by AVIRIS at 10 nm spectral sampling **a**, along with equivalent TM **b** and MSS **c** spectra. In **a** the fine absorption features resulting from atmospheric constituents are shown, along with features normally associated with vegetation spectra

A good discussion of the development of imaging spectrometry may be found in Goetz et al (1985) and Vane and Goetz (1988). Much of the current interest in these devices may lead to the development of similar spaceborne instruments.

Table 1.17 summarises the characteristics of a selection of current aircraft imaging spectrometers, while Fig. 1.10 shows a typical vegetation spectrum recorded by AVIRIS, along with the comparable recordings by Landsat TM and MSS. The fine spectral detail available in imaging spectrometer data is immediately evident.

Owing to the ability of an imaging spectrometer to record the reflectance spectrum of a pixel in enough detail to allow most important spectral absorption and other diagnostic features to be determined, data analysis can extend beyond image analysis and involve procedures for recognition of fine spectral features in recorded spectra of individual pixels. These procedures range from using models based on a scientific understanding of the spectroscopic processes involved in spectrum formation, to efficient library searching techniques that allow identification based on matches to library prototypes.

Owing to the large number of narrow spectral channels and their contiguous nature, there is also a significant degree of redundancy in imaging spectrometry data that can be exploited to reduce data volumes significantly. Data compression methods, therefore, feature prominently in the analytical procedures being developed for these hyperspectral data sets. Data analysis and reduction techniques for hyperspectral data are treated in Chapter 13.

1.5 Image Data Sources in the Microwave Region

1.5.1 Side Looking Airborne Radar and Synthetic Aperture Radar

Remote sensing image data in the microwave range of wavelengths is generally gathered using the technique of side-looking radar, as illustrated in Fig. 1.11. When used with aircraft platforms it is more commonly called SLAR (side looking airborne radar), a technique that requires some modification when used from spacecraft altitudes, as discussed in the following.

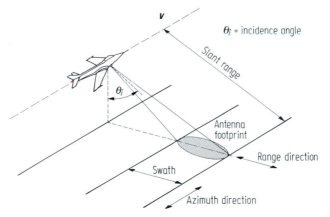

Fig. 1.11. Principle of side looking radar

In SLAR a pulse of electrical energy at the microwave frequency (or wavelength) of interest is radiated to the side of the aircraft at an incidence angle of θ_i. By the same principle as radars used for air navigation and shipping, some of this transmitted energy is scattered from the ground and returned to the receiver on the aircraft. The time delay between transmission and reflection identifies the slant distance to the "target" from the aircraft, while the strength of the return contains information on the so-called scattering coefficient of the target region of the earth's surface. The actual received signal from a single transmitted pulse consists of a continuum of reflections from the complete region of ground actually illuminated by the radar antenna. In Fig. 1.11 this can be identified as the range beamwidth of the antenna. This is chosen at design to give a relation between swath width and altitude, and tends to be rather broad. By comparison the along-track, or so-called azimuth, beamwidth is chosen as small as possible so that the reflections from a single transmitted pulse can be regarded as having come from a narrow strip of terrain broadside to the aircraft. The forward velocity of the aircraft is then arranged so that the next transmitted pulse illuminates the next strip of terrain along the swath. In this manner the azimuth beamwidth of the antenna defines the spatial resolution in the azimuth direction whereas the time resolution possible between echos from two adjacent targets in the range direction defines the spatial resolution in the slant direction.

From an image product viewpoint the slant range resolution is not of interest. Rather it is the projection of this onto the horizontal plane as ground range resolution that is of value to the user. A little thought reveals that the ground range resolution is better at larger incidence angles and thus on the far side of the swath; indeed it can be shown that the ground range size of a resolution element (pixel) is given by

$$r_g = c\tau/2 \sin \theta_i$$

where τ is the length of the transmitted pulse and c is the velocity of light. (Often a simple pulse is not used. Instead a so-called linear chirped waveform is transmitted and signal processing on reception is used to compress this into a narrow pulse. For the present discussion however it is sufficient to consider the transmitted waveform to be a simple pulse or burst of the frequency of interest.)

The azimuth size of a resolution element is related to the length (or aperture) of the transmitting antenna in the azimuth direction, l, the wavelength λ and the range R_0 between the aircraft and the target, and is given by

$$r_a = R_0 \lambda/l$$

This expression shows that a 10 m antenna will yield an azimuth resolution of 20 m at a slant range of 1 km for radiation with a wavelength of 20 cm. However if the slant range is increased to say 100 km – i.e. at low spacecraft altitudes – then a 20 m azimuth resolution would require an antenna of 1 km length, which clearly is impracticable.

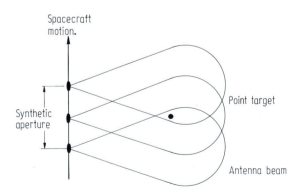

Spacecraft motion.

Synthetic aperture

Point target

Antenna beam

Fig. 1.12. The concept of synthesizing a large antenna by utilizing spacecraft motion along its orbital path. Here a view from above is shown, illustrating that a small real antenna is used to ensure a large real beamwidth in azimuth. As a consequence a point on the ground is illuminated by the full synthetic aperture

Therefore when radar image data is to be acquired from spacecraft, a modification of SLAR, referred to as synthetic aperature radar (SAR) is used. Essentially this utilizes the motion of the space vehicle, during transmission of the ranging pulses, to give an effectively long antenna, or a so-called synthetic aperture. This principle is illustrated in Fig. 1.12, wherein it is seen that an intentionally large azimuth beamwidth is employed to ensure that a particular spot on the ground is illuminated and thus provides reflections over a length of spacecraft travel equivalent to the synthetic aperture required.

A discussion of the details of the synthetic aperture concept and the signal processing required to produce a high azimuth resolution is beyond the scope of this treatment. The matter is pursued further in Ulaby, Moore and Fung (1982), Elachi et al. (1982), Tomiyasu (1978), and Elachi (1988).

1.5.2 The Seasat SAR

The first earth observational space mission to carry a synthetic aperture imaging radar was the Seasat satellite launched in June 1978. Although only short lived it recorded about 126 million square kilometres of image data, including multiple coverage of many regions. Several other remote sensing instruments were also carried, including a radar altimeter, a scatterometer, a microwave radiometer and a visible and infrared imaging radiometer. Relevant characteristics of the satellite and its SAR are summarised in Table 1.18. Polarization referred to in this table relates to the orientation of the electric field vector in the transmitted and received waveforms. Free space propagation of electromagnetic energy, such as that used for radar, takes place as a wave with electric and magnetic field vectors normal to each other and also normal to the direction of propagation. Should the electric field vector be parallel to the earth's surface, the wave is said to be horizontally polarized. Should it be vertical then the wave is said to be vertically polarized. A wavefront with a combination of the two will be either elliptically or circularly polarized. Even though one particular polarization might be adopted for transmission, some rotation can occur when the energy is reflected from the ground. Consequently at the receiver often both vertically and horizontally polarized components are available, each having its own diagnostic properties concerning the earth cover type being sensed. Whether one or the other, or both, are received depends upon the antenna used with the radar. In the case of Seasat, horizontally polarized radiation was transmitted (**H**) and horizontally polarized returns were received (**H**).

Further details on the Seasat SAR will be found in Elachi et al. (1982).

Table 1.18. Characteristics of Seasat SAR, SIR-A, SIR-B, and SIR-C

	Seasat	SIR-A	SIR-B	SIR-C/X-SAR
Altitude	800 km	245 km	225–235 km	225 km
Wavelength	0.235 m	0.235 m	0.235 m	L – 0.235 m, C – 0.058 m, X – 0.031 m
Polarization	*HH*	*HH*	*HH*	HH, HV, VH, VV (only VV at X)
Incidence angle	20°	47°	15–57°	20–55°
Swath width	100 km	50 km	20–50 km	15–90 km
Range resolution	25 m	40 m	58–17 m	13–26 m (L, C), 10–20 (X)
Azimuth resolution	25 m	40 m	25 m	30 m

1.5.3 Spaceborne (Shuttle) Imaging Radar-A (SIR-A)

A modified version of the Seasat SAR was flown as the SIR-A sensor on the second flight of Space Shuttle in November of 1981. Although the mission was shortened to three days, image data of about 10 million square kilometres was recorded. In contrast to Seasat however, in which the final image data was available digitally, the data in SIR-A was recorded and processed optically and thus is available only in film format. For digital processing therefore it is necessary to have areas of interest digitized from film using a device such as a scanning microdensitometer. A summary of SIR-A characteristics is given in Table 1.18, wherein it will be seen that the incidence angle was chosen quite different from that for the Seasat SAR. Interesting features of landform can be brought out by processing the two together.

More details on SIR-A will be found in Elachi et al. (1982) and Elachi (1983).

1.5.4 Spaceborne (Shuttle) Imaging Radar-B (SIR-B)

SIR-B, the second instrument in the NASA shuttle imaging radar program was carried on Space Shuttle mission 41 G in October 1984. Again the instrument was essentially the same as that used on Seasat and SIR-A, however the antenna was made mechanically steerable so that the incidence angle could be varied during the mission. Also about half the data was recorded digitally with the remainder being optically recorded. Details of the SIR-B mission are summarised in Table 1.18; NASA (1984) contains further information on the instrument and experiments planned for the mission. Because of the variable incidence angle both the range resolution and swath width also varied accordingly.

1.5.5 Spaceborne (Shuttle) Imaging Radar-C (SIR-C)/X-band Synthetic Aperture Radar (X-SAR)

SIR-C/X-SAR, the third Shuttle radar mission, was carried out over two 10 day flights in April and September 1994. The SAR carried was the result of cooperation between NASA and DARA, the German Aerospace Agency, and had the characteristics indicated in Table 1.18. Further details of the mission and the SAR will be found in Stofan et al (1995) and Jordan et al (1995).

1.5.6 ERS-1, 2

The European Remote Sensing Satellites ERS-1 and ERS-2 were launched in July 1991 and April 1995 respectively; they carry a number of sensors, one of which is a synthetic aperture radar intended largely for sea state and oceanographic applications. Characteristics of the radar are summarised in Table 1.19.

1.5.7 JERS-1

The Japanese Earth Resources Satellite JERS-1 was launched in February 1992. It carries two imaging instruments; one is an optical sensor and the other an imaging radar. Table 1.19 shows the design characteristics for the radar. The optical sensor, called OPS, has 8 wavebands between $0.52\,\mu m$ and $2.40\,\mu m$ with a swath width of 75 km and a dynamic range of 6 bits. The optical pixel size is 18.3 m (across track) × 24.2 m (along track). Provision is included for stereoscopic imaging.

1.5.8 Radarsat

Canada's Radarsat was launched on 4 November 1995; its SAR is able to operate in the standard and six non-standard modes, one of which will give a 518 km swath (Raney et al, 1991). Table 1.19 lists the characteristics of the Radarsat SAR.

Table 1.19. Characteristics of free flying satellite SAR systems

	ERS-1, 2	JERS-1	Radarsat
Altitude	785 km	568 km	793–821 km
Wavelength*	0.057 m	0.235 m	0.057 m
Polarization	*VV*	*HH*	*HH*
Incidence angle	23°	35°	19–49°
Swath width	100 km	75 km	50–500 km
Range resolution	30 m	18 m	27 m
Azimuth resolution	30 m	18 m	19–24 m

*5.3 GHz for ERS-1, 2 and Radarsat and 1.28 GHz for JERS-1

1.5.9 Aircraft Imaging Radar Systems

Airborne imaging radar systems in SLAR and SAR technologies are also available. As with airborne multispectral scanners these offer a number of advantages over equivalent satellite based systems including flexibility in establishing mission parameters (bearing, incidence angle, spatial resolution etc.) and proprietary rights to data. However the cost of data acquisition is also high.

Table 1.20 summarises the characteristics of four commercial aircraft imaging radars, chosen to illustrate the operating parameter of these devices by comparison to satellite based systems. Note the band: wavelength designations – X: 0.030 m, C: 0.057 m, L: 0.235 m, P: 0.667 m.

Table 1.20. Characteristics of aircraft imaging radars

	MARS LTD AN/APS-94D	MDA LTD IRIS	GOOD-YEAR/GEMS	JPL AIRSAR
Technology	SLAR	SAR	SAR	SAR
Wavelength	X band	X, C, L band	X band	C, L, P band
Range resolution	30 m	5 m to 30 m	~ 10 m	~ 10 m
Azimuth resolution	IFOV = 7.8 mrad	6 m, 10 m	~ 10 m	~ 10 m
Incidence angles	1° to 45°			0° to 70°
Swath width	25 km to 100 km			~23 km

1.6 Spatial Data Sources in General

1.6.1 Types of Spatial Data

The foregoing sections have addressed sources of multispectral digital image data of the earth's surface; each image considered has represented the spatial distribution of energy coming from the earth in one or several wavelength ranges in the electromagnetic spectrum. Other sources of spatially distributed data are also often available for regions of interest. These include simple maps that show topography, land ownership roads and the like, through to more specialised sources of spatial data such as maps of geophysical measurements of the area. Frequently these other spatial data sources contain information not available in multispectral imagery and often judicious combinations of multispectral and other map-like data allow inferences to be drawn about regions on the earth's surface not possible when using a single source on its own. Consequently the image analyst ought to be aware of the range of spatial data available for a region and select that subset likely to assist in the information extraction process.

Table 1.21 is an illustration of the range of spatial data one might expect could be available for a given region. This differentiates the data into three types according as to whether it represents point information, line information or area information. Irrespective of type however, for a spatial data set to be manipulated using the techniques of digital image processing it must share two characteristics with multispectral data of the types discussed in the previous sections. First it must be available in discrete form spatially, and in value. In other words it must consist of, or be able to be converted to, pixels with each pixel describing the properties of a given (small) area on the ground: the value ascribed to each pixel must be expressible in digital form. Secondly it must be in correct geographic relation to a multispectral image data set if the two are to be manipulated together. In situations where multispectral

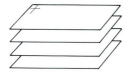

Multispectral source #1
(eg Landsat MSS)

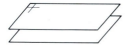

Multispectral source #2
(eg SAR imagery)

Topography
Rainfall

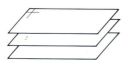

Geophysics
Geochemistry

Fig. 1.13. An integrated spatial data source database

data is not used, the pixels in the spatial data source would normally be arranged to be referenced to a map grid. It is usual however, in digital spatial data handling systems, to have *all* entries in the data set relating to a particular geographical region, mutually registered and referenced to a map base such as the UTM grid system. When available in this manner the data is said to be geocoded. Means by which different data sets can be registered are treated in Sect. 2.5. Such a database is depicted in Fig. 1.13

Table 1.21. Sources of spatial data

Point	Line	Area
Multispectral data	road maps	land ownership
Topography	powerline grids	town plans
Magnetic measurements	pipeline networks	geological maps
Gravity measurements		land use licenses
Radiometric measurements		land use maps
Rainfall		land cover maps
Geochemistry (in ppm)		soil type maps

1.6.2 Data Formats

Not all sources of spatial data are available originally in the pixel oriented digital format depicted in Fig. 1.13. Indeed often the data will be available as analog maps that require digitisation before entry into a digital data base. That is particularly the case with line and area data types, in which case consideration has to be given also to the "value" that will be ascribed to a particular pixel. In line spatial data sources the pixels could be called zero if they were not part of a line and coded to some other number if they formed part of a line of a given type. For a road map, for example, pixels that fall on highways might be given a value of 1 whereas those on secondary roads could be given a value of 2, and so on. On display, using a colour output device, the different numbers could be interpreted and output as different colours. In a similar manner numbers can be assigned to different regions when digitizing area spatial data sources.

Conceptually the digitization process may not be straightforward. Consider the case for example of needing to create a digital topographic map from its analog contour map counterpart. Figure 1.14 illustrates this process. First it is necessary to convert the contours on the paper map to records contained in a computer. This is done by using an input device such as a stylus or cross-hair cursor to mark a series of points on each contour between which the contour is regarded by the computer to be a straight line. Information on a contour at this stage is stored in the computer's memory as a file of points. This format is referred to as *vector format* owing to the vectors that can be drawn from point to point (in principle) to reconstruct a contour on a display or plotter. Some spatial data handling computer systems operate in vector format entirely. However to be able to exploit the techniques of digital image processing the vector formated data has to be turned into a set of pixels arranged on rectangular grid centres. This is referred to as *raster format* (or sometimes grid format); the elevation values for each pixel in the raster form are obtained by a process of interpolation over the points recorded on the contours. The operation is referred to as *vector to raster conversion* and is an essential step in entering map data into a digital spatial data base.

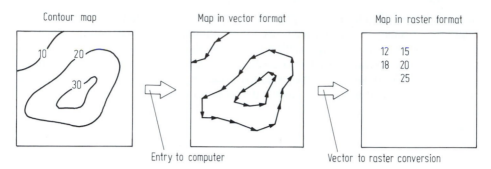

Fig. 1.14. Definition of vector and raster format using the illustration of digitising contour data

Raster format is a natural one for the representation of multispectral image data since data of that type is generated by digitising scanners, is transmitted digitally and is recorded digitally. Moreover many image forming devices such as filmwriters and television monitors operate on a raster display basis, compatible with digital data acquisition and storage. Raster format however is also appealing from a processing point of view since the logical records for the data are the pixel values (irrespective of whether the data is of the point, line or area type) and neighbourhood relationships are easy to establish by means of the pixel addresses. This is important for processing operations that involve near neighbouring groups of pixels. In contrast, vector format does not offer this feature. However an advantage of vector format, often exploited in high quality graphics display devices, is that resolution is not limited by pixel size.

1.6.3 Geographic Information Systems (GIS)

The amount of data to be handled in a database that contains spatial sources such as satellite and aircraft imagery along with maps, as listed in Table 1.21, is enormous, particularly if the data covers a large geographical region. Quite clearly therefore thought has to be given to efficient means by which the data types can be stored and retrieved, manipulated, analysed and displayed. This is the role of the geographic information system (GIS). Like its commercial counterpart, the management information system (MIS), the GIS is designed to carry out operations on the data stored in its database, according to a set of user specifications, without the user needing to be knowledgeable about how the data is stored and what data handling and processing procedures are utilized to retrieve and present the information required. Unfortunately because of the nature and volume of data involved in a GIS many of the MIS concepts developed for data base management systems (DBMS) cannot be transferred directly to GIS design although they do provide guidelines. Instead new design concepts have been needed, incorporating the sorts of operation normally carried out with spatial data, and attention has had to be given to efficient coding techniques to facilitate searching through the large numbers of maps and images often involved.

To understand the sorts of spatial data manipulation operations of importance in GIS one must take the view of the resource manager rather than the data analyst. Whereas the latter is concerned with image reconstruction, filtering, transformation and classification, the manager is interested in operations such as those listed in Table 1.22. These provide information from which management strategies and the like can be inferred. Certainly, to be able to implement many, if not most, of these a substantial amount of image processing may be required. However as GIS technology progresses it is expected that the actual image processing being performed would be transparent to the resource manager; the role of the data analyst will then be in part of the GIS design. A good discussion of the essential issues in GIS will be found in Star and Estes (1990).

A problem which can arise in image data bases of the type encountered in a GIS is the need to identify one image by reason of its similarity to another. In principle, this could be done by comparing the images pixel-by-pixel; however the computational demand in so doing would be enormous for images of any practical size. Instead effort has been

Table 1.22. Some GIS data manipulation operations

Intersection and overlay of data sets (masking)
Intersection and overlay of polygons (grid cells, etc.) with spatial data
Identification of shapes
Identification of points in polygons
Area determination
Distance determination
Thematic mapping
Proximity calculations (shortest route, etc.)
Search by data
Search by location
Search by user-defined attribute
Similarity searching (e. g. of images)

directed to developing codes or signatures for complete images that will allow efficient similarity searching. For example an image histogram could be used (see Sect. 4.2); however as geometric detail is not preserved in a histogram this is rarely a suitable code for an image on its own. One effective possibility that has been explored is the use of image pyramids. A pyramid is created by combining groups of pixels in a neighbour-hood to produce a new composite pixel of reduced resolution, and thus a low resolution image with fewer pixels. This process is repeated on the processed image to form a new image of lower resolution (and fewer pixels) still. Ultimately the image could be reduced to one single pixel that is a global measure of the image's brightness. Since pixels are combined in neighbourhood groups, spatial detail is propagated up through the pyramid, albeit at decreasing resolution. Figure 1.15 illustrates how an image pyramid is constructed by simple averaging of non-overlapping sets of 2×2 pixels. It is a relatively easy matter (see Problem 1.6) to show that the additional memory required to store a complete pyramid, constructed as in the figure, is only 33% more than that required to store just the image itself.

Having developed an image pyramid, signatures that can be used to undertake similarity searching include the histograms computed over rows and colums in the uppermost levels of the pyramid (see Problem 1.7). A little thought shows that this allows an enormous number of images to be addressed, particularly if each pixel is represented by an 8 bit brightness value. As a result very fast searching can be carried out on these reduced representations of images.

Image pyramids are discussed by Rosenfeld (1982) and have been considered in the light of image similarity searching by Chien (1980).

There is sometimes an image processing advantage to be obtained when using a pyramid representation of an image. In edge detection, for example, it is possible to localise edges quickly, without having to search every pixel of an image, by finding apparent edges (regions) in the upper levels of the pyramid. The succeeding lower pixel groupings are then searched to localise the edges better.

Finally the pyramid representation of an image is felt to have some relation to human perception of images. The upper levels contain global features and are therefore not unlike the picture we have when first looking at a scene – generally we take the scene in

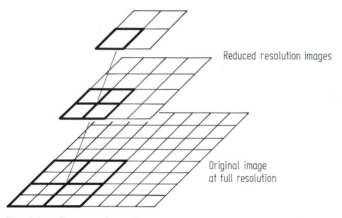

Reduced resolution images

Original image
at full resolution

Fig. 1.15. Construction of an image pyramid by successively averaging groups of 2 × 2 pixels

initially "as a whole" and either miss or ignore detail. Then we focus on regions of interest for which we pay attention to detail because of the information it provides us with.

1.6.4 The Challenge to Image Processing and Analysis

Much of the experience gained with digital image processing and analysis in remote sensing has been with multispectral image data. In principle however any spatial data type in digital format can be processed using the techniques and procedures presented in this book. Information extraction from geophysical data could be facilitated, for example, if a degree of sharpening is applied prior to photointerpretation, while colour density slicing could assist the interpretation of topography. However the real challenge to the image analyst arises when data of mixed types are to be processed together. Several issues warrant comment.

The first relates to differences in resolution, an issue that arises also when treating multi-source satellite data such as Landsat MSS and NOAA AVHRR. The analyst must decide, for example what common pixel size will be used when co-registering the data, since either resolution or coverage will normally be sacrificed. Clearly this decision will be based on the needs of a particular application and is a challenge more to the analyst than the algorithms.

The more important consideration however is in relation to techniques for machine assisted interpretation. There is little doubt that combined multispectral and, say, topographic or land ownership maps can yield more precise thematic (i.e. category of land cover, etc.) information for a particular region than the multispectral data on its own. Indeed the combination of these sources is often employed in photointerpretive studies.

The issue is complicated further when it is recalled that much of the non-spectral, spatial data available is not in numerical point form but rather is in nominal area or line format. With these, image analysis algorithms developed algebraically will not be suitable. Rather some degree of logical processing of labels combined with algebraic processing of arithmetic values (such as pixel brightnesses) is necessary.

Chapter 12 addresses this issue by considering the value of knowledge-based image analysis methods, which lend themselves to handling both numerical and non-numerical data sources.

1.7 A Comparison of Scales in Digital Image Data

Because of IFOV differences the digital images provided by various remote sensing sensors will find application at different scales. As a guide Table 1.23 relates scale to spatial resolution; this has been derived somewhat simplistically by considering an image pixel to be too coarse if it approaches 0.1 mm in size on a photographic product at a given scale. Thus Landsat MSS data is suggested as being suitable for scales smaller than about 1:500,000 whereas NOAA AVHRR data is suitable for scales below 1:10,000,000. Detailed discussions of image quality in relation to scale will be found in Welch (1982), Forster (1985), Woodcock and Strahler (1987) and Light (1990).

Table 1.23. Suggested maximum scales of photographic products as a function of effective ground pixel size (based on 0.1 mm printed pixel).

Scale	Approx. Pixel Size (m)	Sensor (nominal)
1: 50,000	5	aircraft MSS
1: 250,000	25	Spot HRV, Landsat TM
1: 500,000	50	Landsat MSS
1: 5,000,000	500	OCTS
1: 10,000,000	1000	NOAA AVHRR
1: 50,000,000	5000	GMS thermal IR band

References for Chapter 1

More details on the satellite programs treated in this chapter, along with information on sensors and data characteristics can be found in the web sites of the responsible agencies. Some of particular use are:

For weather satellites

http://www.wmo.ch
http://www.ncdc.noaa.gov

For earth observation satellites

http://www.eosat.com
http://www.orbimage.com
http://hdsn.eoc.nasda.go.jp
http://www.spotimage.com.fr

For radar missions

http://www.rsi.ca
http://www.southport.jpl.nasa.gov

For imaging spectrometers

http://www.techexpo.com/WWW/opto-knowledge/

M.T. Chahine, 1983: Interaction Mechanisms within the Atmosphere. In Manual of Remote Sensing, R.N. Colwell (Ed). 2e. American Society of Photogrammetry, Falls Church, Va.

Y.T. Chien, 1980: Hierarchical Data Structures for Picture Storage, Retrieval and Classification. In Pictorial Informations Systems, S.K. Chang and K.S. Fu (Eds.), Springer-Verlag, Berlin.

C. Elachi (Chairman), 1983: Spaceborne Imaging Radar Symposium. Jet Propulsion Laboratory, January 17–20. JPL Publication 83–11.

C. Elachi, T. Bicknell, R.L. Jordan and C. Wu, 1982: Spaceborne Synthetic Aperture Imaging Radars. Applications, Techniques and Technology. Proc. IEEE, 70, 1174–1209.

C. Elachi, 1988: Spaceborne Radar Remote Sensing: Applications and Techniques. N.Y., IEEE.

B.C. Forster, 1985: Mapping Potential of Future Spaceborne Remote Sensing Systems. Institution of Surveyors (Australia) Annual Congress, Alice Springs.

A.F.H. Goetz, G. Vane, T.E. Solomon and B.N. Rock, 1985: Imaging Spectrometry for Earth Remote Sensing, Science, 228, 1147–1153.

R.M. Hoffer, 1978: Biological and Physical Considerations in Applying Computer-Aided Analysis Techniques to Remote Sensor Data. In P.H. Swain and S.M. Davis, Eds., Remote Sensing: The Quantitative Approach, N.Y., McGraw-Hill.

R.L. Jordan, B.L. Huneycutt and M. Werner, 1995: The SIR-C/X-SAR Synthetic Aperture Radar System. IEEE Trans Geoscience and Remote Sensing, 33, 829–839.

A.B. Kahle and A.F.H. Goetz, 1983: Mineralogic Information from a New Airborne Thermal Infra-red Multispectral Scanner, Science, 222, 24–27.

D.L. Light, 1990: Characteristics of Remote Sensors for Mapping and Earth Science Applications. Photogrammetric Engineering and Remote Sensing, 56, 1613–1623.

NASA, 1984: The SIR-B Science Investigations Plan, Jet Propulsion Laboratory Publication 84-3.

R.K. Raney, A.P. Luscombe, E.J. Lanham and S. Ahmed, 1991: Radarsat. Proc. IEEE, 79, 839–849

A. Rosenfeld, 1982: Quadtrees and Pyramids: Hierarchical Representation of Images, Report TR-1171, Computer Vision Laboratory, University of Maryland.

J. Star and J.E. Estes, 1990: Geographic Information Systems, An Introduction. N.J. Prentice-Hall.

E.R. Stofan, D.L. Evans, C. Schmullius, B. Holt, J.J. Plaut, J. van Zyl, S.D. Wall and J. Way, 1995: Overview of Results of Spaceborne Imaging Radar-C, X-Band Synthetic Aperture Radar (SIR-C/X-SAR). IEEE Trans Geoscience and Remote Sensing, 33, 817–828.

K. Tomiyasu, 1978: Tutorial Review of Synthetic-Aperture Radar (SAR) with Applications to Imaging of the Ocean Surface. Proc. IEEE, 66, 563–583.

R. Welch, 1982: Image Quality Requirements for Mapping from Satellite Data. Proc. Int. Soc. Photogrammetry and Remote Sensing, Commission 1. Primary Data Acquisition, Canberra.

F. T Ulaby, R. K. Moore and A. K. Fung, 1981, 1982, 1985: Microwave Remote Sensing, Active and Passive. Vols 1, 2, 3 Reading Mass. Addison-Wesley.

G. Vane and A.F.H. Goetz, 1988: Terrestrial Imaging Spectroscopy. Remote Sensing of Environment, 24, 1–29.

C. E. Woodcock and A. H. Strahler, 1987: The Factor of Scale in Remote Sensing. Remote Sensing of Environment, 21, 311–332.

Problems

1.1 Plot graphs of pixel size in equivalent ground metres as a function of angle from nadir across a swath for

a) Landsat MSS with IFOV of 0.086 mrad, FOV = 11.56°,
b) NOAA AVHRR with IFOV = 1.3 mrad, FOV = 2700 km,
   altitude = 833 km,
c) an aircraft scanner with IFOV = 2.5 mrad, FOV = 80°
   flying at 1000 m AGL (above ground level),

producing separate graphs for the along track and across track dimensions of the pixel.

Replot the graphs to indicate pixel size relative to that at nadir.

1.2 Imagine you have available image data from a multispectral scanner that has two narrow spectral bands. One is centred on 0.65 μm and the other on 1.0 μm wavelength. Suppose the corresponding region on the earth's surface consists of water, vegetation and soil.

Construct a graph with two axes, one representing the brightness of a pixel in the 0.65μm band and the other representing the brightness of the pixel in the 1.0 μm band. In this show where you would expect to find vegetation pixels, soil pixels and water pixels. Note how straight lines could, in principle, be drawn between the three groups of pixels so that if a computer had the equations of these lines stored in its memory it could use them to identify every pixel in the image.

Repeat the exercise for a scanner with bands centred on 0.95 μm and 1.05 μm.

1.3 There are 460 185 km × 185 km frames of Landsat data that cover Australia. Compute the daily data rate (in Gbit/day) for Australia provided by the MSS and TM sensors on Landsat 5, assuming all possible scenes are recorded.

1.4 Assume a "frame" of image data consists of a segment along the track of the satellite, as long as the swath is wide. Compute the data volume of a single frame from each of the following sensors and produce a graph of average data volume per wavelength band versus pixel size.

NOAA	AVHRR
Nimbus	CZCS
Landsat	MSS
Landsat	TM
Spot	HRV (multispectral)

1.5 Determine a relationship between swath width and orbital repeat cycle for a polar orbiting satellite at an altitude of 800 km, assuming that adjacent swaths overlap by 10% at the equator.

1.6 An image pyramid is to be constructed in the following manner: Groups of 2 × 2 pixels are averaged to form single pixels and thereby reduce the number of pixels in the image by a factor of 4, while reducing its resolution as well. Groups of 2 × 2 pixels in the reduced resolution image are then averaged to form a third version of lower resolution still. This process can be continued until the original image is represented by a pyramid of progressively lower resolution images with a single pixel at the top.

Determine the additional memory required to store the complete pyramid by comparison to storing just the image itself. (Hint: Use the properties of a geometric progression.)

Repeat the exercise for the case of a pyramid built by averaging 3 × 3 groups of pixels.

1.7 A particular image data base is to be constructed to allow similarity searching to be performed on sets of binary images – i.e. on images in which pixels take on brightness values of 0 or 1 only. Image pyramids are to be stored in the data base where each succeeding higher level in a pyramid has pixels derived from 3×3 groups in the immediately lower level. The value of the pixel in the higher level is to be that of the majority of pixels in the corresponding lower group. The uppermost level in the pyramid is a 3×3 image.

(i) How much additional storage is required to store the pyramids rather than just the original images?

(ii) The search algorithm to be implemented on the top level of the pyramid is to consist of histogram comparison. In this histograms are taken of the pixels along each row and down each column and a pair of images are 'matched' when all of these histograms are the same for both images. In principle, how many distinct images can be addressed using the top level only?

(iii) An alternative search algorithm to that mentioned in (ii) is to compute just the simple histogram of all the pixels in the top level of the pyramid. How many distinct images could be addressed in this case using the top level only?

(iv) Would you recommend storing the complete pyramid for each image or just the original image plus histogram information for the upper levels of a pyramid?

(v) An alternative means by which the upper levels of the pyramid could be coded is simply by counting and storing the fraction of 1's which occurs in each of the first few upper-most levels.

Suppose this is done for the top three levels. Show how a feature or pattern space could be constructed for the complete image data base, using the 1's fractions for the upper levels in each image, which can then be analysed and searched using pattern classification procedures.

1.8 A particular satellite carries a high resolution optical sensor with 1 m spatial resolution and is at 800 km altitude in a near polar orbit. Orbital period is related to orbital radius by:

$$T = 2\pi \sqrt{\frac{r^3}{\mu}}$$

where $\mu = 3.986 \times 10^{14}$ m^{3}s^{-2}, and orbital radius is given by

$$r = a + h$$

in which $a = 6.378$ Mm and h is altitude.

If the orbit is arranged such that complete earth coverage is possible, how long will that take if there are 2048 pixels per swath? Consequently, what sorts of applications would such a satellite be used for?

Chapter 2
Error Correction and Registration of Image Data

When image data is recorded by sensors on satellites and aircraft it can contain errors in geometry and in the measured brightness values of the pixels. The latter are referred to as radiometric errors and can result from the instrumentation used to record the data, from the wavelength dependence of solar radiation and from the effect of the atmosphere. Image geometry errors can arise in many ways. The relative motions of the platform, its scanners and the earth, for example, can lead to errors of a skewing nature in an image product. Non-idealities in the sensors themselves, the curvature of the earth and uncontrolled variations in the position and attitude of the remote sensing platform can all lead to geometric errors of varying degrees of severity.

When an image is to be utilized it is frequently necessary to make corrections in brightness and geometry if the accuracy of interpretation, either manually or by machine, is not to be prejudiced. For many applications only the major sources of error will require compensation whereas in others more precise correction will be necessary.

It is the purpose of this chapter to discuss the nature of the radiometric and geometric errors commonly encountered in remote sensing images and to develop computational procedures that are used for their compensation. While this is the principal intention, the procedures to be presented find more general application as well, such as in registering together sets of images of the same region but at different times, and in performing operations such as scale changing and zooming (magnification).

Radiometric correction procedures for hyperspectral imagery are treated separately in Chapter 13.

2.1 Sources of Radiometric Distortion

Mechanisms that affect the measured brightness values of the pixels in an image can lead to two broad types of radiometric distortion. First, the relative distribution of brightness over an image in a given band can be different to that in the ground scene. Secondly the relative brightness of a single pixel from band to band can be distorted compared with the spectral reflectance character of the corresponding region on the ground. Both types can result from the presence of the atmosphere as a transmission medium through which radiation must travel from its source to the sensors, and can be a result also of instrumentation effects.

2.1.1 The Effect of the Atmosphere on Radiation

Figure 2.1 depicts the effect the atmosphere has on the measured brightness value of a single pixel for a passive remote sensing system in which the sun is the source of energy, as in the visible and reflective infrared regions. In the absence of an atmosphere the

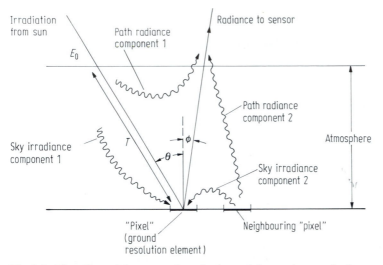

Fig. 2.1. The effect of the atmosphere in determining various paths for energy to illuminate a (equivalent ground) pixel and to reach the sensor

signal measured by the sensor will be a function simply of the level of energy from the sun, actually incident on the pixel, and the reflectance properties of the pixel itself. However the presence of the atmosphere can modify the situation significantly as depicted in the diagram. Before discussing this in detail it is of value to introduce some definitions of radiometric quantities as these will serve to simplify explanations and will allow correction equations to be properly formulated.

Imagine the sun as a source of energy emitting at a given rate of Joules per second, or Watts. This energy radiates through space isotropically in an inverse square law fashion so that at a given distance the sun's emission can be measured as Watts per square metre (given as the power emitted divided by the surface area of a sphere at that distance). This power density is called *irradiance*, a property that can be used to describe the strength of any emitter of electromagnetic energy.

We can measure a level of solar irradiance at the earth's surface. If the surface is perfectly diffuse then this amount is scattered uniformly into the upper hemisphere. The amount of power density scattered in a particular direction is defined by its density per solid angle, since equal amounts are scattered into equal cones of solid angle. This quantity is called *radiance* and has units of Watts per square metre per steradian $(Wm^{-2}sr^{-1})$.

The emission of energy by bodies such as the sun is wavelength dependent, as seen in Fig. 1.2, so that often the term *spectral irradiance* is used to describe how much power density is available incrementally across the wavelength range. Spectral irradiance is typically measured in $Wm^{-2}\mu m^{-1}$.

As an illustration of how these quantities might be used suppose, in the absence of atmosphere, the solar spectral irradiance at the earth is E_λ. If the solar zenith angle (measured from the normal to the surface) is θ as shown in Fig. 2.1 then the spectral irradiance (spectral power density) on the earth's surface is $E_\lambda \cos \theta$. This gives an available irradiance between wavelengths λ_1 and λ_2 of

$$E_{os} = \int_{\lambda_1}^{\lambda_2} E_\lambda \cos \theta \, \mathrm{d}\lambda.$$

In remote sensing the wavebands used ($\Delta\lambda = \lambda_2 - \lambda_1$) are frequently narrow enough to assume

$$E_{os} = E_{\Delta\lambda} \cos \theta \, \Delta\lambda \qquad (2.1)$$

where $E_{\Delta\lambda}$ is the average spectral irradiance in the band $\Delta\lambda$.

Suppose the surface has a reflectance R. This describes what proportion of the incident energy is reflected. If the surface is diffuse then the radiance scattered into the upper hemisphere and available for measurement is

$$L = E_{\Delta\lambda} \cos \theta \, \Delta\lambda \, R/\pi \qquad (2.2)$$

where the divisor π accounts for the upper hemisphere of solid angle. Knowing L it is possible to determine the power detected by a sensor, and indeed the digital count value (or grey level) given in the digital data product from a particular sensor is directly related to the radiance of the scene. If we call the digital value (between 0 and 255 for example) C, then the measured radiance of a particular pixel is

$$L = Ck + L_{min} \qquad (2.3)$$

where $k = (L_{max} - L_{min})/C_{max}$ in which L_{max} and L_{min} are the maximum and minimum measurable radiances of the sensor. These are usually available from the sensor manufacturer or operator.

Equation (2.2) relates to the ideal case of no atmosphere. When an atmosphere is present there are several mechanism that must be taken into account that modify (2.2). These are a result of scattering and absorption by the particles in the atmosphere.

Absorption by atmospheric molecules is a selective process that converts incoming energy into heat. In particular, molecules of oxygen, carbon dioxide, ozone and water attenuate the radiation very strongly in certain wavebands. Sensors commonly used in solid earth and ocean remote sensing are usually designed to operate away from these regions so that the effects are small. Scattering by atmospheric particles is then the dominant mechanism that leads to radiometric distortion in image data (apart from sensor effects).

There are two broadly identified scattering mechanisms. The first is scattering by the air molecules themselves. This is called Rayleigh scattering and is an inverse fourth power function of the wavelength used. The other is called aerosol or Mie scattering and is a result of scattering of the radiation from larger particles such as those associated with smoke, haze and fumes. These particulates are of the order of one tenth to ten wavelengths. Mie scattering is also wavelength dependent, although not as strongly as Rayleigh scattering. When the atmospheric particulates become much larger than a wavelength, such as those common in fogs, clouds and dust the wavelength dependence disappears.

In a clear ideal atmosphere Rayleigh scattering is the only mechanism present. It accounts, for example, for the blueness of the sky. Because the shorter (blue) wavelengths are scattered more than the longer (red) wavelengths we are more likely to see blue when looking in any direction in the sky. Likewise the reddish appearance of sunset is also caused by Rayleigh scattering. This is a result of the long atmospheric path the radiation has to follow at sunset during which most short wavelength radiation is scattered away from direct line of sight by comparison to the longer wavelengths.

In contrast to Rayleigh scattering, fogs and clouds appear white or bluish-white owing to the (near) non-selective scattering caused by the larger particles.

We are now in the position to appreciate the effect of the atmosphere on the radiation that ultimately reaches a sensor. We will do this by reference to Fig. 2.1, commencing with the incoming solar radiation. The effects are identified by title:

Transmittance In the absence of atmosphere transmittance is 100%. However because of scattering and absorption not all of the available solar irradiance reaches the ground. The amount that does, relative to that for no atmosphere, is called the transmittance. Let this be called T_θ, the subscript indicating its dependence on the zenith angle of the source because of the longer path length through the atmosphere. In a similar way there is an atmospheric transmittance T_ϕ to be taken into account between the point of reflection and the sensor.

Sky irradiance Because the radiation is scattered on its travel down through the atmosphere a particular pixel will be irradiated both by energy on the direct path in Fig. 2.1 and also by energy scattered from atmospheric constituents. The path for the latter is undefined and in fact diffuse. A pixel can also receive some energy that has been reflected from surrounding pixels and then, by atmospheric scattering, is again directed downwards. This is the sky irradiance component 2 identified in Fig. 2.1. We will call the sky irradiance at the pixel E_D.

Path radiance Again because of scattering alone, radiation can reach the sensor from adjacent pixels and also via diffuse scattering of the incoming radiation that is actually scattering towards the sensor by the atmospheric constituents before it reaches the ground. These two components are referred to as path radiance and denoted L_p.

Having defined these effects we are now in the position to determine how the radiance measured by the sensor is affected by the presence of the atmosphere. First the total irradiance at the earth's surface now becomes, instead of (2.1)

$$E_G = E_{\Delta\lambda} T_\theta \cos\theta \, \Delta\lambda + E_D$$

where, for simplicity it has been assumed that the diffuse sky irradiance is not a function of wavelength (in the waveband of interest). The radiance therefore due to this global irradiance of the pixel becomes

$$L_T = \frac{R}{\pi} \{ E_{\Delta\lambda} T_\theta \cos\theta \, \Delta\lambda + E_D \}$$

Above the atmosphere the total radiance available to the sensor then becomes

$$L_s = \frac{R T_\phi}{\pi} \{ E_{\Delta\lambda} T_\theta \cos\theta \, \Delta\lambda + E_D \} + L_p \tag{2.4}$$

It is this quantity therefore that should be used in (2.3) to relate the digital count value to measured radiance.

2.1.2 Atmospheric Effects on Remote Sensing Imagery

A result of the scattering caused by the atmosphere is that fine detail in image data will be obscured. Consequently it is important in applications where one is dependent upon the limit of sensor resolution available, such as in urban studies, to take steps to correct for atmospheric effects.

It is important also to consider carefully the effects of the atmosphere on remote sensing systems with wide fields of view in which there will be an appreciable difference in atmospheric path length between nadir and the extremities of the swath. This will be of significance for example with aircraft scanners and satellite missions such as NOAA.

Finally, and perhaps most importantly, because both Rayleigh and Mie scattering are wavelength dependent the effects of the atmosphere will be different in the different wavebands of a given sensor system. In the case of the Landsat multispectral scanner the visible green band (0.5 to 0.6 µm) can be affected appreciably by comparison to the longest infrared band (0.8 to 1.1 µm). This leads to a loss in calibration of the set of brightnesses associated with a particular pixel.

Methods for correcting for the radiometric distortion caused by the atmosphere are discussed in Sect. 2.2.

2.1.3 Instrumentation Errors

Radiometric errors within a band and between bands can also be caused by the design and operation of the sensor system. Band to band errors from this source are normally ignored by comparison to band to band errors from atmospheric effects. However errors within a band can be quite severe and often require correction to render an image product useful.

The most significant of these errors is related to the detector system. An ideal radiation detector should have a transfer characteristic (radiation in, signal out) as shown in Fig. 2.2 a. This should be linear so that there is a proportional increase and decrease of signal with detected radiation level. Real detectors will have some degree of

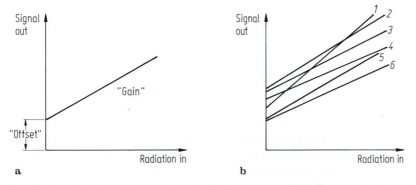

Fig. 2.2. a Transfer characteristic of a radiation detector; **b** Hypothetical mismatches in detector characteristics

nonlinearity (which is ignored here) and will also give a small signal out even when no radiation is being detected. Historically this is known as dark current and is related to the residual electronic noise in the system at any temperature above absolute zero; we will call it an "offset". The slope of the characteristic is frequently called its transfer gain or just simply "gain".

Most remote sensors involve a multitude of detectors. In the case of the Landsat MSS there are 6 per band, for the Landsat TM there are 16 per band and for the SPOT HRV there are 6000 in the panchromatic mode of operation. Each of these detectors will have slightly different transfer characteristics as described by their gains and offsets. The six for one band in Landsat MSS could appear, for example, as shown in Fig. 2.2 b.

In the case of scanners such as the TM and MSS these imbalances will lead to striping in the across swath direction as shown in Fig. 2.4 a. For the HRV longitudinal striping may occur.

2.2 Correction of Radiometric Distortion

In contrast to geometric correction, in which all sources of error are often rectified together, radiometric correction procedures must be specific to the nature of the distortion.

2.2.1 Detailed Correction of Atmospheric Effects

Rectifying image data to remove as much as possible the degrading effects of the atmosphere entails modelling the scattering and absorption processes that take place and establishing how these determine both the transmittances of the various paths and the different components of sky irradiance and path radiance. When available these can be used in (2.3) and (2.4) to relate the digital count values given for the pixels in each band of data C, to the true reflectance R of the surface being imaged. An example of how this can be done is given by Forster (1984) for the case of Landsat MSS data; Forster also gives source material and tables to assist in the computations. Some aspects of his example are given here to establish relative quantities.

Forster considers the case of a Landsat 2 MSS image in the wavelength range 0.8 to 1.1 μm acquired at Sydney Australia on 14 December 1980 at 9:05 a.m. local time. At the time of overpass the atmospheric conditions were

temperature	29 °C	
relative humidity	24%	
atmospheric pressure	1004 mbar	measured at 30 m above sea level.
visibility	65 km	

Based upon the equivalent mass of water vapour in the atmosphere (computed from temperature and humidity measurements) the absorption effect of water molecules was computed. This was the only molecular absorption mechanism considered significant. The measured value for visibility was used to estimate the effect of Mie scattering. Together with the known effect of Rayleigh scattering at that wavelength, these were combined to give the so-called total normal optical thickness of the atmosphere. Its value is for this example

$$\tau = 0.15$$

The transmittance of the atmosphere for an angle of incidence θ of the path is given by

$$T = \exp(-\tau \sec \theta)$$

Thus for a solar zenith angle of 38° (at the time of overpass) and a nadir viewing satellite we have (see Fig. 2.1)

$$T_\theta = 0.827$$
$$T_\phi = 0.861$$

In the waveband of interest Forster notes that the solar irradiance at the earth's surface in the absence of an atmosphere is $E_0 = 256\ \mathrm{Wm}^{-2}$. He further computes that the total global irradiance at the earth's surface is $186.6\ \mathrm{Wm}^{-2}$. Noting from (2.4) that the term in brackets is the global irradiance this leaves the total diffuse sky irradiance as $19.6\ \mathrm{Wm}^{-2}$ – i.e. about 10% of the global irradiance for this example.

Based upon correction algorithms given by Turner and Spencer (1972) which account for Rayleigh and Mie scattering and atmospheric absorption Forster computes the path radiance for this example as

$$L_p = 0.62\ \mathrm{Wm}^{-2}\,\mathrm{sr}^{-1}$$

so that (2.4) becomes

$$L_s = R_7\ 0.274\ (186.6) + 0.62$$
$$\text{i.e.}\quad L_s = 51.5\ R_7 + 0.62 \tag{2.5}$$

where the subscript on R refers to the band.

For the band 7 sensors on Landsat 2 at the time of overpass it can be established in (2.3) that

$$k = (L_{max} - L_{min})/C_{max}$$
$$= (39.1 - 1.1)/63\ \mathrm{Wm}^{-2}\,\mathrm{sr}^{-1}\ \text{per digital value}$$
$$= 0.603$$

so that (2.3) becomes

$$L_s = 0.603\ C_7 + 1.1\ \mathrm{Wm}^{-2}\,\mathrm{sr}^{-1}$$

which when combined with (2.5) gives

$$R_7 = 0.0118\ C_7 + 0.0094$$
$$\text{or}\qquad = 1.18\ C_7 + 0.94\ \%$$

This gives a means by which the % reflectance in band 7 can be computed from the digital count value available in the digital image data. By carrying out similar computations for the other three MSS bands the absolute and differential effects of the atmosphere can be removed. For band 5 for example

$$R_5 = 0.44\ C_5 + 0.5\ \%$$

Note that the effects of the atmosphere (and path radiance in particular) are greater for band 5. This is because of the increasing effect of scattering with decreasing wavelength. If all four MSS bands were considered the effect would be greatest in band 4 and least in band 7.

2.2.2 Bulk Correction of Atmospheric Effects

Frequently, detailed correction for the scattering and absorbing effects of the atmosphere is not required and often the necessary ancilliary information such as visibility and relative humidity is not readily available. In those cases, if the effect of the atmosphere is judged to be a problem in imagery, approximate correction can be carried out in the following manner. First it is assumed that each band of data for a given scene should have contained some pixels at or close to zero brightness value but that atmospheric effects, and especially path radiance, has added a constant value to

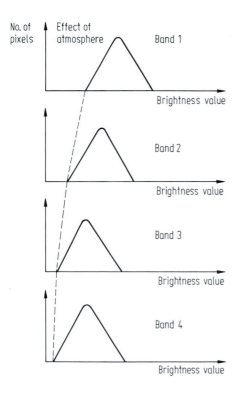

Fig. 2.3. Illustration of the effect of path radiance, resulting from atmospheric scattering, on the four histograms of Landsat MSS image data

each pixel in a band. Consequently if histograms are taken of each band (i.e. graphs of the number of pixels present as a function of brightness value for a given pixel) the lowest significant occupied brightness value will be non-zero as shown in Fig. 2.3. Moreover because path radiance varies as $\lambda^{-\alpha}$ (with α between 0 and 4 depending upon the extent of Mie scattering) the lowest occupied brightness value will be further from the origin for the lower wavelengths as depicted in Fig. 2.3 for the case of Landsat MSS data. Correction amounts first to identifying the amount by which each histogram is "shifted" in brightness away from the origin and then subtracting that amount from each pixel brightness in that band.

It is clear that the effect of atmospheric scattering as implied in the histograms of Fig. 2.3 is to lift the overall brightness value of an image in each band. In the case of a colour composite product (see Sect. 3.2) this will appear as a whitish-bluish haze. Upon correction in the manner just described this haze will be removed and the dynamic range of image intensity will be improved. Consequently the procedure of atmospheric correction outlined in this section is frequently referred to as *haze removal*.

2.2.3 Correction of Instrumentation Errors

Errors in relative brightness such as the within-band line striping referred to in Sect. 2.1.3 and shown in Fig. 2.4a can be rectified to a great extent in the following way. First it is assumed that the detectors used for data aquisition within a band produce signals statistically similar to each other. In other words if the means and standard deviations are computed for the signals recorded by the detectors then they should be the same. This requires the assumption that detail within a band doesn't change significantly over a distance equivalent to that of one scan covered by the set of the detectors (e. g. 474 m for the six scan lines of Landsats 1,2, 3 MSS). This is a reasonable assumption in terms of the mean and standard deviation of the pixel brightness, so that differences in those statistics among the detectors can be attributed to gain and offset mismatches as displayed in Fig. 2.2b. These mismatches can be detected by calculating pixel mean brightness and standard deviation using lines of image data known to come from a single detector. In the case of Landsat MSS this will require the data on every sixth line to be used. In a like manner five other measurements of mean brightness and standard deviation are computed as indications of the performances of the other five MSS detectors. Correction of radiometric mismatches among the detectors can then be effected by adopting one sensor as a standard and adjusting the brightness of all pixels recorded by each other detector so that their mean brightnesses and standard deviations match those of the standard detector. This can be done according to

$$y = \frac{\sigma_d}{\sigma_i} x + m_d - \frac{\sigma_d}{\sigma_i} m_i \qquad (2.6)$$

where x is the old brightness of a pixel and y is its new (destriped) value; m_d and σ_d are the reference values of mean brightness and standard deviation and m_i and σ_i are the mean and standard deviation of the detector under consideration. Alternatively an independent reference mean brightness and standard deviation can be used. This can allow a degree of contrast enhancement to be produced during radiometric correction.

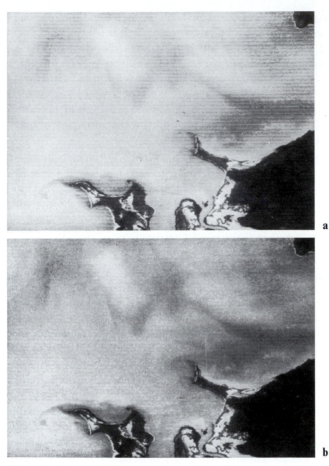

Fig. 2.4. a Landsat MSS visible green image showing severe line striping; **b** The same image after destriping by matching the mean brightnesses and standard deviations of each detector

The method described is frequently referred to as destriping. Figure 2.4 gives an example of destriping a Landsat MSS image in this manner.

The destriping effected by (2.6) is straightforward, but capable only of matching detector responses on the basis of means and standard deviations. A more complete destriping procedure should result if the histograms of the remaining detectors are matched fully to that of the reference detector using the methods of Sect. 4.5. This approach has been used by Weinreb et al. (1989) for destriping weather satellite imagery.

2.3 Sources of Geometric Distortion

There are potentially many more sources of geometric distortion of image data than radiometric distortion and their effects are more severe. They can be related to a number of factors, including

(i) the rotation of the earth during image acquisition,
(ii) the finite scan rate of some sensors,
(iii) the wide field of view of some sensors,
(iv) the curvature of the earth,
(v) sensor non-idealities,
(vi) variations in platform altitude, attitude and velocity, and
(vii) panoramic effects related to the imaging geometry.

It is the purpose of this section to discuss the nature of the distortions that arise from these effects; Sect. 2.4 discusses means by which the distortions can be compensated.

To appreciate why geometric distortion occurs, in some cases it is necessary to envisage how an image is formed from sequential lines of image data. If one imagines that a particular sensor records L lines of N pixels each then it would be natural to form the image by laying the L lines down successively one under the other. If the IFOV of the sensor has an aspect ratio of unity – i.e. the pixels are the same size along and across the scan – then this is the same as arranging the pixels for display on a square grid, such as that shown in Fig. 2.5. The grid intersections are the pixel positions and the spacing between those grid points is equal to the sensor's IFOV.

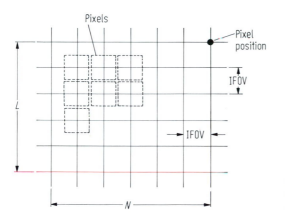

Fig. 2.5. Display grid commonly used to build up an image from the digital data stream of pixels generated by a sensor

2.3.1 Earth Rotation Effects

Line scan sensors such as the Landsat MSS and TM, and the NOAA AVHRR take a finite time to acquire a frame of image data. The same is true of push broom scanners such as the SPOT HRV. During the frame acquisition time the earth rotates from west to east so that a point imaged at the end of the frame would have been further to the west when recording started. Therefore if the lines of image data recorded were arranged for display in the manner of Fig. 2.5 the later lines would be erroneously displaced to the east in terms of the terrain they represent. Instead, to give the pixels their correct positions relative to the ground it is necessary to offset the bottom of the image to the west by the amount of movement of the ground during image acquisition, with all intervening lines displaced proportionately as depicted in Fig. 2.6. The amount by which the image has to be skewed to the west at the end of the frame depends upon the relative velocities of the satellite and earth and the length of the image frame recorded. An example is presented here for Landsats 1, 2, 3.

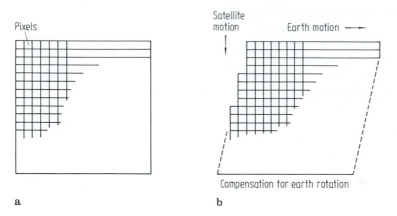

Fig. 2.6. The effect of earth rotation on scanner imagery. **a** Image formed according to Fig. 2.5 in which lines are arranged on a square grid; **b** Offset of successive lines to the west to correct for the rotation of the earth's surface during the frame acquisition time

The angular velocity of the satellite is $\omega_0 = 1.014\,\text{mrad s}^{-1}$ so that a nominal $L = 185$ km frame on the ground is scanned in

$$t_s = L/(r_e\omega_0) = 28.6 \text{ s}$$

where r_e is the radius of the earth (6.37816 Mm).

The surface velocity of the earth is given by

$$v_e = \omega_e r_e \cos\lambda$$

where λ is latitude and ω_e is the earth rotational velocity of 72.72 μrad s^{-1}. At Sydney, Australia $\lambda = 33.8°$ so that

$$v_e = 385.4 \text{ ms}^{-1}$$

During the frame acquisition time the surface of the earth moves to the east by

$$\Delta x_e = v_e t_s = 11.02 \text{ km at } 33.8\,°\text{S latitude}$$

This represents 6% of the frame size. Since the satellite does not pass directly north-south this movement has to be corrected by the inclination angle. At Sydney this is approximately 11° so that the effective sideways movement of the earth is

$$\Delta_x = \Delta x_e \cos 11° = 10.82 \text{ km}$$

Consequently if steps are not taken to correct an image from the first three Landsats for the effect of earth rotation then the image will contain a 6% skew distortion to the east.

2.3.2 Panoramic Distortion

For scanners used on spacecraft and aircraft remote sensing platforms the angular IFOV is constant. As a result the effective pixel size on the ground is larger at the extremities of the scan than at nadir, as illustrated in Fig. 2.7. In particular, if the IFOV is β and the pixel dimension at nadir is p then its dimension in the scan direction at a scan angle of θ as shown is

$$p_\theta = \beta h \sec^2 \theta = p \sec^2 \theta \qquad (2.7a)$$

where h is altitude. Its dimension across the scan line is $p \sec \theta$. For small values of θ these effects are negligible. For example, for Landsats 4 and 5 the largest value of θ is approximately $7.5°$ so that $p_\theta = 1.02\, p$. However for systems with larger fields of view, such as NOAA AVHRR and aircraft scanners, the effect can be quite severe. For an aircraft scanner with FOV $= 80°$ the distortion in pixel size along the scan line is $p_\theta = 1.70\, p$ – i.e. the region on the ground measured at the extremities of the scan is 70% larger laterally than the region sensed at nadir. When the image data is arranged to form an image, as in Fig. 2.5, the pixels are all written as the same size spots on a photographic emulsion or are used to excite the same area of phosphor on a colour

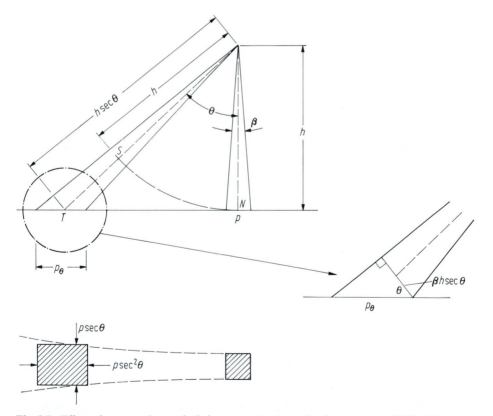

Fig. 2.7. Effect of scan angle on pixel size at constant angular instantaneous field of view

display device. Therefore the displayed pixels are equal across the scan line whereas the equivalent ground areas covered are not. This gives a compression of the image data towards its edges.

There is a second distortion introduced with wide field of view systems and that relates to pixel positions across the scan line. The scanner records pixels at constant angular increments and these are displayed on a grid of uniform centres, as in Fig. 2.5. However the spacings of the effective pixels on the ground increase with scan angle. For example if the pixels are recorded at an angular separation equal to the IFOV of the sensor then at nadir the pixels centres are spaced p apart. At a scan angle θ the pixel centres will be spaced $p\sec^2\theta$ apart as can be ascertained from Fig. 2.7. Thus by placing the pixels on a uniform display grid the image will suffer an across track compression. Again the effect for small angular field of view systems will be negligible in terms of the relative spacings of adjacent pixels. However when the effect is compounded to determine the location of a pixel at the swath edge relative to nadir the error can be significant. This can be determined by computing the arc SN in Fig. 2.7 S being the position to which the pixel at T would appear to be moved if the data is arrayed uniformly. It can be shown readily that $SN/TN = \theta/\tan\theta$ this being the degree of across track scale distortion. In the case of Landsats 1, 2 and 3 $(\theta/\tan\theta)_{\max} = 0.9966$. This indicates that a pixel at the swath edge (92.5 km from the sub-nadir point) will be 314 m out of position along the scan line compared with the ground if the pixel at nadir is in its correct location.

These panoramic effects lead to an interesting distortion in the geometry of large field of view systems. To see this consider the uniform mesh shown in Fig. 2.8 a. Suppose this represents a region on the ground being imaged. For simplicity the cells in the grid could be considered to be features on the ground. Because of the compression in the image data caused by displaying equal-sized pixels on a uniform grid as discussed in the foregoing, the uniform mesh will appear as shown in Fig. 2.8 b. Image pixels are recorded with a constant IFOV and at a constant angular sampling rate. The number of pixels recorded therefore over the outer grid cells in the along scan direction will be smaller than over those near nadir. In the along track direction there is no variation of pixel spacing or density with scan angle as this is established by the forward motion of the platform. Rather pixels near the swath edges will contain information in common owing to the overlapping IFOV.

Linear features such as roads at an angle to the scan direction as shown in Fig. 2.8 will appear bent in the displayed image data because of the along scan compression

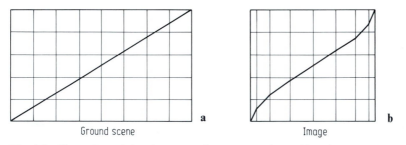

Ground scene   a   Image   b

Fig. 2.8. Illustration of the along scan line compression evident in constant angular IFOV and constant angular scan rate sensors. This leads to so-called S-bend distortion, as shown

effect. Owing to the change in shape caused, the distortion is frequently referred to as *S*-bend distortion and can be a common problem with aircraft line scanners. Clearly, not only linear features are affected; rather the whole image detail near the swath edges is distorted in this manner.

2.3.3 Earth Curvature

Aircraft scanning systems, because of their low altitude (and thus the small absolute swath width of their image data), are not affected by earth curvature. Neither are space systems such as Landsat and SPOT, again because of the narrowness of their swaths. However wide swath width spaceborne imaging systems are affected. For NOAA with a swath width of 2700 km and an altitude of 833 km it can be shown that the deviation of the earth's surface from a plane amounts to about 2.3 % over the swath, which seems insignificant. However it is the inclination of the earth's surface over the swath that causes the greater effect. At the edges of the swath the area of the earth's surface viewed at a given angular IFOV is larger than if the curvature of the earth is ignored. The increase in pixel size can be computed by reference to the geometry of Fig. 2.9. The pixel dimension in the across track direction normal to the direction of the sensor is $\beta[h+r_e(1-\cos\phi)]\sec\theta$ as shown. The geometry of Fig. 2.9 then shows that the effective pixel size on the inclined earth's surface is

$$p_c = \beta[h+r_e(1-\cos\phi)]\sec\theta\sec(\theta+\phi) \tag{2.7b}$$

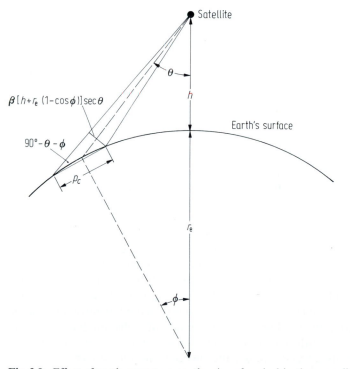

Fig. 2.9. Effect of earth curvature on the size of a pixel in the scan direction (across track)

where βh is the pixel size at nadir and ϕ is the angle subtended at the centre of the earth. Note that this expression reduces to (2.7a) if $\phi = 0$ – i.e. if earth curvature is considered negligible. Using the NOAA satellite as an example $\theta = 54°$ at the edge of the swath and $\phi = 12°$. This shows that the effective pixel size in the along scan direction is 2.89 times larger than that at nadir when earth curvature is ignored, but is 4.94 times that at nadir when the effect of earth curvature is included. This demonstrates that earth curvature introduces a significant additional compressive distortion in the image data acquired by satellites such as NOAA when an image is constructed on a uniform grid such as that in Fig. 2.5. The effect of earth curvature in the along track direction is negligible.

2.3.4 Scan Time Skew

Mechanical line scanners such as the Landsat MSS and TM require a finite time to scan across the swath. During this time the satellite is moving forward leading to a skewing in the along track direction. As an illustration of the magnitude of the effect, the time require to record one MSS scan line of data is 33 ms. During this time the satellite travels forward by 213 m at its equivalent ground velocity of 6.467 km s^{-1}. As a result the end of the scan line is advanced by this amount compared with its start.

2.3.5 Variations in Platform Altitude, Velocity and Attitude

Variations in the elevation or altitude of a remote sensing platform lead to a scale change at constant angular IFOV and field of view; the effect is illustrated in Fig. 2.10a for an increase in altitude with travel at a rate that is slow compared with a frame acquisition time. Similarly, if the platform forward velocity changes, a scale change occurs in the along track direction. This is depicted in Fig. 2.10b again for a change that occurs slowly. For a satellite platform, orbit velocity variations can result from orbit eccentricity and the non-sphericity of the earth.

Platform attitude changes can be resolved into yaw, pitch and roll during forward travel. These lead to image rotation, along track and across track displacement as noted in Fig. 2.10c–e.

While these variations can be described mathematically, at least in principle, a knowledge of the platform ephemeris is required to enable their magnitudes to be computed. In the case of satellite platforms ephemeris information is often telemetered to ground receiving stations. For the Landsat system this is used to apply corrections before the data is distributed.

Attitude variations in aircraft remote sensing systems can potentially be quite significant owing to the effects of atmospheric turbulence. These can occur over a short time, leading to localised distortions in aircraft scanner images. Frequently aircraft roll is compensated for in the data stream. This is made possible by having a data window that defines the swath width; this is made smaller than the complete scan of data over the sensor field of view. A gyro mounted on the sensor is then used to move the position of the data window along the total scan line as the aircraft rolls. Pitch and yaw are generally not corrected unless the sensor is mounted on a three axis stabilized platform.

A comprehensive discussion of the nature and effects of aircraft scanner distortion is given by Silva (1978).

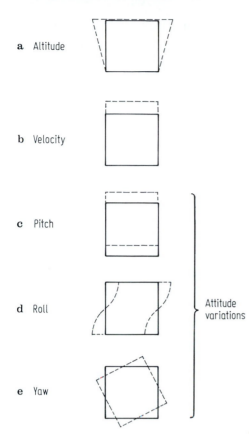

a Altitude

b Velocity

c Pitch

d Roll

Attitude
variations

e Yaw

Fig. 2.10. Effect of platform position and attitude errors on the region of earth being imaged, when those errors occur slowly compared with image acquisition

2.3.6 Aspect Ratio Distortion

The aspect ratio of an image (that is, its scale vertically compared with its scale horizontally) can be distorted by mechanisms that lead to overlapping IFOV's. The most notable example of this occurs with the Landsat multispectral scanner. As discussed in Sect. 1.3.4 samples are taken across a scan line too quickly compared with the IFOV. This leads to pixels having 56 metre centres but sampled with an IFOV of 79 m. Consequently the effective pixel size is 79 m × 56 m and thus is not square. As a result if the pixels recorded by the multispectral scanner are displayed on the square grid of Fig. 2.5 the image will be too wide for its height when related to the corresponding region on the ground. The magnitude of the distortion is $79/56 = 1.411$ so that this is quite a severe error and must be corrected for most applications.

A similar distortion can occur with aircraft scanners if the velocity of the aircraft is not matched to the scanning rate of the sensor. Either underscanning or overscanning can occur leading to distortion in the alongtrack scale of the image.

2.3.7 Sensor Scan Nonlinearities

Line scanners that make use of rotating mirrors, such as the NOAA AVHRR and aircraft scanners, have a scan rate across the swath that is constant, to the extent that the

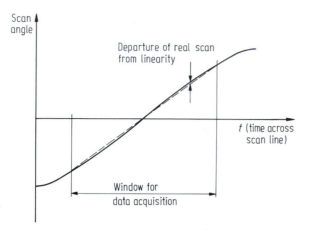

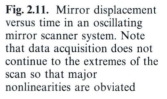

Fig. 2.11. Mirror displacement versus time in an oscillating mirror scanner system. Note that data acquisition does not continue to the extremes of the scan so that major nonlinearities are obviated

scan motor speed is constant. Systems that use an oscillating mirror however, such as the Landsat multispectral scanner, incur some nonlinearity in scanning near the swath edges owing to the need for the mirror to slow down and change directions. This effect is depicted in Fig. 2.11. According to Anuta (1973) this can lead to a maximum displacement in pixel position compared with a perfectly linear scan of about 395 m.

2.4 Correction of Geometric Distortion

There are two techniques that can be used to correct the various types of geometric distortion present in digital image data. One is to model the nature and magnitude of the sources of distortion and use these models to establish correction formulae. This technique is effective when the types of distortion are well characterized, such as that caused by earth rotation. The second approach depends upon establishing mathematical relationships between the addresses of pixels in an image and the corresponding coordinates of those points on the ground (via a map). These relationships can be used to correct the image geometry irrespective of the analyst's knowledge of the source and type of distortion. This procedure will be treated first since it is the most commonly used and, as a technique, is independent of the platform used for data acquisition. Correction by mathematical modelling is discussed later. Before proceeding it should be noted that each band of image data has to be corrected. However since it can often be assumed that the bands are well registered to each other, steps taken to correct one band in an image, can be used on all remaining bands.

2.4.1 Use of Mapping Polynomials for Image Correction

An assumption that is made in this procedure is that there is available a map of the region corresponding to the image, that is correct geometrically. We then define two cartesian coordinate systems as shown in Fig. 2.12. One describes the location of points in the map (x, y) and the other coordinate system defines the location of pixels in the image (u, v). Now suppose that the two coordinate systems can be related via a pair of mapping functions f and g so that

$$u = f(x, y) \qquad (2.8\,a)$$
$$v = g(x, y) \qquad (2.8\,b)$$

If these functions are known then we could locate a point in the image knowing its position on the map. In principle, the reverse is also true. With this ability we could build up a geometrically correct version of the image in the following manner. First we define a grid over the map to act as the grid of pixel centres in the corrected image. This grid is parallel to, or indeed could in fact be, the map coordinate grid itself, described by latitudes and longitudes, UTM coordinates and so on. For simplicity we will refer to this grid as the display grid; by definition this is geometrically correct. We then move over the display grid pixel centre by pixel centre and use the mapping functions above to find the corresponding pixel in the image for each display grid position. Those pixels are then placed on the display grid. At the conclusion of the process we have a geometrically correct image built up on the display grid utilizing the original image as a source of pixels.

While the process is a straightforward one there are some practical difficulties that must be addressed. First we do not know the explicit form of the mapping functions of (2.8). Secondly, even if we did, they may not point exactly to a pixel in the image corresponding to a display grid location; instead some form of interpolation may be required.

2.4.1.1 Mapping Polynomials and Ground Control Points

Since explicit forms for the mapping functions in (2.8) are not known they are generally chosen as simple polynomials of first, second or third degree. For example, in the case of second degree (or order)

$$u = a_0 + a_1 x + a_2 y + a_3 xy + a_4 x^2 + a_5 y^2 \qquad (2.9\,a)$$
$$v = b_0 + b_1 x + b_2 y + b_3 xy + b_4 x^2 + b_5 y^2 \qquad (2.9\,b)$$

Sometimes orders higher than third are used but care must be taken to avoid the introduction of worse errors than those to be corrected, as will be noted later.

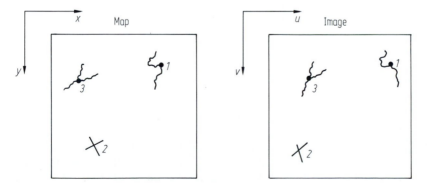

Fig. 2.12. Coordinate systems defined for the image and map, along with the specification of ground control points

If the coefficients a_i and b_i in (2.9) were known then the mapping polynomials could be used to relate any point in the map to its corresponding point in the image as in the foregoing discussion. At present however these coefficients are unknown. Values can be estimated by identifying sets of features on the map that can also be identified on the image. These features, often referred to as *ground control points* (G.C.P's) are well-defined and spatially small and could be road intersections, airport runway intersections, bends in rivers, prominent coastline features and the like. Enough of these are chosen (as pairs – on the map and image as depicted in Fig. 2.12) so that the polynomial coefficients can be estimated by substitution into the mapping polynomials to yield sets of equations in those unknowns. Eqs. (2.9) show that the minimum number required for second order polynomial mapping is six. Likewise a minimum of three is required for first order mapping and ten for third order mapping. In practice however significantly more than these are chosen and the coefficients are evaluated using least squares estimation. In this manner any control points that contain significant positional errors either on the map or in the image will not have an undue influence on the polynomial coefficients.

2.4.1.2 Resampling

Having determined the mapping polynomials explicitly by use of the ground control points the next step is to find points in the image corresponding to each location in the pixel grid previously defined over the map. The spacing of that grid is chosen according to the pixel size required in the corrected image and need not be the same as that in the original geometrically distorted version. For the moment suppose that the points located in the image correspond exactly to image pixel centres. Then those pixels are simply transferred to the appropriate locations on the display grid to build up the rectified image. This is the case in Fig. 2.13.

2.4.1.3 Interpolation

As is to be expected, grid centres from the map-registered pixel grid will not usually project to exact pixel centre locations in the image, as shown in Fig. 2.13, and some decision has to be made therefore about what pixel brightness value should be chosen for placement on the new grid. Three techniques can be used for this purpose.

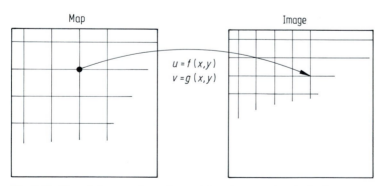

Fig. 2.13. Use of the mapping polynomials to locate points in the image corresponding to display grid positions

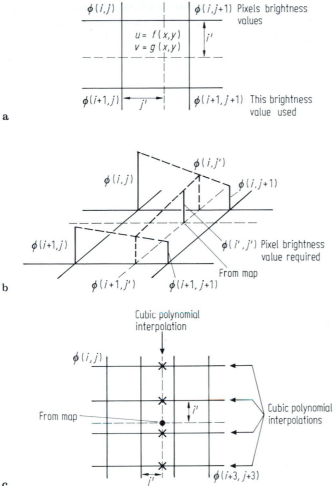

Fig. 2.14. Determining a display grid pixel brightness by **a** nearest neighbour resampling, **b** bilinear interpolation and **c** cubic convolution interpolation; i, j etc. are discrete values of u and v

Nearest neighbour resampling simply chooses the actual pixel that has its centre nearest the point located in the image, as depicted in Fig. 2.14a. This pixel is then transferred to the corresponding display grid location. This is the preferred technique if the new image is to be classified since it then consists of the original pixel brightnesses, simply rearranged in position to give a correct image geometry.

Bilinear interpolation uses three linear interpolations over the four pixels that surround the point found in the image corresponding to a given display grid position. The process is illustrated in Fig. 2.14 b. Two linear interpolations are performed along the scan lines to find the interpolants $\varphi(i, j')$ and $\varphi(i + 1, j')$ as shown. These are given by

$$\phi(i, j') = j' \phi(i, j+1) + (1 - j') \phi(i, j)$$
$$\phi(i+1, j') = j' \phi(i+1, j+1) + (1 - j') \phi(i+1, j)$$

where ϕ is pixel brightness and (i', j') is the position at which an interpolated value for brightness is required. The position is measured with respect to (i, j) and assumes a grid spacing of unity in both directions. The final step is to interpolate linearly over $\phi(i, j')$ and $\phi(i+1, j')$ to give

$$\phi(i', j') = (1 - i')\{j' \phi(i, j+1) + (1 - j') \phi(i, j)\}$$
$$+ i'\{j' \phi(i+1, j+1) + (1 - j') \phi(i+1, j)\} \qquad (2.10)$$

Cubic convolution interpolation uses the surrounding sixteen pixels. Cubic polynomials are fitted along the four lines of four pixels surrounding the point in the image, as depicted in Fig. 2.14c to form four interpolants. A fifth cubic polynomial is then fitted through these to synthesise a brightness value for the corresponding location in the display grid.

The actual form of polynomial that is used for the interpolation is derived from considerations in sampling theory and issues concerned with constructing a continuous function (i.e. interpolating) from a set of samples. These are beyond the scope of this treatment but can be appreciated using the material presented in Chap. 7. An excellent treatment of the problem has been given by Shlien (1979), who discusses several possible cubic polynomials that could be used for the interpolation process and who also demonstrates that the interpolation is a convolution operation. Based on the choice of a suitable polynomial (attributable to Simon (1975)) the algorithm that is used to perform cubic convolution interpolation is (Moik, 1980):

$$\phi(i, j') = j'\{j'(j'[\phi(i, j+3) - \phi(i, j+2) + \phi(i, j+1) - \phi(i, j)]$$
$$+ [\phi(i, j+2) - \phi(i, j+3) - 2\phi(i, j+1) + 2\phi(i, j)])$$
$$+ [\phi(i, j+2) - \phi(i, j)]\}$$
$$+ \phi(i, j+1) \qquad (2.11a)$$

This expression is evaluated for each of the four lines of four pixels depicted in Fig. 2.14c to yield the four interpolants $\phi(i, j')$, $\phi(i+1, j')$, $\phi(i+2, j')$, $\phi(i+3, j')$. These are then interpolated vertically according to

$$\phi(i', j') = i'\{i'(i'[\phi(i+3, j') - \phi(i+2, j') + \phi(i+1, j') - \phi(i, j')]$$
$$+ [\phi(i+2, j') - \phi(i+3, j') - 2\phi(i+1, j') + 2\phi(i, j')])$$
$$+ [\phi(i+2, j') - \phi(i, j')]\}$$
$$+ \phi(i+1, j') \qquad (2.11b)$$

Cubic convolution interpolation, or resampling, yields an image product that is generally smooth in appearance and is often used if the final product is to be treated by photointerpretation. However since it gives pixels on the display grid, with brightnesses that are interpolated from the original data, it is not recommended if classification is to follow since the new brightness values may be slightly different to the actual radiance values detected by the satellite sensors.

2.4.1.4 Choice of Control Points

Enough well defined control point pairs must be chosen in rectifying an image to ensure that accurate mapping polynomials are generated. However care must also be given to

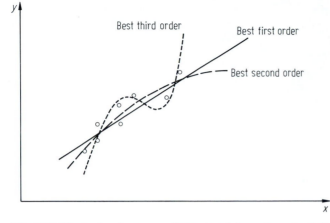

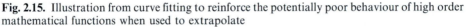

Fig. 2.15. Illustration from curve fitting to reinforce the potentially poor behaviour of high order mathematical functions when used to extrapolate

the locations of the points. A general rule is that there should be a distribution of control points around the edges of the image to be corrected with a scattering of points over the body of the image. This is necessary to ensure that the mapping polynomials are well-behaved over the image. This concept can be illustrated by considering an example from curve fitting. While the nature of the problem is different the undesirable effects that can be generated are similar. In Fig. 2.15 is illustrated a set of data points in a graph through which first order (linear), second order and third order curves are depicted. Note that as the order is higher the curves pass closer to the points. However if it is presumed that the data would have continued for larger values of x with much the same trend as apparent in the points plotted then clearly the linear fit will extrapolate moderately acceptably. In contrast the cubic curve can deviate markedly from the trend when used as an extrapolator. This is essentially true in geometric correction of image data: while the higher order polynomials will be accurate in the vicinity of the control points themselves, they can lead to significant errors, and thus image distortions, for regions of images outside the range of the control points. This is illustrated in the example of Sect. 2.5.4.

2.4.1.5 Example of Registration to a Map Grid

To illustrate the above techniques a small segment of a Landsat MSS image of Sydney, Australia was registered to a map of the region.

It is important that the map has a scale not too different from the scale at which the image data is considered useful. Otherwise the control point pairs may be difficult to establish. In this case a map at 1:250,000 scale was used. The relevant segment is shown reproduced in Fig. 2.16, along with the portion of image to be registered. Comparison of the two reveals the geometric distortion of the image. Eleven control points were chosen for the registration, with the coordinates shown in Table 2.1. Their UTM map coordinates were specified in this exercise by placing the map on a digitizing table, although they could have been read from the map and entered manually. The former

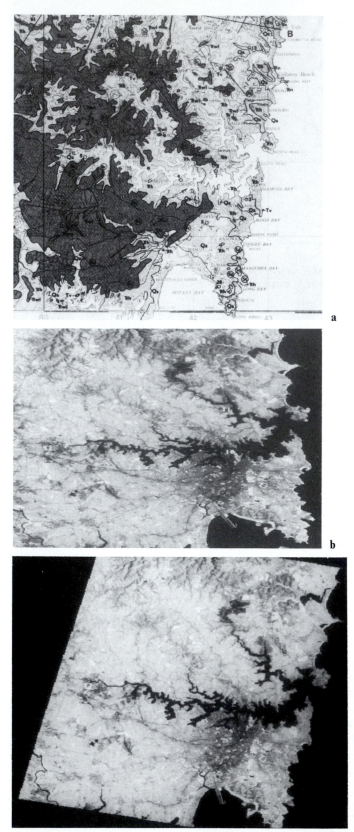

a

b

c

method is substantially more convenient and often more accurate if the digitizing table facility is available.

Using the set of control points, second degree mapping polynomials were generated. To test the effectiveness of these in transferring pixels from the raw image grid to the map display grid, the software system used in the exercise (Dipix Systems Ltd R-STREAM) *computes* the UTM coordinates of the control points from their pixel coordinates in the image. These are compared with the actual UTM coordinates and the differences (residuals) calculated in both directions. A root mean square of all the residuals is then computed in both directions (easting and northing) as shown in Table 2.1, giving an overall impression of the accuracy of the mapping process. In this case the set of control points is seen to lead to an average positional error of 56 m in easting and 63 m in northing, which is smaller than a pixel size in equivalent ground metres and thus would be considered acceptable. At this stage the table can be inspected to see if any individual control point has residuals that are unacceptably high. It could be assumed that this is a result of poor placement; if so it could be re-entered and the polynomial recalculated. If changing that control point leaves the residuals unchanged it may be that there is local distortion in that particular region of the image. A choice has to be made then as to whether the control point should be used to give a degree of correction there, that might also influence the remainder of the image, or whether it should be removed and leave that region in error.

In this example cubic convolution resampling was used in producing an image on a 50 m grid by means of the pair of second order mapping polynomials. The result is shown in Fig. 2.16.

Table 2.1. Control points used in image to map registration example

GCP no.	Image pixel	Image line	Map easting actual	Map easting est.	Map easting residual	Map northing actual	Map northing est.	Map northing residual
1	1909.	1473.	432279.	432230.1	49.4	836471.	836410.1	60.7
2	1950.	1625.	431288.	431418.0	−130.1	822844.	822901.4	−56.9
3	1951.	1747.	428981.	428867.9	112.6	812515.	812418.2	96.8
4	1959.	1851.	427164.	427196.9	−33.2	803313.	803359.4	−46.7
5	1797.	1847.	417151.	417170.3	−18.9	805816.	805759.3	57.1
6	1496.	1862.	397860.	397871.6	−11.2	808128.	808187.2	−59.6
7	1555.	1705.	404964.	404925.8	38.6	821084.	820962.6	121.6
8	1599.	1548.	411149.	411138.5	10.5	833796.	833857.3	−61.1
9	1675.	1584.	415057.	415129.0	−72.4	829871.	829851.1	19.8
10	1829.	1713.	422019.	421986.6	32.7	816836.	816884.5	−48.1
11	1823.	1625.	423530.	423507.8	22.0	824422.	824504.8	−83.2

Standard error in easting = 55.92 m
Standard error in northing = 63.06 m

Fig. 2.16. a Map and **b** Landsat MSS image segment to be registered. The result obtained from second order mapping polynomials is shown in **c**

2.4.2 Mathematical Modelling

If a particular distortion in image geometry can be represented mathematically then the mapping functions in (2.8) can be specified explicitly. This obviates the need to choose arbitary polynomials as in (2.9) and to use control points to determine the polynomial coefficients. In this section some of the more common distortions are treated from this point of view. However rather than commence with expressions that relate image coordinates (u, v) to map coordinates (x, y) it is probably simpler conceptually to start the other way around, i.e. to model what the true (map) positions of pixels should be given their positions in an image. This expression can then be inverted if required to allow the image to be resampled on to the map grid.

2.4.2.1 Aspect Ratio Correction

The easiest source of distortion to model is that caused by the 56 m equivalent ground spacing of the 79 m × 79 m equivalent pixels in the Landsat multispectral scanner. As noted in Sect. 2.3.6 this leads to an image that is too wide for its height by a factor of $79/56 = 1.411$. Consequently to produce a geometrically correct image either the vertical dimension has to be expanded by this amount or the horizontal dimension must be compressed. We will consider the former. This requires that the pixel axis horizontally be left unchanged (i.e. $x = u$), but that the axis vertically be scaled (i.e. $y = 1.411\,v$). This can be expressed conveniently in matrix notation as

$$\begin{bmatrix} x \\ y \end{bmatrix} = \begin{bmatrix} 1 & 0 \\ 0 & 1.411 \end{bmatrix} \begin{bmatrix} u \\ v \end{bmatrix}. \tag{2.12}$$

One way of implementing this corection would be to add extra lines of pixel data to expand the vertical scale. This could be done by duplicating four lines in every ten. Alternatively, and more precisely, (2.12) can be inverted to give

$$\begin{bmatrix} u \\ v \end{bmatrix} = \begin{bmatrix} 1 & 0 \\ 0 & 0.709 \end{bmatrix} \begin{bmatrix} x \\ y \end{bmatrix}. \tag{2.13}$$

Thus, as with the techniques of the previous section, a display grid is defined over the map (with coordinates (x, y)) and (2.13) is used to find the corresponding location in the image (u, v). The interpolation techniques of Sect. 2.4.1.3 are then used to generate brightness values for the display grid pixels.

2.4.2.2 Earth Rotation Skew Correction

To correct for the effect of earth rotation it is necessary to implement a shift of pixels to the left that is dependent upon the particular line of pixels measured with respect to the top of the image. Their line addresses as such (v) are not affected. Using the results of Sect. 2.3.1, these corrections are implemented by

$$\begin{bmatrix} x \\ y \end{bmatrix} = \begin{bmatrix} 1 & \alpha \\ 0 & 1 \end{bmatrix} \begin{bmatrix} u \\ v \end{bmatrix}$$

with $\alpha = -0.058$ for Sydney, Australia. Again this can be implemented in an approximate sense by making one pixel shift to the left every 17 lines of image data measured down from the top, or alternatively the expression can be inverted to give

$$\begin{bmatrix} u \\ v \end{bmatrix} = \begin{bmatrix} 1 & -\alpha \\ 0 & 1 \end{bmatrix} \begin{bmatrix} x \\ y \end{bmatrix} = \begin{bmatrix} 1 & 0.058 \\ 0 & 1 \end{bmatrix} \begin{bmatrix} x \\ y \end{bmatrix} \tag{2.14}$$

which again is used with interpolation procedures from Sect. 2.4.1.3 to generate display grid pixels.

2.4.2.3. Image Orientation to North-South

Although not strictly a geometric distortion it is an inconvenience to have an image that is corrected for most major effects but is not oriented vertically in a north-south direction. It will be recalled for example that the Landsat orbits in particular are inclined to the north-south line by about 9°. (This of course is different with different latitudes). To rotate an image by an angle ζ in the counter – or anticlockwise direction (as required in the case of Landsat) it is easily shown that (Foley and Van Dam 1982)

$$\begin{bmatrix} x \\ y \end{bmatrix} = \begin{bmatrix} \cos\zeta & \sin\zeta \\ -\sin\zeta & \cos\zeta \end{bmatrix} \begin{bmatrix} u \\ v \end{bmatrix}$$

so that

$$\begin{bmatrix} u \\ v \end{bmatrix} = \begin{bmatrix} \cos\zeta & -\sin\zeta \\ \sin\zeta & \cos\zeta \end{bmatrix} \begin{bmatrix} x \\ y \end{bmatrix} \tag{2.15}$$

2.4.2.4 Correction of Panoramic Effects

The discussion in Sect. 2.3.2 makes note of the pixel positional error that results from scanning with a fixed IFOV at a constant angular rate. In terms of map and image coordinates, the distortion can be described by

$$\begin{bmatrix} x \\ y \end{bmatrix} = \begin{bmatrix} \tan\theta/\theta & 0 \\ 0 & 1 \end{bmatrix} \begin{bmatrix} u \\ v \end{bmatrix}$$

where θ is the instantaneous scan angle, which in turn can be related to x or u, viz. $x = h\tan\theta$, $u = h\theta$, where h is altitude. Consequently resampling can be carried out according to

$$\begin{bmatrix} u \\ v \end{bmatrix} = \begin{bmatrix} \theta\cot\theta & 0 \\ 0 & 1 \end{bmatrix} \begin{bmatrix} x \\ y \end{bmatrix} = \begin{bmatrix} h\tan^{-1}\left(\frac{x}{h}\right)/x & 0 \\ 0 & 1 \end{bmatrix} \begin{bmatrix} x \\ y \end{bmatrix} \tag{2.16}$$

2.4.2.5 Combining the Corrections

Clearly any exercise in image correction usually requires several distortions to be rectified. Using the techniques in Sect. 2.4.1 it is assumed that all sources are rectified simultaneously. When employing mathematical modelling, a correction matrix has to be devised for each source considered important, as in the preceding sub-sections, and the set of matrices combined. For example if the aspect ratio of a Landsat MSS image is corrected first, followed by correction of the effect of earth rotation, then the following single linear transformation can be established for resampling.

$$\begin{bmatrix} x \\ y \end{bmatrix} = \begin{bmatrix} 1 & \alpha \\ 0 & 1 \end{bmatrix} \begin{bmatrix} 1 & 0 \\ 0 & 1.411 \end{bmatrix} \begin{bmatrix} u \\ v \end{bmatrix}$$
$$= \begin{bmatrix} 1 & 1.411\alpha \\ 0 & 1.411 \end{bmatrix} \begin{bmatrix} u \\ v \end{bmatrix}$$

which for $\alpha = -0.058$ (at Sydney) gives

$$\begin{bmatrix} u \\ v \end{bmatrix} = \begin{bmatrix} 1 & 0.058 \\ 0 & 0.709 \end{bmatrix} \begin{bmatrix} x \\ y \end{bmatrix}$$

2.5 Image Registration

2.5.1 Georeferencing and Geocoding

Using the correction techniques of the preceding sections an image can be registered to a map coordinate system and therefore have its pixels addressable in terms of map coordinates (eastings and northings, or latitudes and longitudes) rather than pixel and line numbers. Other spatial data types, such as geophysical measurements, image data from other sensors and the like, can be registered similarly to the map thus creating a georeferenced integrated spatial data base of the type used in a geographic information system.

Expressing image pixel addresses in terms of a map coordinate base is often referred to as geocoding; ultimately it would be anticipated that remote sensing image data could be purchased according to bounds expressed in map coordinates rather than in scenes or frames.

2.5.2 Image to Image Registration

Many applications of remote sensing image data require two or more scenes of the same geographical region, acquired at different dates, to be processed together. Such a situation arises for example when changes are of interest, in which case registered images allow a pixel by pixel comparison to be made.

Two images can be registered to each other by registering each to a map coordinate base separately, in the manner demonstrated in the previous section. Alternatively, and particularly if georeferencing is not important, one image can be chosen as a master to which the other, known as the slave, is to be registered. Again the techniques of Sect. 2.4 are used, however the coordinates (x, y) are now the pixel coordinates in the master image rather than the map coordinates. As before (u, v) are the coordinates of the image to be registered (i.e. the slave). An advantage in image to image registration is that only one registration step is required in comparison to two if both are taken back to a map base. Furthermore an artifice known as a sequential similarity detection algorithm can be used to assist in accurate co-location of control point pairs.

2.5.3 Sequential Similarity Detection Algorithm

A correlation procedure is of value in locating the position of a control point in the master image having identified it in the slave. Known as a sequential similarity detection algorithm (SSDA), the technique has several variations, as treated by Bernstein (1983). Only one specific implementation is considered here to illustrate the nature of the method.

Suppose a control point has been chosen in the slave image and it is necessary to determine its counterpart in the master image. In principle a rectangular sample of pixels surrounding the control point in the slave image can be extracted as a window to be correlated with the master image. Because of the spatial properties of the pair of images near the control points a high correlation should occur when the slave window is located over its exact counterpart region in the master, and thus the master location of the control point is identified. Obviously it is not necessary to move the slave window over the complete master image since the user knows approximately where the control point should occur in the master. Consequently it is only necessary to specify a search region in the neighbourhood of the approximate location. Software systems that provide this option allow the user to choose both the size of the window of pixels from the slave image control point neighbourhood and the size of the search region in the master image over which the window of slave pixels is moved to detect an acceptable correlation.

The correlation measure used need not be sophisticated. Indeed a simple similarity check that can be used is to compute the sum of the absolute differences of the slave and master image pixel brightnesses over the window, for each possible location of the window in the search region. The location that gives the smallest absolute difference defines the control point position as that pixel at the current centre of the window. Obviously the sensitivity of the method will be reduced if there is a large average difference in brightness between the two images — such as that owing to seasonal variations. A refinement therefore is to compute the summed absolute difference of the pixel brightnesses relative to their respective means in the search window.

Clearly the use of techniques such as these to locate control points depends upon there not being major changes of an uncorrelated nature between the scenes in the vicinity of a control point being tested. For example a vegetation flush due to rainfall in part of the search window can lead to an erroneous location. Nevertheless with a judicious choice of window size and search region, measures such as SSDA can give very effective guidance to the user, especially when available on an interactive image processing facility.

2.5.4 Example of Image to Image Registration

To illustrate image to image registration, but more particularly to see clearly the effect of control point distribution and the significance of the order of the mapping polynomials to be used for registration, two segments of Landsat MSS infrared image data in the northern suburbs of Sydney were chosen. One was acquired on December 29, 1979 and was used as the master. The other was acquired on December 14, 1980 and was used as the slave image. These are shown in Fig. 2.17 wherein careful inspection shows the differences in image geometry.

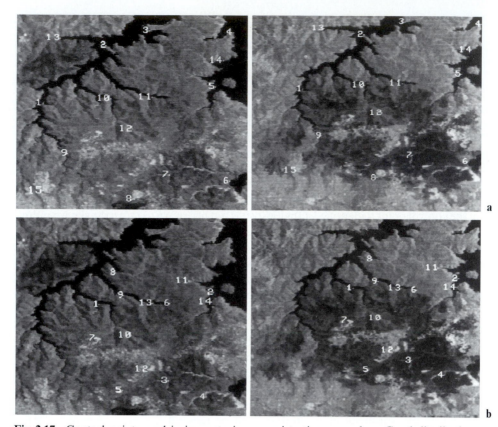

Fig. 2.17. Control points used in image to image registration example. **a** Good distribution; **b** Poor distribution

Two sets of control points were chosen. In one the points were distributed as nearly as possible in a uniform manner around the edge of the image segment as shown in Fig. 2.17 a, with some points located across the centre of the image. This set would be expected to give a reasonable registration of the images. The second set of control points was chosen injudiciously, closely grouped around one particular region, to illustrate the resampling errors that can occur. These are shown in Fig. 2.17 b. In both cases the control point pairs were co-located with the assistance of a sequential similarity detection algorithm. This worked well particularly for those control points around the coastal and river regions where the similarity between the images is unmistakable. To minimise tidal influences on the location of control points, those on water boundaries were chosen as near as possible to be on headlands, and certainly were never chosen at the end of inlets.

For both sets of control points third degree mapping polynomials were used along with cubic convolution resampling. As expected the first set of points led to an acceptable registration of the images whereas the second set gave a good registration in the immediate neighbourhood of the points but beyond them produced gross distortion.

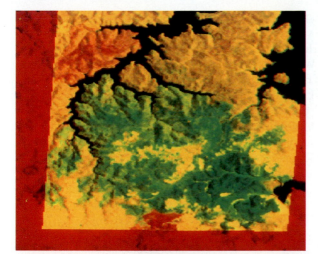

a

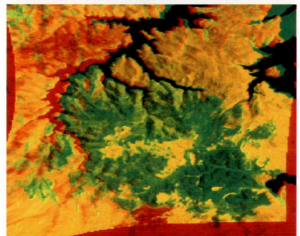

b

c

Fig. 2.18. a Registration of 1980 image (green) with 1979 image (red) using the control points of Fig. 2.17a, with third order mapping polynomials; **b** Third order mapping of 1980 image (green) to 1979 image (red) using the control points of Fig. 2.17b; **c** As for **b** but using first order mapping polynomials

The adequacy of the registration process can be assessed visually if the master and resampled slave images can be superimposed in different colours using a colour display system. Fig. 2.18a and 2.18b show the master image in red with the resampled slave image superimposed in green. Where good registration has been achieved the resultant is yellow (with the exception of regions of gross dissimilarity in pixel brightness – in this case associated with fire burns). However misregistration shows quite graphically by a red-green separation. This is particularly noticeable in Fig. 2.18b where the poor extrapolation obtained with third order mapping is demonstrated.

The exercise using the poor set of control points (Fig. 2.17b) was repeated. However this time only first order mapping polynomials were used. While these obviously will not remove non-linear differences between the images and will give poorer matches at the control points themselves, they are well behaved in extrapolation beyond the vicinity of the control points and lead to an acceptable registration as shown in Fig. 2.18c.

2.6 Miscellaneous Image Geometry Operations

While the techniques of Sect. 2.4 and 2.5 have been devised for treating errors in image geometry and for registering sets of images, and images to maps, they can be exploited also for performing intentional changes to image geometry. Image rotation and scale changing are chosen here as illustrations.

2.6.1 Image Rotation

Rotation of image data by an angle about the pixel grid can be useful for a number of applications. Most often it is used to align the pixel grid, and thus the image, to a north-south orientation as treated in Sect. 2.4.2.3. However the transformation in (2.15) is perfectly general and can be used to rotate any image in an anticlockwise sense by any specified angle ζ.

2.6.2 Scale Changing and Zooming

The scales of an image in both the vertical and horizontal directions can be altered by the transformation.

$$\begin{bmatrix} x \\ y \end{bmatrix} = \begin{bmatrix} a & 0 \\ 0 & b \end{bmatrix} \begin{bmatrix} u \\ v \end{bmatrix}$$

where a and b are the desired scaling factors. To resample the scaled image onto the display grid we use the inverse operation, as before, to locate pixel positions in the original image corresponding to each display grid position, viz

$$\begin{bmatrix} u \\ v \end{bmatrix} = \begin{bmatrix} 1/a & 0 \\ 0 & 1/b \end{bmatrix} \begin{bmatrix} x \\ y \end{bmatrix}.$$

Again interpolation is used to establish actual pixel brightnesses to use, since u, v will not normally fall on exact grid locations.

Frequently $a = b$ so that the image is simply magnified (although different magnification factors could be used in each direction if desired). This is often called zooming, particularly if the process is implemented in an interactive image display system. If the nearest neighbour interpolation procedure is used in the resampling process the zoom implemented is said to occur by pixel replication and the image will look progressively blocky for larger zoom factors. If cubic convolution interpolation is used there will be a change in magnification but the image will not take on the blockly appearance. Often this process is called interpolative zoom. Both pixel replication zoom and interpolative zoom can also be implemented in hardware to allow the process to be performed in real time.

References for Chapter 2

Good discussions on the effect of the atmosphere on image data in the visible and infrared wavelength ranges will be found in Slater (1980) and Forster (1984). Forster gives a detailed set of calculations to illustrate how correction procedures for compensating radiometric distortion caused by the atmosphere are applied. Definitions and derivations of radiometric quantities are covered comprehensively by Slater.

Extensive treatments of geometric distortion and means for geometric correction are covered by Anuta (1973), Billingsley (1983) and Bernstein (1983). Discussions of resampling interpolation techniques in particular, and the optimum distribution of control points will be found in Shlien (1979) and Orti (1981).

An interesting account of geometrical transformations in general, but especially as related to computer graphics, is found in Foley and Van Dam (1982).

P. E. Anuta, 1973: Geometric Correction of ERTS-1 Digital MSS Data. Information Note 103073, Laboratory for Applications of Remote Sensing, Purdue University, West Lafayette, Indiana.

R. Bernstein, 1983: Image Geometry and Rectification. In R. N. Colwell (Ed.) Manual of Remote Sensing, 2e, Chapter 21, Falls Church, Va. American Society of Photogrammetry.

F. C. Billingsley, 1983: Data Processing and Reprocessing in R. N. Colwell (Ed.) Manual of Remote Sensing, 2e, Chapter 17, Falls Church, Va. American Society of Photogrammetry.

J. D. Foley and A. Van Dam, 1982: Fundamentals of Interactive Computer Graphics, Philippines, Addison-Wesley.

B. C. Forster, 1984: Derivation of Atmospheric Correction Procedures for Landsat MSS with Particular Reference to Urban Data. Int. J. Remote Sensing, 5, 799–817.

T. G. Moik, 1980: Digital Processing of Remotely Sensed Images, Washington, NASA.

F. Orti, 1981: Optimal Distribution of Control Points to Minimize Landsat Image Registration Errors. Photogrammetric Engineering and Remote Sensing, 47, 101–110.

S. Shlien, 1979: Geometric Correction, Registration and Resampling of Landsat Imagery. Canadian J. Remote Sensing, 5, 74–89.

L. F. Silva, 1978: Radiation and Instrumentation in Remote Sensing. In P. H. Swain & S. M. Davis (Eds.) Remote Sensing: The Quantitative Approach, N. Y., Mc-Graw-Hill.

K. Simon, 1975: Digital Reconstruction and Resampling for Geometric Manipulation. Proc. Symp. on Machine Processing of Remotely Sensed Data, Purdue University, June 3–5.

P. N. Slater, 1980: Remote Sensing: Optics and Optical System, Reading, Mass., Addison-Wesley.

R. E. Turner and M. M. Spencer, 1972: Atmospheric Model for Correction of Spacecraft Data. Proc. 8th Int. Symp. on Remote Sensing of the Environment, Ann Arbor, Michigan 895–934.

M. P. Weinreb, R. Xie, I. H. Lienesch and D. S. Crosby, 1989: Destriping GOES Images by Matching Empirical Distribution Functions. Remote Sensing of Environment, 29, 185–195.

Problems

2.1 (a) Consider a (hypothetical) region on the ground consisting of a square grid. For simplicity suppose the grid "lines" are 79 m in width and the grid spacing is 790 m. Sketch how the region would appear in Landsat multispectral scanner imagery, before any geometric correction has been applied. Include only the effect of earth rotation and the effect of 56 m horizontal spacing of the 79 m × 79 m ground resolution elements.

(b) Develop a pair of linear (first order) mapping polynomials that will correct the image data of part (a). Assume the "lines" on the ground have a brightness of 100 and the background brightness is 20. Resample onto a 50 m grid and use a nearest neighbour interpolation. You will not want to compute all the resampled pixels unless a small computer program is used for the purpose. Instead you may wish simply to consider some significant pixels in the resampling to illustrate the accuracy of the geometric correction.

2.2 A sample of pixels from each of three cover types present in the Landsat MSS scene of Sydney, Australia, acquired on 14 December, 1980 is given in Table 2.2(a). Only the brightnesses (digital count values) in the visible red band (0.6 to 0.7 μm) and the second of the infrared bands (0.8 to 1.1μm) are given.

Table 2.2. Pixels from three cover types in wavelength bands 5 (0.6 to 0.7 μm) and 7 (0.8 to 1.1 μm) (on a scale of 0 to 255)

(a) Landsat MSS image of Sydney, 14 December 1980

Water		Vegetation		Bare	
Band 5	Band 7	Band 5	Band 7	Band 5	Band 7
20	11	60	142	74	66
23	7	53	130	103	82
21	8	63	140	98	78
21	7	52	126	111	86
22	7	34	92	84	67
19	3	38	120	76	67
17	1	38	151	72	67
20	4	38	111	98	71
24	8	31	81	99	80
19	4	50	158	108	71

(b) Landsat MSS image of Sydney, 8 June 1980

Water		Vegetation		Bare	
Band 5	Band 7	Band 5	Band 7	Band 5	Band 7
11	2	19	41	43	27
13	5	24	45	43	34
13	2	20	44	40	30
11	1	22	30	27	19
9	1	15	22	34	23
14	4	14	26	36	26
13	4	21	27	34	27
15	5	17	38	70	50
12	4	24	37	37	30
15	4	20	27	44	30

For this image Forster (1984) has computed the following relations between reflectance (R) and the digital count values (C) measured in the image data, where the subscript 7 refers to the infrared data and the subscript 5 refers to the visible red data:

$$R_5 = 0.44 C_5 + 0.5$$
$$R_7 = 1.18 C_7 + 0.9 .$$

Table 2.2 (b) shows samples of MSS digital count values for Sydney acquired on 8 June 1980. For this scene, Forster has determined

$$R_5 = 3.64\,C_5 - 1.6$$
$$R_7 = 1.52\,C_7 - 2.6\,.$$

Compute mean values of the digital count values for each cover type in each scene and plot these (along with bars that indicate standard deviation) in a multispectral space. This has two axes; one is the pixel digital value in the infrared band and the other is the value in the visible red band.

Instead of plotting the mean and standard deviations of the digital count values, convert the data to reflectances first. Comment on the effect correction of the raw digital count values to reflectance data (in which atmospheric effects have been removed) has on the apparent spectral separation of the three cover types.

2.3 Aircraft line scanners acquire image data using a mirror that sweeps out lines of data at right angles to the fuselage axis. In the absence of a cross wind, scanning therefore will be orthogonal to the aircraft ground track, as is the case for satellite scanning. However aircraft data is frequently recorded in the presence of a cross wind. The aircraft fuselage then maintains an angle to the ground track so that scanning is no longer orthogonal to the effective forward motion. Discuss the nature of the distortion, referred to as "crabbing", that this causes in the image data as displayed on a rectangular grid. Remember to take account also of the finite scan time across a line.

Push broom scanners, employing linear arrays of charge coupled device sensors, are also used on aircraft. What is the nature of the geometric distortion incurred with those devices in the presence of a cross wind?

It is technically feasible to construct two dimensional detector arrays for use as aircraft sensors, with which frames of images data would be recorded in a "snapshot fashion" much like the Landsat RBV. What geometric distortion would be incurred with this device as a result of a cross-wind?

2.4 Compute the skew distortion resulting from earth rotation in the case of Landsats 4 and 5, and Spot.

2.5 For a particular application suppose it was necessary to apply geometric correction procedures to an image prior to *classification*. (See Chap. 3 for an overview of classification). What interpolation technique would you see in the resampling process? Why?

2.6 Destriping of Landsat multispectral scanner images is often performed by computing six (modulo-6 line) histograms and then either (i) matching all six to a standard histogram or (ii) choosing one of the six as a reference and matching the other five to it. Which method is to be preferred if the image is to be analysed by photointerpretation or by classification (see Chap. 3)?

2.7 In a particular problem you have to register five Landsat images to a map. Would you register each image to the map separately, register one image to the map and then the other four images to that one, or image 1 to the map, image 2 to image 1, image 3 to image 2 etc?

2.8 (Requires a background in digital signal processing and sampling theory).

Remote sensing digital images are simply two dimensional uniform samples of the ground scene. In particular one line of image data is a regular sequence of samples. The spatial frequency spectrum of the line data will therefore be periodic as depicted in Fig. 2.19; it is assumed here there

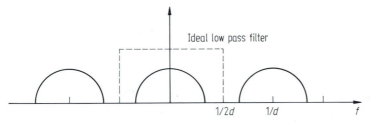

d = pixel spacing

Fig. 2.19. Spatial frequency spectrum of the line data

is no aliasing. From sampling theory it is well known that the original data can be recovered by low pass filtering the spectrum, using the ideal filter as indicated in the figure. Multiplication of the spectrum by this ideal filter is equivalent to convolving the original line samples by the inverse Fourier transform of the filter function. From the theory of the Fourier transform, the inverse of the filter function is

$$s(x) = \frac{2d}{\pi} \frac{\sin x}{x}$$ with $x = \zeta/2d$ in which ζ is a spatial variable along lines of data, and d is the inter-

pixel spacing. This is known generally as an interpolating function. Determine some cubic polynomial approximations to this function. These could be determined from a simple Taylor series expansion or could be derived from cubic splines. For a set of examples see Shlien (1979).

2.9 A multispectral scanner has been designed for aircraft operation. It has a field of view (FOV) of $\pm 35°$ about nadir and an instantaneous field of view (IFOV) of 2mrad. The sensor is designed to operate at a flying height of 1000 m.

    (i) Determine the pixel size, in metres, at nadir.

    (ii) Determine the pixel size at the edge of a swath compared with that at nadir.

    (iii) Discuss the nature of the distortion in the image geometry encountered if the pixels across a scan line are displayed on uniform pixel centre.

2.10 Determine the maximum angle of the field of view for an airborne optical sensor with a constant instantaneous field of view (IFOV), so that the pixel dimension along the scan line at the extremes is less than 1.5 times that at nadir (ignore earth curvature effect).

2.11 Consider the panoramic along scan line distortion of an airborne optical remote sensing system with a constant instantaneous field of view (IFOV); sketch the image formed for the ground scene shown in Fig. 2.20 below, and explain why it appears as you have sketched it.

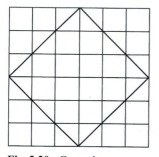

Fig. 2.20. Ground scene

Chapter 3
The Interpretation of Digital Image Data

3.1 Approaches to Interpretation

When image data is available in digital form, spatially quantised into pixels and radiometrically quantised into discrete brightness levels, several approaches are possible in endeavouring to extract information. One involves the use of a computer to examine each pixel in the image individually with a view to making judgement about pixels specifically based upon their attributes. This is referred to as *quantitative analysis* since pixels with like attributes are often counted to give area estimates. Means for doing this are described in Sect. 3.4. Another approach involves a human analyst/interpreter extracting information by visual inspection of an image composed from the image data. In this he or she notes generally large scale features and is often unaware of the spatial and radiometric digitisations of the data. This is referred to as *photointerpretation* or sometimes *image interpretation*; its success depends upon the analyst exploiting effectively the spatial, spectral and temporal elements present in the composed image product. Information spatially, for example, is present in the qualities of shape, size, orientation and texture. Roads, coastlines and river systems, fracture patterns, and lineaments generally, are usually readily identified by their spatial disposition. Temporal data, such as the change in a particular object or cover type in an image from one date to another can often be used by the photointerpreter as, for example, in discriminating deciduous or ephemeral vegetation from perennial types. Spectral clues are utilised in photointerpretation based upon the analyst's foreknowledge of, and experience with, the spectral reflectance characteristics of typical ground cover types, and how those characteristics are sampled by the sensor on the satellite or aircraft used to acquire the image data.

Those two approaches to image interpretation have their own roles and often these are complementary. Photointerpretation is aided substantially if a degree of digital image processing is applied to the image data beforehand, whilst quantitative analysis depends for its success on information provided at key stages by an analyst. This information very often is drawn from photointerpretation.

A comparison of the attributes of photointerpretation and quantitative analysis is given in Table 3.1. From this it can be concluded that photointerpretation, involving direct human interaction and therefore high level decisions, is good for spatial assessment but poor in quantitative accuracy. Area estimates by photointerpretation, for instance, would involve planimetric measurement of regions identified visually; in this, boundary definition errors will prejudice area accuracy. By contrast, quantitative analysis, requiring little human interaction and low level software decisions, has poor spatial ability but high quantitative accuracy. Its high accuracy comes from the ability of a computer, if required, to process every pixel in a given image and to take account of the full range of spectral, spatial and radiometric detail present. Its poor spatial properties come from the difficulty with which decisions about shape, size, orientation and texture can be made using standard sequential computing techniques. Developments in parallel computing and multi-resolution image analysis may obviate this in due course.

Table 3.1. A comparison of photointerpretation and quantitative analysis

Photointerpretation (by a human analyst/interpreter)	Quantitative analysis (by computer)
On a scale large relative to pixel size	At individual pixel level
Inaccurate area estimates	Accurate area estimates possible
Only limited multispectral analysis	Can perform true multispectral (multidimensional) analysis
Can assimilate only a limited number of distinct brightness levels (say 16 levels in each feature)	Can make use quantitatively of all available brightness levels in all features (e.g. 256, 1024, 4096)
Shape determination is easy	Shape determination involves complex software decisions
Spatial information is easy to use in a qualitative sense	Limited techniques available for making use of spatial data

With hyperspectral data sets pixel identification is also possible using knowledge of the spectroscopic properties of earth surface materials. In a sense, by operating at the pixel level, such an approach is analogous to classification. However, in the classification approaches normally adopted in remote sensing the pixel is identified through patterns in the properties of the recorded data. In contrast, because of the high spectral resolutions available in hyperspectral imagery, as illustrated in Fig. 1.10, it is possible for a skilled analyst to identify and label a pixel because of its spectroscopic characteristics. Also, because the spectroscopic nature of the pixel is so well defined, and provided the data has been properly calibrated, data bases of previously recorded pixel spectra, or spectra recorded in the laboratory, can be established to aid in pixel labelling.

3.2 Forms of Imagery for Photointerpretation

Multispectral or single dimensional image data can be procured either in photographic form or as digital data on storage media. The latter is usually more expensive although more flexible since first, photographic products can be created from the digital data and secondly, the data can be processed digitally for enhancement before hard copies are produced.

There are two fundamental types of photographic product. The first is a black and white reproduction of each band in the image data. If produced from the raw digital data then black will correspond to a digital brightness value of 0 whereas white will correspond to the highest digital value. This is usually 63, 127, 255 or 4095 (for 6 bit, 7 bit, 8 bit and 12 bit data respectively).

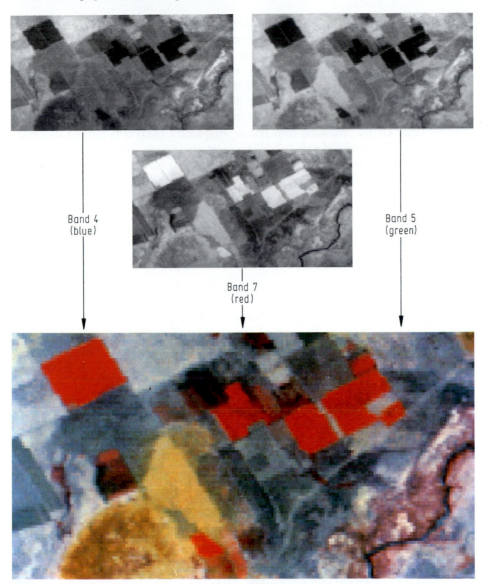

Fig. 3.1. Formation of a Landsat multispectral scanner false colour composite by displaying band 7 as red, band 5 as green and band 4 as blue

The second photographic product is a colour composite in which selected features or bands in multispectral data are chosen to be associated with the three additive colour primaries in the display device which produces the colour product. When the data consists of more than three features a judgement has to be made as to how to discard all but three, or alternatively, a mapping has to be invented that will allow all the features to be combined suitably into the three primaries. One possible mapping is the principal components transformation developed in Chap. 6. Usually this approach is not

adopted since the three new features are synthetic and the analyst is therefore not able
to call upon experience of spectral reflectance characteristics. Instead a subset of original
bands is chosen to form the colour composite. When the data available consists of a large
number of bands (such as produced by aircraft scanners of by imaging spectrometers) only experience, and the area of application, tell which three bands should
be combined into a colour product. For data with limited spectral bands however
the choice is more straightforward. An example of this is Landsat multispectral
scanner data. Of the available four bands, frequently band 6 is simply discarded
since it is highly correlated with band 7 for most cover types and also is more highly
correlated with bands 4 and 5 than is band 7. Bands 4, 5 and 7 are then associated with
the colour primaries in the order of increasing wavelength:

Landsat MSS Bands		Colour primaries
4 (green)		blue
5 (red)		green
7 (infrared)		red

increasing
wavelength

    An example of the colour product obtained by such a procedure is seen in Fig. 3.1.
This is often referred to as a false colour composite or sometimes, by association with
colour infrared film, a colour infrared image. In this, vegetation shows as variations in
red (owing to the high infrared response associated with vegetation), soils show as

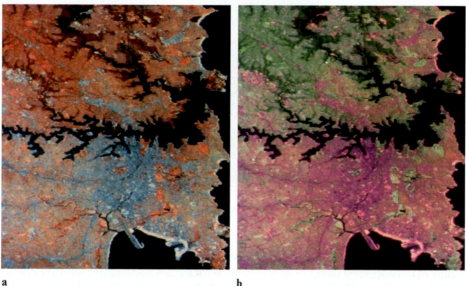

a   b

Fig. 3.2. Standard Landsat multispectral scanner false colour composite **a** compared with a
product in which band 7 has been displayed as green, band 5 as red and band 4 as blue **b**. Finer
detail is more apparent in the second product owing to the sensitivity of the human vision system
to yellow-green hues. The image segment shown is Sydney, the capital of New South Wales,
Australia, acquired on December 14, 1980

blue, green and sometimes yellow and water as black or deep blue. These colour associations are easily determined by reference to the spectral reflectance characteristics of earth surface cover types in Sect. 1.1; it is important also to take notice of the spectral imbalance created by computer enhancement of the brightness range in each wavelength band as discussed in Sect. 3.3 following.

It is of interest to note that the correlation matrix for the image of Fig. 3.1 is

	Band 4	Band 5	Band 6	Band 7
Band 4	1.00			
Band 5	0.85	1.00		
Band 6	0.31	0.39	1.00	
Band 7	−0.09	−0.07	0.86	1.00

wherein the redundancy present in band 6 can be seen.

In many ways the choice of colour assignments for the Landsat multispectral scanner bands is an unfortunate one since this yields, for most scenes, an image product substantially composed of reds and blues. These are the hues in which the human visual system is least sensitive to detail. Instead it would have been better to form an image in which yellows and greens predominate since then many fine details become more apparent. An illustration of this is given in Fig. 3.2.

Colour composite products for other remote sensing image data sets are created, similarly, by associating bands with display colour primaries in a wavelength monotonic fashion.

3.3 Computer Processing for Photointerpretation

When image data is available in digital form it can be processed before an actual image or photograph is written in order to ensure that the clues used for photointerpretation are enhanced. Little can be done about temporal clues, but judicious processing makes spectral and spatial data more meaningful. This processing is of two types. One deals with the radiometric (or brightness value) character of the image and is termed radiometric enhancement. The other has to do with the image's perceived spatial or geometric character and is referred to as geometric enhancement. The latter normally involves such operations as smoothing noise present in the data, enhancing and highlighting edges, and detecting and enhancing lines. Radiometric enhancement is concerned with altering the contrast range occupied by the pixels in an image. From the point of view of computation, radiometric enhancement procedures involve determining a new brightness value for a pixel (by some specified algorithm) from its existing brightness value. They are often referred to therefore as point operations and can be effectively implemented using look up tables. These are two-column tables (either in software or hardware) that associate a set of new brightness values with the set of old brightnesses. Specific radiometric enhancement techniques are treated in Chap. 4.

Geometric enhancement procedures involve establishing a new brightness value for a pixel by using the existing brightnesses of pixels over a specified neighbourhood of pixels. These cannot be implemented in look up table form and are usually time-consuming to evaluate. The range of geometric enhancement techniques commonly

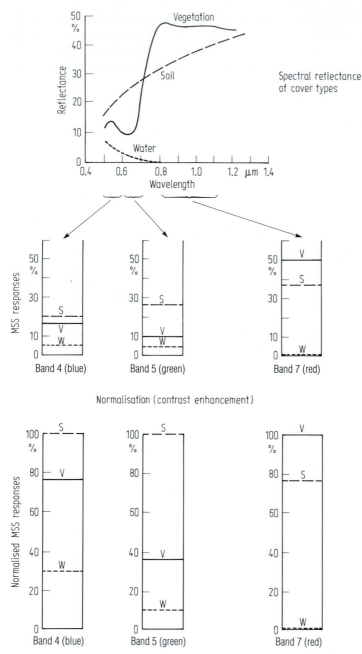

Fig. 3.3. Indication of how contrast enhancement can distort the feature-to-feature or band-to-band relativity (and thus colour relativity) in an image. Without contrast enhancement both soil and vegetation cover types would have a reddish appearance, whereas after enhancement the soil takes on its characteristic bluish tones. The bands indicated correspond to the Landsat MSS; the same effect will occur with similar band combinations from other sensors (eg. SPOT bands 1, 2 and 3)

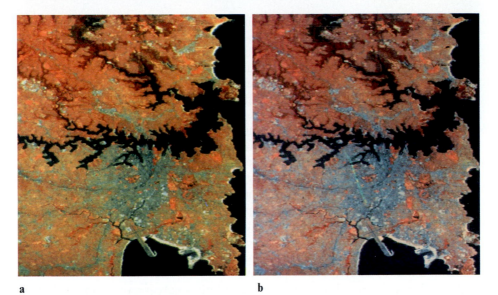

a   b

Fig. 3.4. Image in which each band has identical contrast enhancement before colour composition **a** compared to that in which each band has been enhanced independently to cover its full brightness range **b**. This causes a loss of band to band relativity and thus gives a different range of hues

used in the treatment of remote sensing image data is given in Chap. 5. Both radiometric and geometric enhancement can be of value in highlighting spatial information. It is generally only radiometric or contrast enhancement, however, that amplifies an image's spectral character. A word of caution however is in order here. When contrast enhancement is utilised, each feature in the data is generally treated independently. This can lead to a loss of feature-to-feature relativity and thus, in the case of colour composites, can lead to loss of colour relativity. The reason for this is depicted in Fig. 3.3, and the effect is illustrated in Fig. 3.4.

3.4 An Introduction to Quantitative Analysis — Classification

Identification of features in remote sensing imagery by photointerpretation is effective for global assessment of geometric characteristics and general appraisal of ground cover types. It is, however, impracticable to apply at the pixel level unless only a handful of pixels is of interest. As a result it is of little value for determining accurate estimates of the area in an image corresponding to a particular ground cover type, such as the hectarage of a crop. Moreover as noted in Sect. 3.2, since photointerpretation is based upon the ability of the human analyst-interpreter to assimilate the available data, only three or so of the complete set of spectral components of an image can be used readily. Yet there are four bands available in Landsat multispectral scanner imagery, seven for Landsat thematic mapper data and many for imaging spectrometer data. It is not that all of these would necessarily need to be used in the identification of a pixel; rather, should they all require consideration or evaluation, then the photointerpretive approach is clearly limited. Furthermore the human analyst is unable to discriminate to the limit of the radiometric resolution available in scanner and other forms of imagery. By comparison if a computer can be used for analysis, it could conceivably do so at the pixel level and could examine and identify as many pixels as required. In addition, it should be possible for computer analysis of remote sensing image data to take full account of the multidimensional aspect of the data, and its radiometric resolution.

Computer interpretation of remote sensing image data is referred to as quantitative analysis because of its ability to identify pixels based upon their numerical properties and owing to its ability for counting pixels for area estimates. It is also generally called classification, which is a method by which labels may be attached to pixels in view of their spectral character. This labelling is implemented by a computer by having trained it beforehand to recognise pixels with spectral similarities.

Clearly the image data for quantitative analysis must be available in digital form. This is an advantage with image data types, such as that from Landsat, SPOT, IRS, etc, as against more traditional aerial photographs. The latter require digitisation before quantitative analysis can be performed.

Detailed procedures and algorithms for quantitative analysis are the subject of Chap. 8, 9 and 10; Chap. 11 is used to show how these are developed into classification methodologies for effective quantitative analysis. The remainder of this chapter however is used to provide an outline of the essential concepts in classification. As a start it is necessary to devise a model with which to represent remote sensing multispectral image data in a form amenable to the development of analytical procedures.

3.5 Multispectral Space and Spectral Classes

The most effective means by which multispectral data can be represented in order to formulate algorithms for quantitative analysis is to plot them in a pattern space, or multispectral vector space, with as many dimensions as there are spectral components. In this space, each pixel of an image plots as a point with co-ordinates given by the brightness value of the pixel in each component. This is illustrated in Fig. 3.5 for a simple two dimensional band 5, band 7 Landsat MSS space. Provided the spectral bands have been designed to provide good discrimination it is expected that pixels would form groups in multispectral space corresponding to various ground cover types, the sizes and shapes of the groups being dependent upon varieties of cover type, systematic noise and topographic effects. The groups or clusters of pixel points are referred to as *information classes* since they are the actual classes of data which a computer will need to be able to recognise.

In practice the information class groupings may not be single clusters as depicted in Fig. 3.5. Instead it is not unusual to find several clusters for the same region of soil, for the same apparent type of vegetation and so on for other cover types in a scene. These are not only as a result of specific differences in types of cover but also result from differences in moisture content, soil types underlying vegetation and topographic influences. Consequently, a multispectral space is more likely to appear as shown in Fig. 3.6 wherein each information class is seen to be composed of several *spectral classes*.

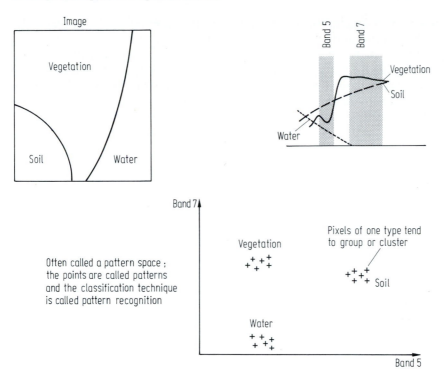

Fig. 3.5. Illustration of a two dimensional multispectral space showing its relation to the spectral reflectance characteristics of ground cover types

In many cases the information classes of interest do not form distinct clusters or groups of clusters but rather are part of a continuum of data in the multispectral space. This happens for example when, in a land systems exercise, there is a gradation of canopy closure with position so that satellite or aircraft sensors might see a gradual variation in the mixture of canopy and understory. The information classes here might correspond to nominated percentage mixtures rather than to sets of well defined sub-classes as depicted in Fig. 3.6. It is necessary in situations such as this to determine

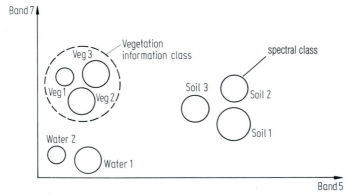

Fig. 3.6. Representation of information classes by sets of spectral classes

appropriate sets of spectral classes that represent the information classes effectively. This is demonstrated in the exercises chosen in Chap. 11.

In quantitative analysis it is the spectral classes that a computer will be asked to work with since they are the "natural" groupings or clusters in the data. After quantitative analysis is complete the analyst simply associates all the relevant spectral classes with the one appropriate information class. Shortly, spectral classes will be seen to be unimodal probability distributions and information classes as possible multimodal distributions. The latter need to be resolved into sets of single modes for convenience and accuracy in analysis.

3.6 Quantitative Analysis by Pattern Recognition

3.6.1 Pixel Vectors and Labelling

Recognition that image data exists in sets of spectral classes, and identification of those classes as corresponding to specific ground cover types, is carried out using the techniques of mathematical pattern recognition or pattern classification and their more recent neural network counterparts. The patterns are the pixel themselves, or strictly the mathematical *pixel vectors* that contain the sets of brightness values for the pixels arranged in column form:

$$x = \begin{bmatrix} x_1 \\ x_2 \\ \vdots \\ x_N \end{bmatrix}$$

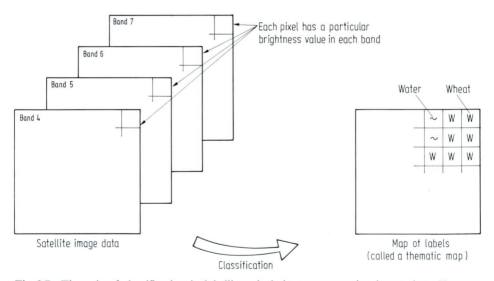

Fig. 3.7. The role of classification in labelling pixels in remote sensing image data. Here an illustration from Landsat multispectral scanner data is depicted

where x_1 to x_N are the brightnesses of the pixel x in bands 1 to N respectively. It is simply a mathematical convention that these are arranged in a column and enclosed in an extended square bracket. A summary of essential results from the algebra used for describing and manipulating these vectors is given in Appendix C.

Classification involves labelling the pixels as belonging to particular spectral (and thus information) classes using the spectral data available. This is depicted as a mapping in Fig. 3.7. In the terminology of statistics this is more properly referred to as allocation rather than classification. However throughout this book, classification, categorization, allocation and labelling are generally used synonomously.

There are two broad classes of classification procedure and each finds application in the analysis of remote sensing image data. One is referred to as supervised classification and the other unsupervised classification. These can be used as alternative approaches but are often combined into hybrid methodologies as demonstrated in Chap. 11.

3.6.2 Unsupervised Classification

Unsupervised classification is a means by which pixels in an image are assigned to spectral classes without the user having foreknowledge of the existence or names of those classes. It is performed most often using clustering methods. These procedures can be used to determine the number and location of the spectral classes into which the data falls and to determine the spectral class of each pixel. The analyst then identifies those classes *a posteriori*, by associating a sample of pixels in each class with available reference data, which could include maps and information from ground visits. Clustering procedures are generally computationally expensive yet they are central to the analysis of remote sensing imagery. While the information classes for a particular exercise are known, the analyst is usually totally unaware of the spectral classes, or sub-classes as they are sometimes called. Unsupervised classification is therefore useful for determining the spectral class composition of the data prior to detailed analysis by the methods of supervised classification.

The range of clustering algorithms frequently used for determination of spectral classes and for unsupervised classification is treated in Chap. 9.

3.6.3 Supervised Classification

Supervised classification procedures are the essential analytical tools used for the extraction of quantitative information from remotely sensed image data. There are a number of specific techniques and these are treated in detail in Chap. 8. It is the role of this section to introduce the framework of the approach.

An important assumption in supervised classification usually adopted in remote sensing is that each spectral class can be described by a probability distribution in multispectral space: this will be a multivariable distribution with as many variables as dimensions of the space. Such a distribution describes the chance of finding a pixel belonging to that class at any given location in multispectral space. This is not unreasonable since it would be imagined that most pixels in a distinct cluster or spectral class would lie towards the centre and would decrease in density for positions away from the class centre, thereby resembling a probability distribution. The distribution

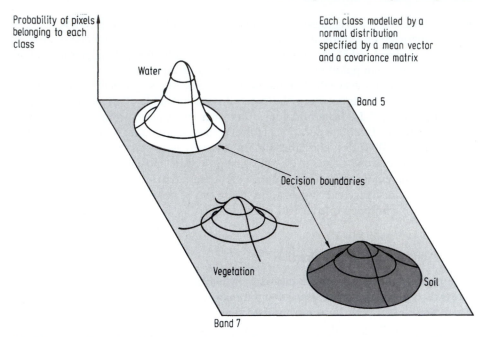

Fig. 3.8. Two dimensional multispectral space with the spectral classes represented by Gaussian probability distributions. Here the space is defined in terms of the Landsat multispectral scanner bands 5 and 7

found to be of most value is the normal or Gaussian distribution. It gives rise to tractable mathematical descriptions of the supervised classification process, and is robust in the sense that classification accuracy is not overly sensitive to violations of the assumptions that the classes are normal. A two dimensional multispectral space with the spectral classes so modelled is depicted in Fig. 3.8. The *decision boundaries* shown in the figure represent those points in multispectral space where a pixel has equal chance of belonging to two classes. The boundaries therefore partition the space into regions associated with each class; this is developed further in Sect. 8.2.4.

A multidimensional normal distribution is described as a function of a vector location in multispectral space by:

$$p(x) = \frac{1}{(2\pi)^{N/2} |\Sigma|^{1/2}} \exp\left\{ -\tfrac{1}{2}(x-m)^t \, \Sigma^{-1}(x-m) \right\}$$

where x is a vector location in the N dimensional pixel space: m is the mean position of the spectral class – i.e. the position x at which a pixel from the class is most likely to be found, and Σ is the covariance matrix of the distribution, which describes its spread directionally in the pixel space. Equation (6.2) shows how this matrix is defined; Appendix D summarises some of the important properties of this distribution.

The multidimensional normal distribution is specified completely by its mean vector and its covariance matrix. Consequently, if the mean vectors and covariance matrices are known for each spectral class then it is possible to compute the set of probabilities that describe the relative likelihoods of a pattern at a particular location belonging to each of those classes. It can then be considered as belonging to the class which indicates the highest probability. Therefore if m and Σ are known for every spectral class in an image, every pixel in the image can be examined and labelled corresponding to the most likely class on the basis of the probabilities computed for the particular location for a pixel. Before that classification can be performed however m and Σ are estimated for each class from a representative set of pixels, commonly called *a training set*. These are pixels which the analyst knows as coming from a particular (spectral) class. Estimation of m and Σ from training sets is referred to as supervised learning. Supervised classification consists therefore of three broad steps. First a set of training pixels is selected for each spectral class. This may be done using information from ground surveys, aerial photography, topographic maps or any other source of reference data. The second step is to determine m and Σ for each class from the training data. This completes the learning phase. The third step is the classification phase, in which the relative likelihoods for each pixel in the image are computed and the pixel labelled according to the highest likelihood.

The view of supervised classification adopted here has been based upon an assumption that the classes can be modelled by probability distributions and, as a consequence, are described by the parameters of those distributions. Other supervised techniques also exist, in which neither distribution models nor parameters are relevant. These are referred to as non-parametric methods and are discussed extensively in Nilsson (1965). More recently, neural network non-parametric classification methods have been shown to offer promise in remote sensing applications, as demonstrated in Sect. 8.9.

References for Chapter 3

An overview of the essential issues in image interpretation, including the role of computer processing for improving the impact of imagery and for quantitative analysis, will be found in Estes et al. (1983). A good summary, with extensive references, of the spectral reflectance characteristics of common earth surface cover types has been given by Hoffer (1978). Material of this type is important in photointerpretation. Landgrebe (1981) and Hoffer (1979) have provided good general discussions on computer classification of remote sensing image data.

More recent quantitative treatments will be found in Schowengerdt (1997) and Mather (1987). Schott (1997) has treated remote sensing data flow from a systems perspective.

J. E. Estes, E. J. Hajic & L. R. Tinney, 1983: Fundamentals of Image Analysis: Analysis of Visible and Thermal Infrared Data in D. S. Simonett (ed.), Manual of Remote Sensing, Vol. 1, 2e, American Society of Photogrammetry, Falls Church, Virginia.

R. M. Hoffer, 1978: Biological and Physical Considerations in Applying Computer-Aided Analysis Techniques to Remote Sensing Data, in P. H. Swain & S. M. Davis, Eds.: Remote Sensing: The Quantitative Approach, McGraw-Hill, N. Y.

R. M. Hoffer, 1979: Computer Aided Analysis Techniques for Mapping Earth Surface Features, Technical Report 020179, Laboratory for Applications of Remote Sensing, Purdue University, West Lafayette, Indiana.

D. A. Landgrebe, 1981: Analysis Technology for Land Remote Sensing, Proc. IEEE, 69, 628–642.

P. M. Mather, 1987: Computer Processing of Remotely Sensed Images, Wiley, Chichester.

N. J. Nilsson, 1965: Learning Machines, McGraw-Hill, N. Y.

J. R. Schott, 1997: Remote Sensing: The Image Chain Approach, Oxford UP, N. Y.

R. A. Schowengerdt, 1997: Remote Sensing Models and Methods for Image Processing, 2e, Academic, MA.

Problems

3.1 For each of the following applications would photointerpretation or quantitative analysis be the most appropriate analytical technique? Where necessary, assume spectral discrimination is possible.

   (i) Lithological mapping in geology

  (ii) Structural mapping in geology

 (iii) Assessment of forest condition

 (iv) Mapping movements of floods

  (v) Crop area determination

 (vi) Crop health assessment

 (vii) Bathymetric charting

(viii) Soil mapping

 (ix) Mapping drainage patterns

  (x) Land system mapping

3.2 Can contrast enhancing image data beforehand improve its discrimination for machine analysis?

3.3 Prepare a table comparing the attributes of supervised and unsupervised classification. You may care to consider the issues of training data, cost (see Chap. 11), analyst interaction and spectral class determination.

3.4 A problem with using probability models to describe classes in multispectral space is that atypical pixels can be erroneously classified. For example, a pixel with high band 5 and band 7 brightness in Fig. 3.8 would be classified as vegetation even though it is more reasonably soil. This is a result of the positions of the decision boundaries shown. Suggest a means by which this situation can be avoided. (This is taken up in Sect. 8.2.5).

3.5 The collection of the four brightness values for a pixel in a Landsat multispectral scanner image is often called a vector. Each of the four components in such a vector can take either 128 or 64 different values, depending upon the band. How many distinct pixel vectors are possible? How many are there for Landsat thematic mapper image data?

    It is estimated that the human visual system can discriminate about 20,000 colours (J. O Whittaker, Introduction to Psychology, Saunders, Philadelphia, 1965).

    Comment on the radiometric handling capability of machine analysis, compared to colour discrimination by a human analyst/interpreter.

3.6 Information classes are resolved into so-called spectral classes prior to classification. These are pixel groups amenable to modelling by single multivariate Gaussian or normal distribution functions. Why are more complex distributions not employed to obviate the need to establish spectral classes? (Hint: How much is known about *multi*-variate distributions other than Gaussian?)

Chapter 4
Radiometric Enhancement Techniques

4.1 Introduction

4.1.1 Point Operations and Look Up Tables

Image analysis by photointerpretation is often facilitated when the radiometric nature of the image is enhanced to improve its visual impact. Specific differences in vegetation and soil types, for example, may be brought out by increasing the contrast of an image. In a similar manner subtle differences in brightness value can be highlighted either by contrast modification or by assigning quite different colours to those levels. The latter method is known as colour density slicing.

It is the purpose of this chapter to present a variety of radiometric modification procedures often used with remote sensing image data. The range of techniques treated is characterised by the common feature that a new brightness value for a pixel is generated only from its existing value. Neighbouring pixels have no influence, as they do in the geometric enhancement procedures that are the subject of Chap. 5. Consequently, radiometric enhancement techniques are sometimes referred to as point or pixel-specific operations.

All of the techniques to be covered in this chapter can be represented either as a graph or as a table that expresses the relationship between the old and new brightness values. In tabular form this is referred to as a look up table (LUT) since a particular procedure can be implemented usually with simple arrays in software or with hardwired look up tables (sometimes called function memories) in hardware.

4.1.2 Scalar and Vector Images

Two particular image types require consideration when treating image enhancement. The first could be referred to as a *scalar* image, in which each pixel has only a single brightness value associated with it. Such is the case for a simple black and white image. The second type is a *vector* image, wherein each pixel is represented by a vector of brightness values, which might be the blue, green and red components of the pixel in a colour scene or, for a remote sensing multispectral image, may be the various spectral response components for the pixel. Most image enhancement techniques relate to scalar images and also to the scalar components of vector imagery. Such is the case with all techniques given in this chapter. Enhancement methods that relate particularly to vector imagery tend to be transformation oriented. Those are treated in Chap. 6.

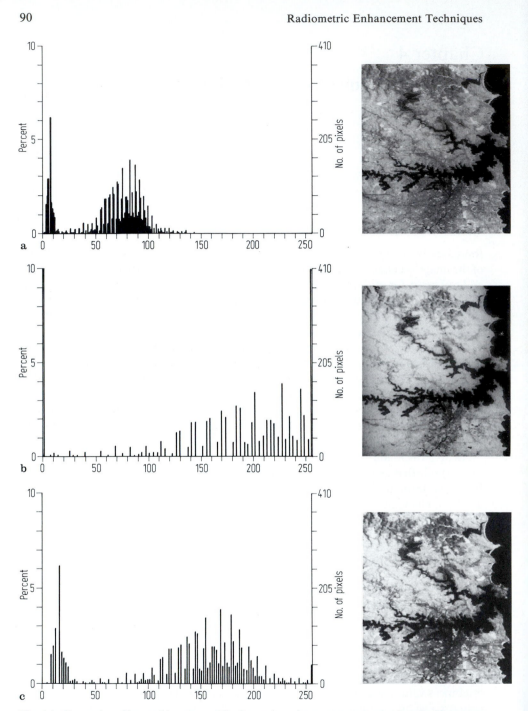

Fig. 4.1. Examples of image histograms. The image in **a** shows poor contrast since its histogram utilizes a restricted range of brightness value. The image in **b** is very contrasty with saturation in the black and white regions resulting in some loss of discrimination of bright and dull features. The image in **c** makes optimum use of the available brightness levels and shows good contrast. Its histogram shows a good spread of bars but without the large bars at black and white indicative of the saturation in image **b**

4.2 The Image Histogram

Consider a spatially quantised scalar image such as that corresponding to one of the Landsat multispectral scanner bands; in this case the brightness values are also quantised. If each pixel in the image is examined and its brightness value noted, a graph of number of pixels with a given brightness versus brightness value can be constructed. This is referred to as the histogram of the image. The tonal or radiometric quality of an image can be assessed from its histogram as illustrated in Fig. 4.1. An image which makes good use of the available range of brightness values has a histogram with occupied bins (or bars) over its full range, but without significantly large bars at black or white.

An image has a unique histogram but the reverse is not true in general since a histogram contains only radiometric and no spatial information. A point of some importance is that the histogram can be viewed as a discrete probability distribution since the relative height of a particular bar indicates the chance of finding a pixel with that particular brightness value somewhere in the image. Associated with this concept is the cumulative histogram, or cumulative discrete probability distribution, which is a graph of total number of pixels below a brightness level threshold versus the brightness level threshold as illustrated in Fig. 4.2.

It is useful in what is to follow if the histogram is regarded as a continuous function of a continuously varying brightness value. While this is not the case for digital imagery, such a concept will simplify some derivations and will lead to results that are easily interpreted in digital form. A continuous histogram function $h(x)$ is illustrated in Fig. 4.3 where x is the continuous variable representing brightness value, and h represents the value of the discrete histogram at that brightness. Strictly the height of a bar in the discrete histogram will be $h(x)dx$ where dx is the brightness value increment. When the number of brightness values is very large $dx \simeq 1$. However in general it can be shown that $dx = (L-1)/L$ where L is the total number of brightness values.

In statistical terms $h(x)$ is a probability density function and $\int_0^x h(x)\,dx$ is the cumulative probability function.

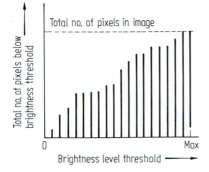

Fig. 4.2. Illustration of a cumulative histogram

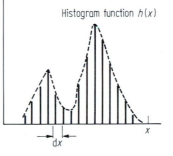

Fig. 4.3. Description of the continuous histogram function $h(x)$ and its relation to the discrete histogram when $dx = 1$

4.3 Contrast Modification in Image Data

4.3.1 Histogram Modification Rule

Suppose one has available a digital image with poor contrast, such as that in Fig. 4.1 a, and it is desired to improve its contrast to obtain an image with a histogram that has a good spread of bars over the available brightness range, resembling that in Fig. 4.1 c. In other words, a so-called contrast stretching of the image data is required. Often the degree of stretching desired is apparent. For example the original histogram may occupy brightness values between 40 and 75 and it might be wished to expand this range to the maximum possible, say 0 to 255. Even though the modification is somewhat obvious it is necessary to express it in mathematical terms in order to relegate it to a computer. Contrast modification is a mapping of brightness values, in that the brightness value of a particular histogram bar is respecified more favourably. The bars themselves though are not altered in size, although in some cases some bars may be mapped to the same new brightness value and will be superimposed. In general, however, the new histogram will have the same number of bars as the old. They will simply be at different locations.

The mapping of brightness values associated with contrast modification can be described as

$$y = f(x)$$

where x is the old brightness value of a particular bar in the histogram and y is the corresponding new brightness value. Let $h_i(x)$ be the histogram function of the original image and $h_o(y)$ be the histogram function of the contrast modified image. The

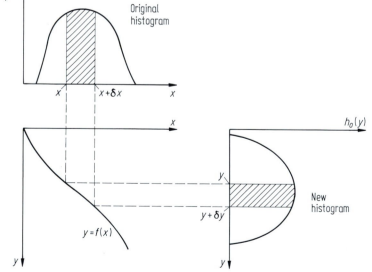

Fig. 4.4. Diagrammatic representation of contrast modification by the brightness value mapping function $y = f(x)$

subscripts i and o here are meant to connote the "input" histogram to a contrast modification process, and the resulting "output" histogram. The design of such a process and indeed an observation of the results of the radiometric modification of an image, follows from an answer to the query: knowing $h_i(x)$ and $y = f(x)$ what is the form of $h_o(y)$? The answer can be obtained by reference to Fig. 4.4, which has been adapted from Castleman (1996).

In Fig. 4.4 the number of pixels represented by the range y to $y + \delta y$ in the modified histogram must, by definition in the diagram, be equal to the number of pixels represented in the range x to $x + \delta x$ in the original histogram. Given that $h_i(x)$ and $h_o(y)$ are strictly density functions, this implies

$$h_i(x)\, \delta x = h_o(y)\, \delta y$$

so that in the limit as $\delta x, \delta y \to 0$

$$h_o(y) = h_i(x) \frac{\mathrm{d}x}{\mathrm{d}y} \tag{4.1}$$

Since $y = f(x)$, $x = f^{-1}(y)$ and (4.1) becomes

$$h_o(y) = h_i(f^{-1}(y)) \frac{\mathrm{d}(f^{-1}(y))}{\mathrm{d}y} \tag{4.2}$$

which is an analytical expression for the output histogram. It should be noted, mathematically, that this requires the inverse $x = f^{-1}(y)$ to exist. For the contrast modification procedures used with remote sensing image data this is generally the case, particularly for those situations in which (4.2) would be used. Should the inverse not exist – for example, if $y = f(x)$ is not monotonic – Castleman (1996) recommends treating the original brightness value range x as a set of contiguous subranges over each of which $y = f(x)$ is monotonic. These are then treated separately and the resulting histograms added.

4.3.2 Linear Contrast Enhancement

As an illustration of contrast modification, consider the simple linear variation described by

$$y = f(x) = ax + b$$

so that $\quad x = f^{-1}(y) = (y - b)/a,$

and $\quad \mathrm{d}f^{-1}(y)/\mathrm{d}y = 1/a.$

The modified histogram therefore from (4.2) is

$$h_o(y) = \frac{1}{a} h_i\left(\frac{y - b}{a}\right)$$

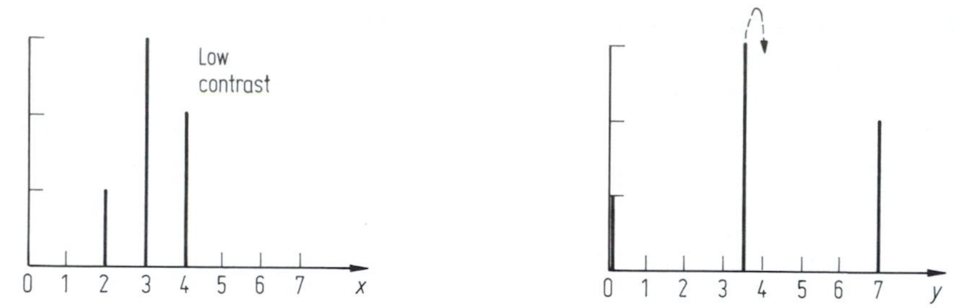

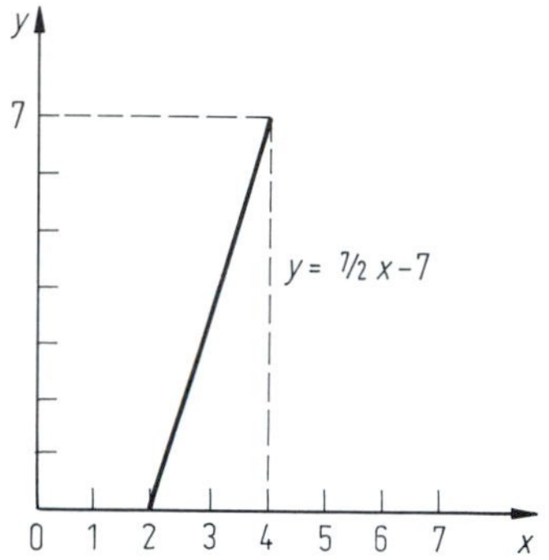

LUT	
x	y
0	0
1	0
2	0
3	4
4	7
5	7
6	7
7	7

Fig. 4.5. Simple numerical example of linear contrast modification. The available range of discrete brightness values is 0 to 7. Note that a non-integral output brightness value might be indicated. In practice this is rounded to the nearest integer

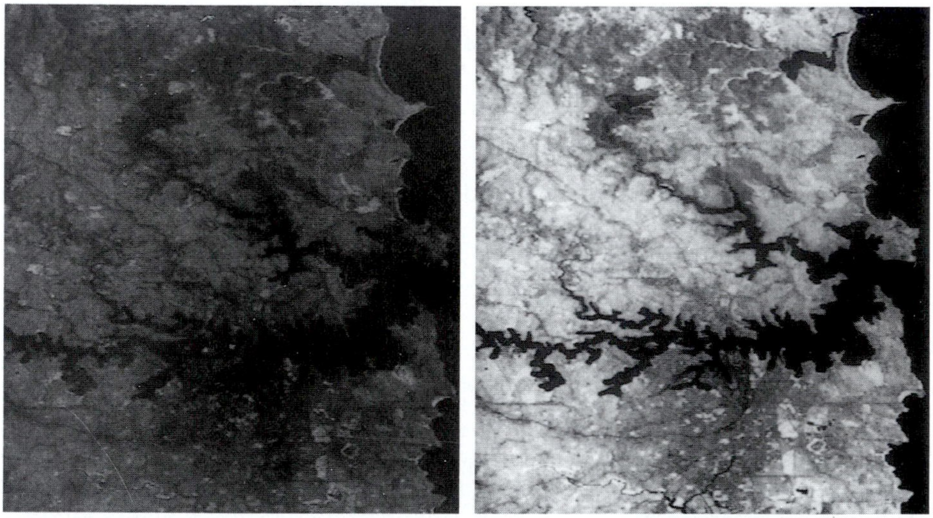

a **b**

Fig. 4.6. Linear contrast modification of the image in **a** to produce the visually better product in **b**

Relative to the original histogram, the modified version is shifted owing to the effect of b, is spread or compressed, depending on whether a is greater or less than 1 and is modified in amplitude. The last effect only relates to the continuous histogram function and cannot happen for discrete brightness value data. A numerical example of linear contrast modification is shown in Fig. 4.5, whereas a poorly contrasting image that has been radiometrically enhanced by linear contrast stretching is shown in Fig. 4.6.

The look-up table for the particular linear stretch in Fig. 4.5 has been included in the figure. In practice this would be used by a computer routine to produce the new image. This is done by reading the brightness values of the original version, pixel by pixel, substituting these into the left hand side of the table and then reading the new brightness value for a pixel from the corresponding entry on the right hand side of the table. It is important to note in digital image handling that the new brightness values, just as the old, must be discrete, and cover usually the same range of brightnesses. Generally this will require some rounding to integer form of the new brightness values calculated from the mapping function $y = f(x)$. A further point to note in the example of Fig. 4.5 is that the look-up table is undefined outside the range 2 to 4 of inputs. To do so would generate output brightness values that are outside the range valid for this example. In practice, linear contrast stretching is generally implemented as the saturating linear contrast enhancement technique in Sect. 4.3.3 following.

4.3.3 Saturating Linear Contrast Enhancement

Frequently a better image product is given when linear contrast enhancement is used to give some degree of saturation at the black and white ends of the histogram. Such is the case, for example, if the darker regions in an image correspond to the same ground cover type within which small radiometric variations are of no interest. Similarly, a particular region of interest in an image may occupy a restricted brightness value range; saturating linear contrast enhancement is then employed to expand that range

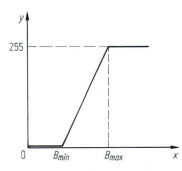

Fig. 4.7. Saturating linear contrast mapping

to the maximum possible dynamic range of the display device with all other regions being mapped to either black or white. The brightness value mapping function $y = f(x)$ for saturating linear contrast enhancement is shown in Fig. 4.7, in which B_{max} and B_{min} are the user-determined maximum and minimum brightness values that are to be expanded to the lowest and highest brightness levels supported by the display device.

4.3.4 Automatic Contrast Enhancement

Most remote sensing image data is too low in brightness and poor in contrast to give an acceptable image product if displayed directly in raw form. This is a result of the need to have the dynamic range of satellite and aircraft sensors so adjusted that a variety of cover types over many images can be detected without leading to saturation of the detectors or without useful signals being lost in noise. As a consequence a single typical image will contain a restricted set of brightnesses.

Image display systems frequently implement an automatic contrast stretch on the raw data in order to give a product with good contrast. Such a procedure is also of value when displaying image data on other output devices as well, at least for a first look at the data.

Typically the automatic enhancement procedure is a saturating linear stretch. The cut-off and saturation limits B_{min} and B_{max} are chosen by determining the mean brightness of the raw data and its standard deviation and then making B_{min} equal to the mean less three standard deviations and B_{max} equal to the mean plus three standard deviations.

4.3.5 Logarithmic and Exponential Contrast Enhancement

Logarithmic and exponential mappings of brightness values between original and modified images are useful for enhancing dark and light features respectively. The mapping functions are depicted in Fig. 4.8, along with their mathematical expressions. It is particularly important with these that the output values be scaled to lie within the range of the device used to display the product (or the range appropriate to files used for storage in a computer memory) and that the output values be rounded to allowed, discrete values.

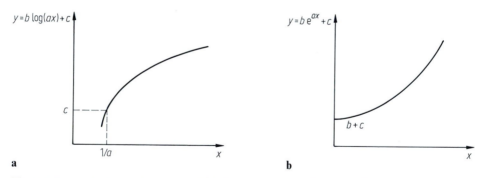

Fig. 4.8. Logarithmic **a** and exponential **b** brightness mapping functions. The parameters a, b and c are usually included to adjust the overall brightness and contrast of the output product

4.3.6 Piecewise Linear Contrast Modification

A particularly useful and flexible contrast modification procedure is the piecewise linear mapping function shown in Fig. 4.9. This is characterised by a set of user specified break points as shown. Generally the user can also specify the number of break points.

This method has particular value in implementing some of the contrast matching procedures in Sect. 4.4 and 4.5 following.

It should be noted that this is a more general version of the saturating linear contrast stretch of Sect. 4.3.3.

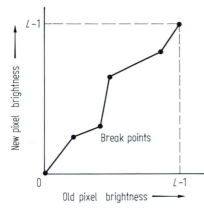

Fig. 4.9. Piecewise linear contrast modification function, characterised by the break points shown. These are user specified (as new, old pairs). It is clearly important that the function commence at 0,0 and finish at $L-1, L-1$ as shown, where L is the total number of brightness levels

4.4 Histogram Equalization

4.4.1 Use of the Cumulative Histogram

The foregoing sections have addressed the task of simple expansion (or contraction) of the histogram of an image. In many situations however it is desirable to modify the contrast of an image so that its histogram matches a preconceived shape, other than a simple closed form mathematical modification of the original version. A particular and important modified shape is the uniform histogram in which, in principle, each bar has the same height. Such a histogram has associated with it an image that utilises the available brightness levels equally and thus should give a display in which there is good representation of detail at all brightness values. In practice a perfectly uniform histogram cannot be achieved for digital image data; the procedure following however produces a histogram that is quasi-uniform on the average. The method of producing a uniform histogram is known generally as histogram equilization.

As before, let $h_i(x)$ be the histogram function of the original image and $h_o(y)$ represent the modified histogram, which is to be uniform. If the image contains a total of N pixels and there are L histogram bins or brightness values, then each of the brightness values in the modified histogram should have a bar of N/L pixels associated with it. Recall also that the bars in a discrete histogram have the values $h_o(y)\,\mathrm{d}y$. In the case of L available brightness values, $\mathrm{d}y = (L-1)/L$ so that for a uniform histogram

$$h_o(y)\ (L-1)/L = N/L$$

giving $h_o(y) = N/(L-1)$. From (4.1) therefore

$$\frac{\mathrm{d}y}{\mathrm{d}x} = \frac{\mathrm{d}}{\mathrm{d}x}\{f(x)\} = \frac{L-1}{N}\,h_i(x)$$

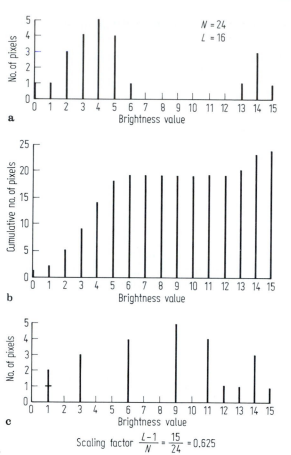

Fig. 4.10. Example of histogram equalisation. **a** Original histogram; **b** Cumulative histogram used to produce the look up table in Table 4.1; **c** The resulting quasi-uniform histogram

in which $y = f(x)$ is the sought-for mapping or transformation of brightness values that takes the original histogram of an image into a uniform histogram. Consequently,

$$y = f(x) = \frac{L-1}{N} \int h_i(x)\, dx \tag{4.3}$$

The histogram equalization transform therefore is the integral of the original histogram function times a scaling factor. The integral is just the continuous cumulative histogram; this can be replaced by the discrete cumulative histogram in the case of imagery with quantised brightness values, in which case (4.3) will yield a look-up table that can be used to move histogram bars to new brightness value locations. To illustrate the concept, consider the need to "flatten" the simple histogram shown in Fig. 4.10a. This corresponds to a hypothetical image with 24 pixels, each of which can take on one of 16 possible brightness values. The corresponding cumulative histogram is shown in Fig. 4.10b, and the scaling factor in (4.3) is $(L-1)/N = 15/24 = 0.625$. According to (4.3) the new brightness value location of a histogram bar is given by finding its original location on the abcissa of the cumulative histogram (x) and then reading its unscaled new location (y) from the ordinate. Multiplication by the scaling factor then produces the required new value. It is likely, however, that this may not be one of the discrete brightness values available (for the output display device) in which case the associated bar is moved to the nearest available brightness value. This procedure is summarised, for the example at hand, in Table 4.1, and the new, quasi-uniform histogram is given in Fig. 4.10c. It is important to emphasise that additional brightness values cannot be created nor can pixels from a single brightness value in an original histogram be distributed over several brightness values in the modified version. All that can be done is to re-map the brightness values to give a histogram that is as uniform as possible. Sometimes this entails some bars from the original histogram being moved to the same new location and thereby being superimposed, as is observed in the example.

In practice, the look up table created in Table 4.1 would be applied to every pixel in the image by feeding into the table the original brightness value for the pixel and reading from the table the new brightness value.

When the integral in (4.3) is interpreted as the cumulative histogram of an image, a simpler interpretation is possible. In such a case the maximum value of the integral is N, the total number of pixels in the image. The effect therefore of the denominator N in (4.3) is to normalise the height of the cumulative histogram to unity. The $L-1$ multiplier in (4.3) then scales the vertical axis of the cumulative histogram to a maximum of $L-1$. This is the same as that on the horizontal axis ($L-1$ is the largest brightness of L values starting at zero and incrementing by 1) so that the resulting cumulative histogram is a true brightness value mapping function, correctly scaled.

Figure 4.11 shows an example of an image with a simple linear contrast modification compared to the same image but in which contrast modification by histogram

Table 4.1. Look up table generation for histogram equalization example

Original brightness value	Unscaled new value	Scaled new value	Nearest available brightness value
0	1	0.63	1
1	2	1.25	1
2	5	3.13	3
3	9	5.63	6
4	14	8.75	9
5	18	11.25	11
6	19	11.88	12
7	19	11.88	12
8	19	11.88	12
9	19	11.88	12
10	19	11.88	12
11	19	11.88	12
12	19	11.88	12
13	20	12.50	13
14	23	14.40	14
15	24	15.00	15

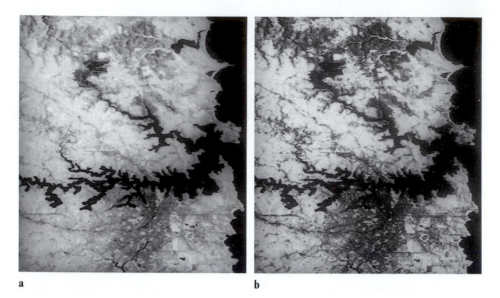

a   b

Fig. 4.11. Image with linear contrast stretch **a** compared with the same image enhanced with a stretch from histogram equalization **b**

equalization has been implemented. Many of these subtle contrast changing techniques only give perceived improvement of detail on some image types and sometimes require all components of a colour composite image to be so processed before an "improvement" is noticeable.

It is not necessary to retain the same number of distinct brightness values in an equalized histogram as in the original. Sometimes it is desirable to have a smaller

Table 4.2. Look up table for histogram equalization using 8 output brightnesses from 16 input brightnesses

Original brightness value	Unscaled new value	Scaled new value	Nearest available brightness value
0	1	0.29	0
1	2	0.58	1
2	5	1.46	1
3	9	2.63	3
4	14	4.08	4
5	18	5.25	5
6	19	5.54	6
7	19	5.54	6
8	19	5.54	6
9	19	5.54	6
10	19	5.54	6
11	19	5.54	6
12	19	5.54	6
13	20	5.83	6
14	23	6.70	7
15	24	7.00	7

output set and thereby produce a histogram with (fewer) bars that are closer in height than would otherwise be the case. This is implemented by redefining L in (4.3) to be the new total number of bars. Repeating the example of Table 4.1 and Fig. 4.10 for the case of $L = 8$ (rather than 16) gives the look up table of Table 4.2. Such a strategy would be an appropriate one to adopt when using an output device with a small number of brightness values (grey levels).

4.4.2 Anomalies in Histogram Equalization

Images with extensive homogeneous regions will give rise to histograms with large bars at the corresponding brightness values. A particular example is a Landsat multispectral scanner infrared image with a large expanse of water. Because histogram equalization creates a histogram that is uniform on the average by grouping smaller bars together, the equalized version of an image such as that just described will have poor contrast and little detail – quite the opposite to what is intended. The reason for this can be seen in the simple illustration of Fig. 4.12. The cumulative histogram used as the look-up table for the enhancement is dominated by the large bar at brightness value 0. The resulting image would be mostly grey and white with little grey level discrimination.

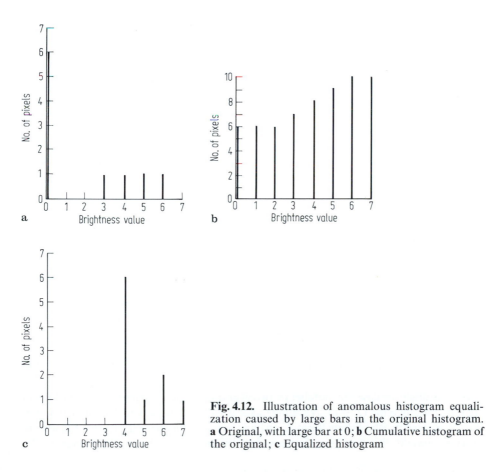

Fig. 4.12. Illustration of anomalous histogram equalization caused by large bars in the original histogram. **a** Original, with large bar at 0; **b** Cumulative histogram of the original; **c** Equalized histogram

A similar situation happens when the automatic contrast enhancement procedure of Sect. 4.3.4 is applied to images with large regions of constant brightness. This can give highly contrasting images on colour display systems; an acceptable display may require some manual adjustment of contrast taking due regard of the abnormally large histogram bars.

To avoid the anomaly in histogram equalization caused by the types of image discussed it is necessary to reduce the significance of the dominating bars in the image histograms. This can be done simply by arbitarily reducing their size when constructing the look up table, remembering to take account of this in the scale factor of (4.3). Another approach is to produce the cumulative histogram and thus look-up table on a subset of the image that does not include any, or any substantial portion, of the dominating region. Hogan (1981) has also provided an alternative procedure, based upon accumulating the histogram over "buckets" of brightness value. Once a bucket is full to a prespecified level, a new bucket is started.

4.5 Histogram Matching

4.5.1 Principle of Histogram Matching

Frequently it is desirable to match the histogram of one image to that of another image and in so doing make the apparent distribution of brightness values in the two images as close as possible. This would be necessary for example when a pair of contiguous images are to be joined to form a mosaic. Matching their histograms will minimise the brightness value variations across the join. In another case, it might be desirable to match the histogram of an image to a pre-specified shape, other than the uniform distribution treated in the previous section. For example, it is often found of value in photointerpretation to have an image whose histogram is a Gaussian function of brightness, in which most pixels have mid-range brightness values with only a few in the extreme white and black regions. The histogram matching technique, to be derived now, allows both of these procedures to be implemented.

The process of histogram matching is best looked at as having two stages, as depicted in Fig. 4.13. Suppose it is desired to match the histogram of a given image, $h_i(x)$, to the histogram $h_o(y)$; $h_o(y)$ could be a pre-specified mathematical expression or the histogram of the second image. Then the steps in the process are to equalize the histogram $h_i(x)$ by the methods of the previous section to obtain an intermediate histogram $h^*(z)$, which is then modified to the desired shape $h_o(y)$.

If $z = f(x)$ is the transformation that flattens $h_i(x)$ to produce $h^*(z)$ and $z = g(y)$ is the operation that would flatten the reference histogram $h_o(y)$ then the overall mapping of brightness values required to produce $h_o(y)$ from $h_i(x)$ is

$$y = g^{-1}(z), \quad z = f(x) \quad \text{or} \quad y = g^{-1}\{f(x)\}. \tag{4.4}$$

If, as is often the case, the number of pixels and brightness values in $h_i(x)$ and $h_o(y)$ are the same, then the $(L-1)/N$ scaling factor in (4.3) will cancel in (4.4) and can therefore be ignored in establishing the look up table which implements the contrast matching process. Should the number of pixels be different, say N_1 in the image to be

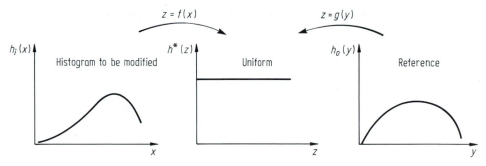

Fig. 4.13. The stages in histogram matching

modified and N_2 in the reference image then a scaling factor of N_2/N_1 will occur in (4.4). All scaling considerations can be bypassed however if the cumulative histograms are always scaled to some normalised value such as unity, or 100% (of the total number of pixels in an image).

4.5.2 Image to Image Contrast Matching

Figure 4.14 illustrates the steps implicit in (4.4) in matching source and reference histograms. In this case the reference histogram is that of a second image. Note that the procedure is to use the cumulative histogram of the source image to obtain new brightness values in the manner of the previous section by reading ordinate values corresponding to original brightness values entered on the abcissa. The new values are then entered into the *ordinate* of the cumulative reference histogram and the final brightness values (for the bars of the source histogram) are read from the *abcissa*; i.e. the cumulative reference histogram is used in reverse as indicated by the g^{-1} operation in (4.4). The look up table for this example is shown in Table 4.3. Again, note that some

Table 4.3. Look up table generation for contrast matching

Source histogram brightness values x	Intermediate (equalized) values z	Modified values y	Nearest available brightness values
0	0	0	0
1	0	0	0
2	0	0	0
3	0	0	0
4	0	0	0
5	1	1	1
6	3	1.8	2
7	5	2.6	3
8	6	3	3
9	7	4	4
10	8	8	5
11	8	8	5

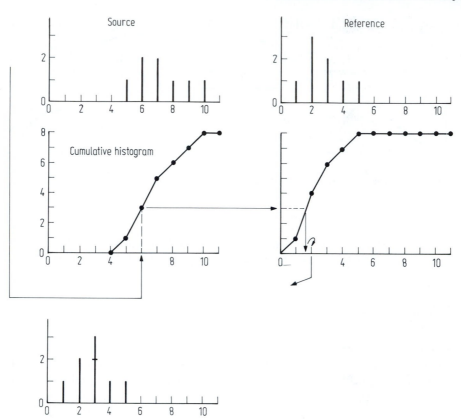

Fig. 4.14. An illustration of the steps in histogram matching

of the new brightness values produced may not be in the available range; as before, they are adjusted to the nearest acceptable value.

An example using a pair of contiguous image segments is shown in Fig. 4.15. Because of seasonal differences the contrasts are quite different. Using the cumulative histograms an acceptable matching is achieved. Such a process, as noted earlier, is an essential step in producing a mosaic of separate contiguous images. Another step is to ensure geometric integrity of the join. This is done using the geometric registration procedures of Sect. 2.5.

4.5.3 Matching to a Mathematical Reference

In some applications it is of value to pre-specify the desired shape of an image histogram to give a modified image with a particular distribution of brightness values. To implement this it is necessary to take an existing image histogram and modify it according to the procedures of Sect. 4.5.1. The reference is a mathematical function that describes the desired shape. A particular example is to match an image histogram to a Gaussian or normal shape. Often this is referred to as applying a "gaussian stretch" to an image; it yields a modified version with few black and white regions and in which

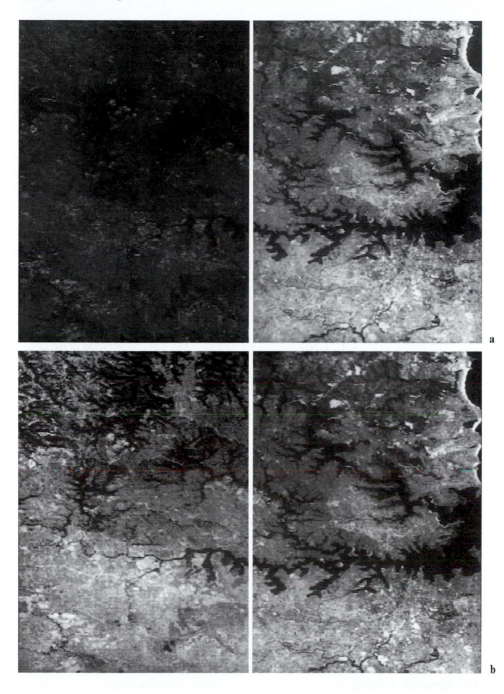

Fig. 4.15. a Contiguous Landsat multispectral scanner images showing contrast and brightness differences resulting from seasonal effects. The left hand image is an autumn scene and that on the right a summer scene, both of the northern suburbs of Sydney, Australia. **b** The same image pair but in which the histogram of the autumn scene has been matched to that of the summer scene

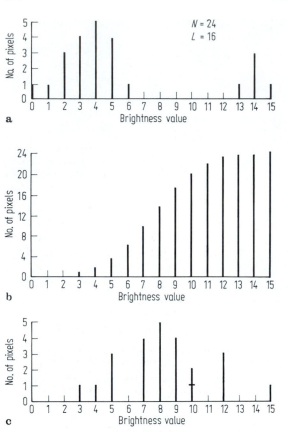

a

b

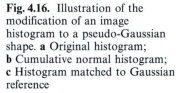

c

Fig. 4.16. Illustration of the modification of an image histogram to a pseudo-Gaussian shape. **a** Original histogram; **b** Cumulative normal histogram; **c** Histogram matched to Gaussian reference

most detail is contained in the mid-grey range. This requires a reference histogram in the form of a normal distribution. However since a cumulative version of the reference is to be used, it is really a cumulative normal distribution that is required. Fortunately cumulative normal tables and curves are readily available. To use such a table in the contrast matching situation requires its ordinate to be adjusted to the total number of pixels in the image to be modified and its abcissa to be chosen to match the maximum allowable brightness range in the image. The latter requires consideration to be given to the number of standard deviations of the Gaussian distribution to be contained in the total brightness value range, having in mind that the Gaussian function is continuous to $\pm \infty$. The mean of the distribution is placed usually at the mid-point of the brightness scale and commonly the standard deviation is chosen such that the extreme black and white regions are three standard deviations from the mean. A simple illustration is shown in Fig. 4.16.

4.6 Density Slicing

4.6.1 Black and White Density Slicing

A point operation often performed with remote sensing image data is to map *ranges* of brightness value to particular shades of grey. In this way the overall discrete number of

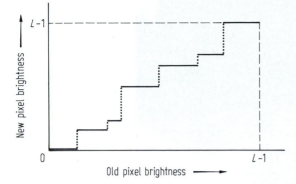

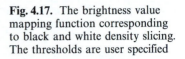

Fig. 4.17. The brightness value mapping function corresponding to black and white density slicing. The thresholds are user specified

brightness values used in the image is reduced and some detail is lost. However the effect of noise can also be reduced and the image becomes segmented, or sometimes contoured, in sections of similar grey level, in which each segment is represented by a user specified brightness. The technique is known as density slicing and finds value, for example, in highlighting bathymetry in images of water regions when penetration is acceptable. When used generally to segment a scalar image into significant regions of interest it is acting as a simple one dimensional parallelepiped classifier (see Sect. 8.4). The brightness value mapping function for density slicing is as illustrated in Fig. 4.17. The thresholds in such a function are entered by the user. An image in which the technique has been used to highlight bathymetry is shown in Fig. 4.19. Here differences in Landsat multispectral scanner visible imagery, at brightnesses too low to be discriminated by eye, have been mapped to new grey levels to make the detail apparent.

4.6.2 Colour Density Slicing and Pseudocolouring

A simple yet lucid extension of black and white density slicing is to use colours to highlight brightness value ranges, rather than simple grey levels. This is known as

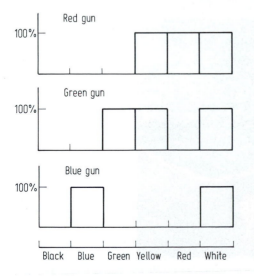

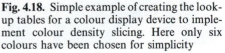

Fig. 4.18. Simple example of creating the look-up tables for a colour display device to implement colour density slicing. Here only six colours have been chosen for simplicity

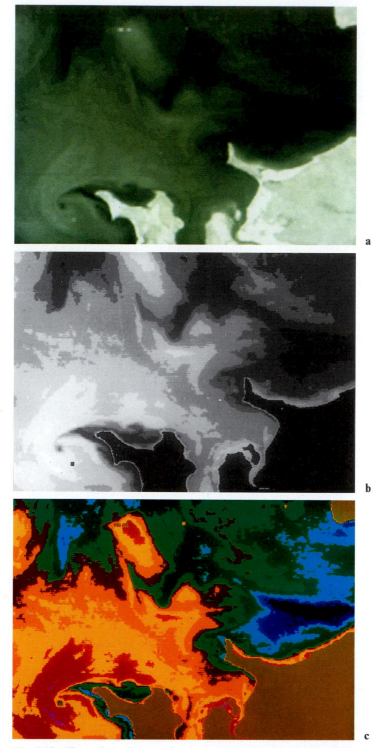

Fig. 4.19. Illustration of contouring in water detail using density slicing. **a** The image used is a band 5 + band 7 composite Landsat multispectral scanner image, smoothed to reduce line striping and then density sliced; **b** Black and white density slicing; **c** Colour density slicing

colour density slicing. Provided the colours are chosen suitably, it can allow fine detail to be made immediately apparent. It is a particularly simple operation to implement on a display system by establishing three brightness value mapping functions in the manner depicted in Fig. 4.18. Here one function is applied to each of the colour primaries used in the display device. An example of the use of colour density slicing, again for bathymetric purposes, is given in Fig 4.19.

This technique is also used to give a colour rendition to black and white imagery. It is then usually called pseudocolouring. Where possible this uses as many distinct hues as there are brightness values in the image. In this way the contours introduced by density slicing are avoided. Moreover it is of value in perception if the hues used are graded continuously. For example, starting with black, moving from dark blue, mid blue, light blue, dark green, etc. through to oranges and reds will give a much more acceptable pseudocoloured product than one in which the hues are chosen arbitarily.

References for Chapter 4

Much of the material on contrast enhancement and contrast matching treated in this chapter will be found also in Castleman (1996) and Gonzalez and Woods (1992) but in more mathematical detail. Passing coverages are also given by Moik (1980) and Hord (1982). More comprehensive treatments will be found in Schowengerdt (1997), Jensen (1986), Mather (1987) and Harrison and Jupp (1990).

The papers by A. Schwartz (1976) and J. M. Soha et al. (1976) give examples of the effect of histogram equalization and of Gaussian contrast stretching. Chavez et al. (1979) have demonstrated the performance of multicycle contrast enhancement, in which the brightness value mapping function $y = f(x)$ is cyclic. Here, several sub-ranges of input brightness value x are each mapped to the full range of output brightness value y. While this destroys the radiometric calibration of an image it can be of value in enhancing structural detail.

K. R. Castleman, 1996: Digital Image Processing, 2e, N. J., Prentice-Hall.
P. S. Chavez, G. L. Berlin and W. B. Mitchell, 1979: Computer Enhancement Techniques of Landsat MSS Digital Images for Land Use/Land Cover Assessment. Private Communication, US Geological Survey, Flagstaff, Arizona.
R. C. Gonzalez and R. E. Woods, 1992: Digital Image Processing, Mass., Addison-Wesley.
B. A. Harrison and D. L. B. Jupp, 1990: Introduction to Image Processing, Canberra, CSIRO.
A. Hogan, 1981: A Piecewise Linear Contrast Stretch Algorithm Suitable for Batch Landsat Image Processing. Proc. 2nd Australasian Conf. on Remote Sensing, Canberra, 6.4.1–6.4.4.
R. M. Hord, 1982: Digital Image Processing of Remotely Sensed Data, N.Y., Academic.
J. R. Jensen, 1986: Introductory Digital Image Processing — a Remote Sensing Perspective. N. J., Prentice-Hall.
P. M. Mather, 1987: Computer Processing of Remotely-Sensed Images. Suffolk, Wiley.
J. G. Moik, 1980: Digital Processing of Remotely Sensed Images, Washington, NASA.
A. Schwartz, 1976: New Techniques for Digital Image Enhancement, in Proc. Caltech/JPL Conf. on Image Processing Technology, Data Sources and Software for Commercial and Scientific Applications, California, Nov. 3–5, 2.1–2.12.
R. A. Schowengerdt, 1997: Remote Sensing Models and Methods for Image Processing, 2e, New York, Academic.
J. M. Soha, A. R. Gillespie, M. J. Abrams and D. P. Madura, 1976: Computer Techniques for Geological Applications; in Proc. Caltech/JPL Conf. on Image Processing Technology, Data Sources and Software for Commercial and Scientific Applications, Nov. 3–5, 4.1–4.21.

Problems

4.1 One form of histogram modification is to match the histogram of an image to a Gaussian or normal function. Suppose a raw image has the histogram indicated in Fig. 4.20. Produce the look-up table that describes how the brightness values of the image should be changed if the histogram is to be mapped, as nearly as possible, to a Gaussian histogram with a mean of 8 and a standard deviation of 2 brightness values. *Note that the sum of counts in the Gaussian reference histogram must be the same as that in the raw data histogram.*

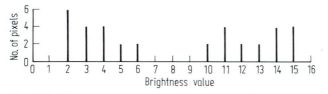

Fig. 4.20. Histogram

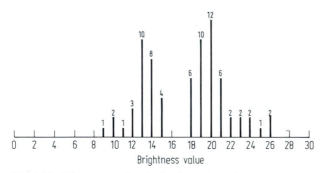

Fig. 4.21. Histogram of a single dimensional image

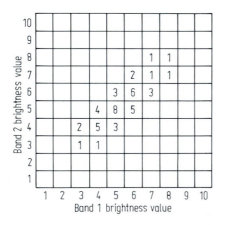

Fig. 4.22. Two-dimensional histogram

4.2 The histogram of a particular image is shown in Fig. 4.21. Produce the modified version that results from:

(i) a simple linear contrast stretch which makes use of the full range of brightness values
(ii) a simple piecewise linear stretch that maps the range (12, 23) to (0, 31) and
(iii) histogram equalization (i. e. producing a quasi-uniform histogram).

4.3 A two-dimensional histogram for a particular two band image data is shown in Fig. 4.22. Determine the histogram that results from a simple linear contrast stretch on each band individually.

4.4 Determine, algebraically, the contrast mapping function that equalizes the contrast of an image which has a Gaussian histogram at the centre of the brightness value range, with the extremities of the range being three standard deviations from the mean.

4.5 What is the shape of the cumulative histogram of an image that has been contrast (histogram) equalized? Can this be used as a figure of merit in histogram equalization?

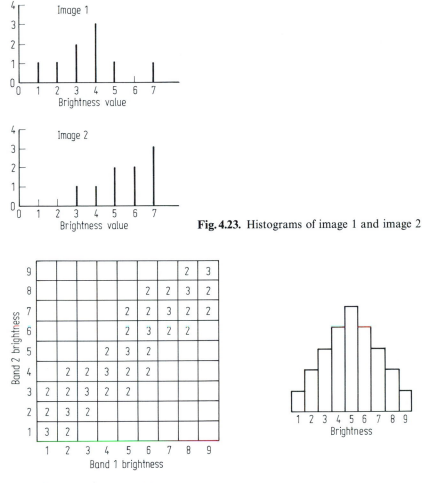

Fig. 4.23. Histograms of image 1 and image 2

Fig. 4.24. Two dimensional histogram

4.6 Clouds and large regions of clear, deep water frequently give histograms for near infrared imagery that have large high brightness level or low brightness value bars respectively. Sketch histograms of these types. Qualitatively, equalize the histograms using the material of Sect. 4.4 and comment on the undesirable appearance of the corresponding contrast enhanced images. Show that the situation can be rectified somewhat by artificially limiting the large bars to values not greatly different to the heights of other bars in the histogram, provided the accompanying cumulative histograms are normalised to correspond to the correct number of pixels in the image. A similar, but more effective procedure has been given in A. Hogan (1981).

4.7 Two Landsat images are to be joined side by side to form a mosaic for a particular application. To give the new, combined image a uniform appearance it is decided that the range and distribution of brightness levels in the first image should be made to match those of the second image, before they are joined. This is to be carried out by matching the histogram of image 1 to that of image 2. The original histograms are shown in Fig. 4.23. Produce a look up table that can be used to transform the pixel brightness values of image 1 in order to match the histograms as nearly as possible. Use the look-up table to modify the histogram of image 1 and comment on the degree to which contrast matching has been achieved.

4.8 (a) Contrast enhancement is frequently carried out on remote sensing image data. Describe the advantages in doing so, if the data is to be analysed by

(i) photointerpretation

(ii) quantitative computer methods.

(b) A particular two band image has the two dimensional histogram shown in Fig. 4.24. It is proposed to enhance the contrast of the image by matching the histograms in each band to the triangular profile shown. Produce look-up tables to enable each band to be enhanced, and from these produce the new two-dimensional histogram for the image.

4.9 Plot the equilized histogram for the example of Table 4.2. Compare it with Fig. 4.10 and comment on the effect of restricting the range of output brightnesses. Repeat the exercise for the cases of 4 and 2 output brightness values.

Chapter 5
Geometric Enhancement Using Image Domain Techniques

5.1 Neighbourhood Operations

This chapter presents methods by which the geometric detail in an image may be modified and enhanced. The specific techniques covered are applied to the image data directly and could be called image domain techniques. These are alternatives to procedures used in the spatial frequency domain which require Fourier transformation of the image beforehand. Those are treated in Chap. 7.

In contrast to the point operations used for radiometric enhancement, techniques for geometric enhancement are characterised by operations over neighbourhoods. The procedures still determine modified brightness values for an image's pixels; however, the new value for a given pixel is derived from the brightnesses of a set of the surrounding pixels. It is this spatial interdependence of the pixel values that leads to variations in the perceived image geometric detail. The neighbourhood influence will be apparent readily in the techniques of this chapter; for the Fourier transformation methods of Chap. 7 it will be discerned in the definition of the Fourier operation.

5.2 Template Operators

Geometric enhancements of most interest in remote sensing generally relate to smoothing, edge detection and enhancement, and line detection. Enhancement of edges and lines leads to image sharpening. Each of these operations is considered in the following sections. Most of the methods to be presented are, or can be expressed as, template techniques in which a template, box or window is defined and then moved

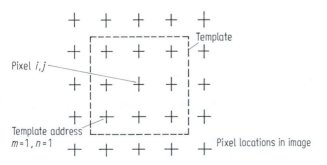

Fig. 5.1. A 3×3 template positioned over a group of nine image pixels, showing the relative locations of pixels and template entry addresses

over the image row by row and column by column. The products of the pixel brightness values, covered by the template at a particular position, and the template entries are taken and summed to give the template response. This response is then used to define a new brightness value for the pixel currently at the centre of the template. When this is done for every pixel in the image, a radiometrically modified image is produced that enhances or smooths geometric features according to the specific numbers loaded into the template. A 3×3 template is illustrated in Fig. 5.1. Templates of any size can be defined, and for an M by N pixel sized template, the response for image pixel i, j is

$$r(i, j) = \sum_{m=1}^{M} \sum_{n=1}^{N} \phi(m, n) \, t(m, n) \tag{5.1}$$

where $\phi(m, n)$ is the pixel brightness value, addressed according to the template position and $t(m, n)$ is the template entry at that location. Often the template entries collectively are referred to as the 'kernel' of the template and the template technique generally is called convolution, in view of its similarity to time domain convolution in linear system theory. This concept is developed in Sect. 5.3 below wherein the template approach is justified.

5.3 Geometric Enhancement as a Convolution Operation

This section presents a brief linear system theory basis for the use of the template expression of (5.1). It contains no results essential to the remainder of the chapter and can be safely passed over by the reader satisfied with (5.1) from an intuitive viewpoint.

Consider a signal in time represented as $x(t)$. Suppose this is passed through a system of some sort to produce a modified signal $y(t)$ as depicted in Fig. 5.2. The system here could be an intentional one such as an amplifier or filter, inserted to change the signal in a predetermined way; alternatively it could represent unintentional modification of the signal such as by distortion or the effect of noise. The properties of the system can be described by a function of time $h(t)$. This is called its impulse response (or sometimes its transfer function, although that term is more properly used for the Fourier transform of the impulse response, as noted in Chap. 7).

The relationship between $y(t)$ and $x(t)$ is described by the convolution operation. This can be expressed as an integral

$$y(t) = \int_{-\infty}^{\infty} x(\tau) \, h(t - \tau) \, d\tau \overset{\Delta}{=} x(t) * h(t) \tag{5.2}$$

as shown in McGillem and Cooper (1984). McGillem and Cooper, Castleman (1996) and Brigham (1974, 1988) all give comprehensive accounts of the properties of convolution and the characteristics of linear systems derived from the operation of convolution.

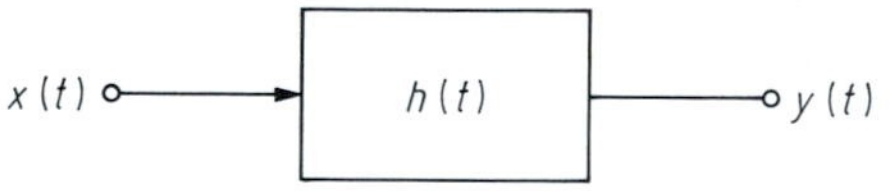

Fig. 5.2. Signal model of a linear system

A similar mathematical description applies when images are used in place of signals in (5.2) and Fig. 5.2. The major difference is that the image has two independent variables (its i and j pixel position indices, or address) whereas the signal $x(t)$ in Fig. 5.2 has only one – time. Consequently the transfer function of a system that operates on an image is also two dimensional, and the processed image is given by a two dimensional version of the convolution integral in (5.2). In this case the system can represent any process that modifies the image. It could, for example, account for degradation brought about by the finite point spread function of an image acquisition instrument or an image display device. It could also represent the effect of intentional image processing such as that used in geometric enhancement. In both cases if the new and old versions of the image are described by $r(x, y)$ and $\phi(x, y)$ respectively, where x and y are continuous position variables that describe the locations of points in a continuous image domain, then the two dimensional convolution operation is described as

$$r(x, y) = \int\limits_{-\infty}^{\infty} \int\limits_{-\infty}^{\infty} \phi(u, v)\, t'(x - u, y - v)\, du\, dv \tag{5.3}$$

where $t'(x, y)$ is the two dimensional system transfer function (impulse response). It will also be called the system function here.

Even though, in principle, $\phi(x, y)$ and $t'(x, y)$ are both defined over the complete range of x and y, in practice they are both limited. Clearly the image itself must be finite in extent spatially; the system function $t'(x, y)$ is also generally quite limited. Should it represent the point spread function of an imaging device it would be significantly non-zero over only a small range of x and y. (If it were an impulse it can be shown that (5.3) yields $r(x, y) = \phi(x, y)$ as would be expected).

In order to be applicable to digital image data it is necessary to modify (5.3) so that the discrete natures of x and y are made explicit and, consequently, the integrals are replaced by suitable summations. If we let i, j represent discrete values of x, y and similarly μ, v represent discrete values of the integration variables u, v then (5.3) can be written

$$r(i, j) = \sum_{\mu} \sum_{v} \phi(\mu, v)\, t'(i - \mu, j - v) \tag{5.4}$$

which is a digital form of the two dimensional convolution integral. The sums are taken over all values of μ, v for which a non-zero result exists.

To see how (5.4) would be used in practice it is necessary to interpret the sequence of operations implied. For clarity, assume the non-zero range of $t'(i, j)$ is quite small compared with that for the image $\phi(i, j)$. Also assume $t'(i, j)$ is a square array of samples – for example 3×3. These assumptions in no way prejudice the generality of what follows.

In (5.4) the negative signs on μ and v in $t'(i - \mu, j - v)$ imply a reflection through both axes. This is tantamount to a rotation of the system function through $180°$ before any further operations take place. Let the rotated form be called $t(\mu - i, v - j)$.

Equation (5.4) implies that a brightness value for the response image at pixel location i, j – viz. $r(i, j)$ is given by taking the non-zero products of the original version of the image and the rotated system function and adding these together. In so doing, note that the origin of the μ, v co-ordinates is the same as for the i, j co-ordinates just as the dummy and real variable co-ordinates in (5.2) and (5.3) are the same. Also note that the effect of $\mu - i, v - j$ in $t(\mu - i, v - j)$ is to shift the origin of the rotated system function to the location i, j – the current pixel address for which a new brightness value is to be calculated. These two points are illustrated in Fig. 5.3. The need to produce brightness values for pixels in the response image at every i, j means that the origin of the rotated system function must be moved progressively, implying that a different set of products between the original image and rotated system function is taken every time.

The sequence of operations described between the rotated system function and the original image are the same as those noted in Sect. 5.2 in regard to (5.1). The only difference in fact between (5.1) and (5.4) lies in the definitions of the indices m, n and μ, v. In (5.1) the pixel addresses are referred to an origin defined at the bottom left hand corner of the template, with the successive shifts mentioned in the accompanying description. This is a simple way to describe the template and readily allows any template size to be defined. In (5.4) the shifts are incorporated into the expression by defining the image and system function origins correctly.

The templates of Sect. 5.2 are equivalent to the rotated system functions of this section. Consequently any image modification operation that can be modelled by convolution, and described in principle in a manner similar to that in Fig. 5.2, can also be expressed in template form. For example, if the point spread function of a display device is known, then an equivalent template can be devised, noting that the 180° rotation is important if the system function is not symmetric. In a like manner intentional modifications of an image – such as smoothing and sharpening – can also be implemented using templates. The actual template entries to be used can often be developed intuitively, having careful regard to the desired results. Alternatively the system function $t'(i, j)$ necessary to implement a particular desired filtering operation can be defined first in the spatial frequency domain, using the material from Chap. 7, and then transformed back to the image domain. Rotation by 180° then gives the required template.

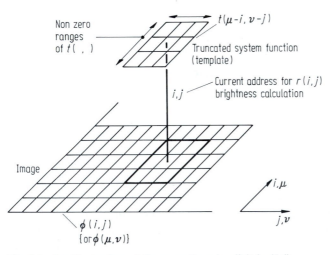

Fig. 5.3. An illustration of the operations implicit in (5.4)

5.4 Image Domain Versus Fourier Transformation Approaches

Most geometric enhancement procedures can be implemented using either the Fourier transform approach of Chap. 7 or the image domain procedures of this chapter. Which option to use depends upon several factors such as available software, familiarity with each method including its limitations and idiosynchrasies, and ease of use. A further consideration relates to computer processing time. This last issue is pursued here in order to indicate, from a cost viewpoint, when one method should be chosen in favour of the other.

Both the Fourier transform, frequency domain process and the template approach consist only of sets of multiplications and additions. No other mathematical operations are involved. It is sufficient, therefore, from the point of view of cost, to make a comparison based upon the number of multiplications and number of additions necessary to achieve a result. Here we will ignore the additions since they are generally faster than multiplications for most computers and also since they are comparable in number to the multiplications involved.

For an image of $K \times K$ pixels (only a square image is considered for simplicity) and a template of size $M \times N$ the total number of multiplications necessary to evaluate (5.1) for every image pixel (ignoring any difficulties with the edges of the image) is

$$N_C = MNK^2 \tag{5.5a}$$

From the material presented in Sect. 7.8.4 it can be seen that the number of (complex) multiplications required in the frequency domain approach is,

$$N_F = 2K^2 \log_2 K + K^2 \tag{5.5b}$$

A cost comparison therefore is

$$\frac{N_C}{N_F} = MN/(2\log_2 K + 1) \tag{5.6}$$

When this figure is less than unity it is more economical to use the template operator approach. Otherwise the Fourier transformation procedure is more cost-effective. Clearly this does not take into account program overheads (such as the bit shuffling required in the frequency domain approach, how data is buffered into computer memory from disc for processing) and the added cost of complex multiplications; however it is a reasonable starting point in choosing between the methods.

Table 5.1 contains a number of values of N_C/N_F for various image and template sizes, from which it is seen that, provided a 3×3 template will implement the enhancement required, then it is always more cost-effective than enhancement based upon Fourier transformation. Similarly, a non isotropic 3×5 template is more cost-effective for practical image sizes. However the spatial frequency domain technique will be economical if very large templates are needed, although only marginally so for large images.

As a final comment in this comparison it should be remarked that the frequency domain method is able to implement processes not possible (or at least not viable) with

Table 5.1. Time comparison of geometric enhancement by template operation compared with the Fourier transformation approach – based upon multiplication count comparison, described by (5.6) in which the added cost of complex multiplication is ignored

Template size Image size	3×3	3×5	5×5	5×7	7×7
16×16	1.00	1.67	2.78	3.89	5.44
64×64	0.69	1.15	1.92	2.69	3.77
128×128	0.60	1.00	1.67	2.33	3.27
256×256	0.53	0.88	1.47	2.06	2.88
512×512	0.47	0.79	1.32	1.84	2.58
1024×1024	0.43	0.71	1.19	1.67	2.33
2048×2048	0.39	0.65	1.09	1.52	2.13
4096×4096	0.36	0.60	1.00	1.40	1.96

template operators. Removal of periodic noise is one example. This is particularly simple in the spatial frequency domain but requires unduly complex templates or even nonlinear operators (such as median filtering) in the image domain. Notwithstanding these remarks the template approach is a popular one since often 3×3 and 5×5 templates are sufficient to achieve desired results.

5.5   Image Smoothing (Low Pass Filtering)

5.5.1 Mean Value Smoothing

Images can contain random noise superimposed on the pixel brightness values owing to noise generated in the transducers which acquire the image data, systematic quantisation noise in the signal digitising electronics and noise added to the video signal during transmission. This will show as a speckled 'salt and pepper' pattern on the image in regions of homogeneity; it can be removed by the process of low pass filtering or smoothing, unfortunately usually at the expense of some high frequency information in the image. To smooth an image a uniform template in (5.1) is used with entries

$$t(m, n) = 1/MN \quad \text{for all} \quad m, n$$

so that the template response is a simple average of the pixel brightness values currently within the template, viz

$$r(i, j) = \frac{1}{MN} \sum_{m=1}^{M} \sum_{n=1}^{N} \phi(m, n) \tag{5.7}$$

The pixel at the centre of the template is thus represented by the average brightness level in a neighbourhood defined by the template dimensions. This is an intuitively obvious template for smoothing and is equivalent to using running averages for smoothing time series information.

It is evident that high frequency information such as edges will also be averaged and lost. This loss of high frequency detail can be circumvented somewhat if a threshold is applied to the template response in the following manner.
Let

$$\varrho(i,j) = \frac{1}{MN} \sum_{m=1}^{M} \sum_{n=1}^{N} \phi(i,j)$$

then

$$r(i,j) = \varrho(i,j) \quad \text{if} \quad |\phi(i,j) - \varrho(i,j)| < T$$
$$= \phi(i,j) \quad \text{otherwise}$$

where T is a prespecified threshold. T could be determined *a priori* based upon knowledge of or an estimate of scene signal to noise ratio.

Eliason and McEwan (1990) recommend choosing the threshold as a multiple of the standard deviation of brightness within the template window. This provides better noise removal in homogeneous regions while allowing better preservation of edges and other valid high spatial frequency detail.

A simple illustration of image smoothing by averaging over a template, both with and without the application of a threshold, is given in Fig. 5.4. For clarity this is based upon a hypothetical one dimensional image, or alternatively a single line of image data, with which a 3×1 template is used. In this manner the actual numerical modification of pixel brightness values can be observed.

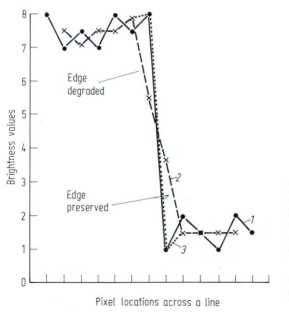

Pixel locations across a line

Fig. 5.4. Illustration of the effect of 3×1 averaging across a single line of image data with and without thresholding. Note, thresholding preserves edges whilst reducing noise.
1 original image,
2 3×1 smoothing,
3 3×1 smoothing with threshold of 1

In principle, templates of any shape and size can be used. Larger templates give more smoothing (and greater loss of high frequency detail) whereas horizontal rectangular templates will smooth horizontal noise but leave noise and high frequency detail in the

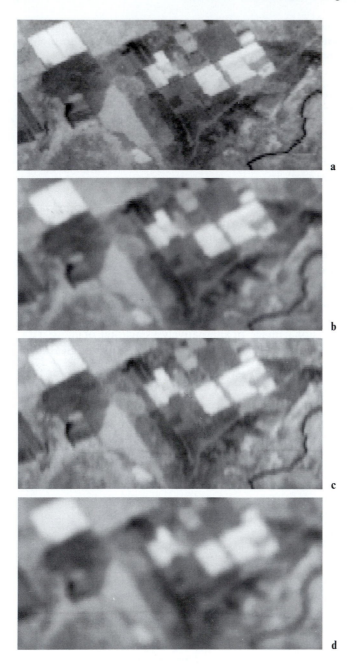

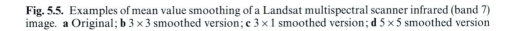

Fig. 5.5. Examples of mean value smoothing of a Landsat multispectral scanner infrared (band 7) image. **a** Original; **b** 3×3 smoothed version; **c** 3×1 smoothed version; **d** 5×5 smoothed version

vertical direction relatively unaffected by comparison. In Fig. 5.5 several different smoothing templates have been applied to a Landsat multispectral scanner infrared image.

Commonly, smoothing by template methods is referred to as box car filtering. When based upon (5.7) it is also called mean value smoothing, or averaging.

5.5.2 Median Filtering

Disadvantages of the thresholding method for avoiding edge deterioration are that it adds to the computational cost of the smoothing operation and T must be determined. An alternative technique for smoothing in which the edges in an image are maintained is that of median filtering. In this the pixel at the centre of the template is given the median brightness value of all the pixels covered by the template — i. e. that value which has as many values higher and lower. (For example, the median of 4, 6, 3, 7, 9, 2, 1, 8, 8 is 6, whereas the mean is 5.3). Figure 5.6 shows the effect of median filtering on a single line of image data compared with simple box car averaging, which uses the mean of pixel brightness values. Again, it can be seen that most of the original edge is preserved.

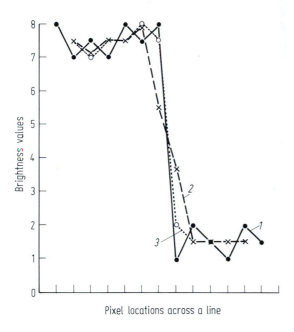

Fig. 5.6. Comparison of simple averaging and median filtering of a single line of image data. *1* original image, *2* 3×1 smoothing, *3* 3×1 median filtering

An application for which median filtering is well suited is the removal of impulse-like noise. This is because pixels corresponding to noise spikes are atypical in their neighbourhood and will be replaced by the most typical pixel in that neighbourhood. Figure 5.7 gives an example of median filtering on an image with added black and white impulsive noise.

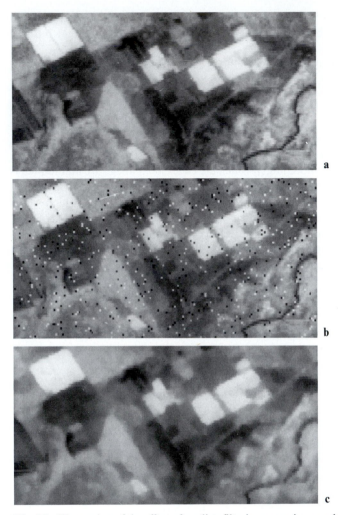

Fig. 5.7. Illustration of the effect of median filtering on an image which contains impulsive noise.
a Original image; **b** Image with noise; **c** Filtered image

Finally it should be noted that median filtering is not a linear function of the
brightness values of the image pixels. Consequently it is not a convolution operation in
the sense described in Sect. 5.3.

5.6 Edge Detection and Enhancement

Edge enhancement is a particularly simple and effective means for increasing geometric
detail in an image. It is performed by first detecting edges and then either adding these
back into the original image to increase contrast in the vicinity of an edge, or
highlighting edges using saturated (black, white or colour) overlays on borders.

There are essentially three economical techniques for detecting edges using image domain techniques. These are

(i) by using an edge detecting template,
(ii) by calculating spatial derivatives, or
(iii) by subtracting a smoothed image from its original.

These three approaches are treated in the following sections.

5.6.1 Linear Edge Detecting Templates

A 3×3 template that detects vertical edges in image data is

$$t(m, n) = \begin{array}{|c|c|c|} \hline -1 & 0 & +1 \\ \hline -1 & 0 & +1 \\ \hline -1 & 0 & +1 \\ \hline \end{array}$$

(5.8 a)

As can be inferred from its structure it computes a value for the central pixel under the template that is the accumulated difference horizontally between pixels on three adjacent rows. To see this, consider a region of an image which is basically dull (brightness value 2) into which protrudes a bright object (brightness value 8) as depicted in Fig. 5.8 a. Application of the template yields the responses shown in Figure 5.8 b, wherein the vertical edge between the object and background has been detected but not the horizontal edge. Note that the edge is defined by two columns of pixels, one on either side of the true edge position. A threshold would normally be applied to the template response (say 9 in the case of Fig. 5.8) to define the edge pixels.

```
2 2 2 2 8 8 8 8          .   .   .    .   .   .   .   .
2 2 2 2 8 8 8 8          .  0   0   18  18   0   0  .
2 2 2 2 8 8 8 8          .  0   0   18  18   0   0  .
2 2 2 2 8 8 8 8          .  0   0   12  12   0   0  .
2 2 2 2 2 2 2 2          .  0   0    6   6   0   0  .
2 2 2 2 2 2 2 2          .  0   0    0   0   0   0  .
2 2 2 2 2 2 2 2          .  0   0    0   0   0   0  .
2 2 2 2 2 2 2 2          .   .   .    .   .   .   .   .
```

dull background   vertical edge 2 pixels wide

a   **b**

Fig. 5.8. Image **a** and edges detected by a vertically sensitive template **b**; Dots indicate indeterminate edge responses for this example

Templates for detecting edges in other orientations are:

$$\begin{array}{|c|c|c|} \hline -1 & -1 & -1 \\ \hline 0 & 0 & 0 \\ \hline +1 & +1 & +1 \\ \hline \end{array} \qquad \begin{array}{|c|c|c|} \hline 0 & +1 & +1 \\ \hline -1 & 0 & +1 \\ \hline -1 & -1 & 0 \\ \hline \end{array} \qquad \begin{array}{|c|c|c|} \hline +1 & +1 & 0 \\ \hline +1 & 0 & -1 \\ \hline 0 & -1 & -1 \\ \hline \end{array}$$

horizontal   diagonal   (5.8 b)

Clearly all four 3×3 templates have to be applied to an image to detect its edges in all orientations. This requires four passes over the image data, computing each template response for each pixel. At the completion of all processing the four template responses for each pixel are compared and the pixel labelled (as an edge in a particular direction) according to the largest template response provided that the response is also above a user specified threshold. Choosing a threshold too low will lead to many false edge counts. These contribute to noise in the processed image. Conversely, if the threshold is set too high, there will be little continuity in the detected edges.

5.6.2 Spatial Derivative Techniques

If an image consists of a continuous brightness function of a pair of continuous co-ordinates, x and y, say $\phi(x, y)$, then a vector gradient can be defined in the image according to

$$\nabla \phi(x, y) = \frac{\partial}{\partial x} \phi(x, y) \, i + \frac{\partial}{\partial y} \phi(x, y) \, j \tag{5.9}$$

where i, j are a pair of unit vectors. The direction of the vector gradient is the direction of maximum upward slope and its amplitude is the value of the slope. For edge detection operations usually only the magnitude of the gradient, defined by

$$|\nabla| = \sqrt{\nabla_1^2 + \nabla_2^2} \tag{5.10a}$$

is retained, in which

$$\nabla_1 = \frac{\partial}{\partial x} \phi(x, y) \quad \nabla_2 = \frac{\partial}{\partial y} \phi(x, y) \tag{5.10b}$$

The direction of the gradient is usually of interest only in contouring applications or in determining aspect in digital terrain models.

5.6.2.1 The Roberts Operator

For digital image data, in which x and y are discretised, the continuous derivatives in (5.10) are replaced by differences. For example, it is possible to define

$$\nabla_1 = \phi(i, j) - \phi(i + 1, j + 1) \tag{5.11a}$$

and

$$\nabla_2 = \phi(i + 1, j) - \phi(i, j + 1) \tag{5.11b}$$

which are the discrete components of the vector derivative at the point $i + \frac{1}{2}, j + \frac{1}{2}$, in the diagonal directions. This estimate of gradient is called the Roberts operator, and is by definition associated with the pixel i, j.

Application of the Roberts operator to the model image of Fig. 5.8a yields the results shown in Fig. 5.9a, in which it will be seen that both horizontal and vertical edges are detected, as will be diagonal edges. Since this procedure computes a local gradient it is necessary to choose a threshold value above which edge gradients are said to occur.

This is usually chosen with experience of a particular image. Frequently however it is useful to produce gradient maps in which pixels, for which the local gradient lies within prespecified upper and lower bounds, are displayed. Conventionally, the responses are placed to the left and upper sides of the edges.

```
0  0  0  8.5   0    0    0   .              .   .    .    .    .    .   .
0  0  0  8.5   0    0    0   .              .  0  0  24  24   0   0   .
0  0  0  8.5   0    0    0   .              .  0  0  24  24   0   0   .
0  0  0  6.0  8.5  8.5  8.5  .              .  0  0  19  25  24  24  .
0  0  0   0    0    0    0   .              .  0  0   8  19  24  24  .
0  0  0   0    0    0    0   .              .  0  0   0   0   0   0   .
0  0  0   0    0    0    0   .              .  0  0   0   0   0   0   .
.  .   .   .    .    .    .                  .   .    .    .    .    .   .

a                                           b
```

Fig. 5.9. Response of **a** the Robert's operator and **b** the Sobel operator to the model image data of Fig. 5.8 a. Dots are indeterminate responses from edge pixels

5.6.2.2 The Sobel Operator

A better edge estimator than the Roberts operator is the Sobel operator, which computes discrete gradient in the horizontal and vertical directions *at* the pixel location i, j. For this, which is clearly more costly to evaluate, the orthogonal components of gradient are

$$\nabla_1 = \{\phi(i-1, j+1) + 2\phi(i-1, j) + \phi(i-1, j-1)\} \tag{5.12a}$$
$$- \{\phi(i+1, j+1) + 2\phi(i+1, j) + \phi(i+1, j-1)\}$$

and

$$\nabla_2 = \{\phi(i-1, j+1) + 2\phi(i, j+1) + \phi(i+1, j+1)\} \tag{5.12b}$$
$$- \{\phi(i-1, j-1) + 2\phi(i, j-1) + \phi(i+1, j-1)\}$$

Applying this to the example of Fig. 5.8a produces the responses shown in Fig. 5.9 b. Again, both horizontal and vertical edges are detected as will be edges on a diagonal slope. As before, a threshold on the responses is generally chosen to allow an edge map to be produced in which small responses, resulting from noise or minor gradients, are suppressed. Also gradient maps can be produced illustrating regions in which the local slope lies within user specified bounds.

   It can be seen that the Sobel operator is equivalent to simultaneous application of the templates:

$$\nabla_1 = \begin{array}{|c|c|c|} \hline 1 & 2 & 1 \\ \hline 0 & 0 & 0 \\ \hline -1 & -2 & -1 \\ \hline \end{array} \qquad \nabla_2 = \begin{array}{|c|c|c|} \hline -1 & 0 & 1 \\ \hline -2 & 0 & 2 \\ \hline -1 & 0 & 1 \\ \hline \end{array}$$

5.6.3 Thinning, Linking and Border Responses

Should an edge map be of interest (or indeed a line map using the methods of Sect. 5.7) then the product resulting from using the above procedures is likely to contain many double width, or wider lines, such as those seen in Figs. 5.8 and 5.9 and, in addition, may have lines with many breaks. Such a map can be tidied up by thinning edges or lines that are too thick and by linking together segments that appear to belong to the same edge but are separated by a break. Thinning and linking are not commonly employed in remote sensing image analysis. However should they require consideration available techniques will be found in Babu and Nevatia (1980) and Paul and Shanmugan (1982).

In the examples of Figs. 5.8 and 5.9 border pixels for which detector responses could not be determined were simply left unprocessed. Since images encountered in remote sensing are frequently much larger than 100×100 pixels, this is a common practice as the loss of borders is not all that significant. A more elegant means for treating edge pixels however is to create artificial borders of pixels around the image. These are used in the generation of edge pixel responses but are not themselves replaced by a template response. The values given to the artificial border pixels can be taken simply from the adjacent image pixels or, more acceptably from a theoretical viewpoint, they can be taken from the pixels on the extreme opposite edge of the image if only small templates are used. This is based upon the concept, drawn from digital signal processing, that the image, being spatially discretised or sampled, should be regarded as one period both horizontally and vertically of an infinite periodic replication of the array of pixels.

5.6.4 Edge Enhancement by Subtractive Smoothing (Sharpening)

While treated in the context of edge enhancement this technique really leads to the enhancement of all high spatial frequency detail in an image including edges, lines and points of high gradient. It is probably better regarded therefore as a sharpening technique.

A smoothed image retains all low spatial frequency information but has its high frequency features, such as edges and lines, attenuated (unless edge preservation procedures such as thresholding are employed). Consequently, if a smoothed image is subtracted from its original the resultant difference image will have only the edges and lines substantially remaining. This is illustrated for a single line of image data in Fig. 5.10. After the edges are determined in this manner, the difference image can be added back to the original (in varying proportions) to give an edge enhanced image. This is also illustrated in Fig. 5.10.

The difference operation to create a high spatial frequency image can give negative brightness values as seen in Fig. 5.10b. Provided the image is not displayed, this produces no problems. For display however it is common to scale the difference image such that a zero difference is displayed as mid-grey with positive differences towards white and negative differences towards black. When the difference image is added back to the original, negative brightnesses can again result. Again, this can be handled by level shifting or scaling, or simply by setting negative brightness values to zero.

Figure 5.11 shows the sharpening technique of subtractive smoothing applied to bands 4, 5 and 7 of a Landsat multispectral scanner image and the effect this has on the colour composite formed from these bands. As noted the sharpened image has clearer high frequency detail; however there is a tendency for noise to be enhanced, as might be expected.

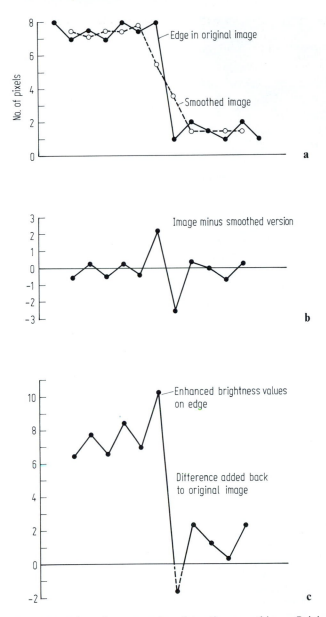

Fig. 5.10. Edge enhancement by subtractive smoothing. **a** Original line of image data, along with smoothed version; **b** Original line of data minus the smoothed version to leave 'edges' detected; **c** Addition of 'edges' (general high frequency detail) back to the original image to provide a sharpened version

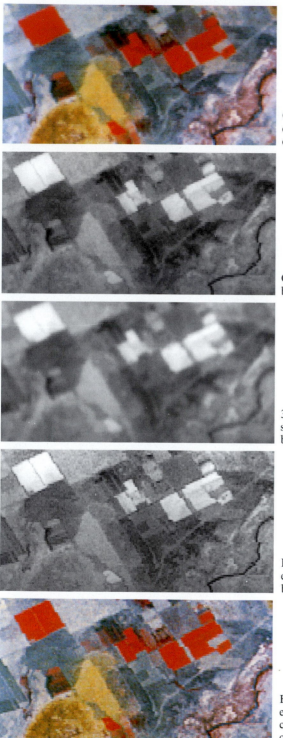

Original
colour
composite

Original
band 7

3 × 3
smoothed
band 7

Edge
enhanced
band 7

Edge
enhanced
colour
composite

Fig. 5.11. Illustration of subtractive smoothing as an image sharpening procedure

5.7 Line Detection

5.7.1 Linear Line Detecting Templates

Line features such as rivers and roads in satellite images can be detected as pairs of edges if they are more than one pixel wide or alternatively, if they are a single pixel in width, they can be detected using the following line detecting templates:

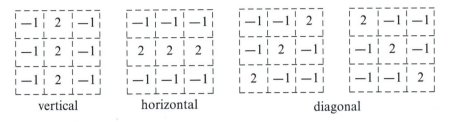

<table>
<tr><td>−1</td><td>2</td><td>−1</td></tr>
<tr><td>−1</td><td>2</td><td>−1</td></tr>
<tr><td>−1</td><td>2</td><td>−1</td></tr>
</table>
vertical

<table>
<tr><td>−1</td><td>−1</td><td>−1</td></tr>
<tr><td>2</td><td>2</td><td>2</td></tr>
<tr><td>−1</td><td>−1</td><td>−1</td></tr>
</table>
horizontal

<table>
<tr><td>−1</td><td>−1</td><td>2</td></tr>
<tr><td>−1</td><td>2</td><td>−1</td></tr>
<tr><td>2</td><td>−1</td><td>−1</td></tr>
</table>

<table>
<tr><td>2</td><td>−1</td><td>−1</td></tr>
<tr><td>−1</td><td>2</td><td>−1</td></tr>
<tr><td>−1</td><td>−1</td><td>2</td></tr>
</table>
diagonal

These templates seem not to have been used to any great extent in remote sensing image processing since lines, in addition to edges, are enhanced using the gradient and subtractive smoothing techniques of Sect. 5.6. Moreover, with sensor resolutions available up to 1982, not many single pixel width linear features have been apparent in imagery. With resolutions in the range of 10 m to 30 m however, cultural features such as roads, could be amenable to detection using line related templates.

5.7.2 Non-linear and Semi-linear Line Detecting Templates

The line detecting templates of Sect. 5.7.1 are regarded as linear since their convolution with image data is a linear mathematical operation. Some nonlinear line detecting template operations have also been proposed. To describe these it is of value to denote a 3×3 neighbourhood of pixels in an image as

$$A_1 \quad B_1 \quad C_1$$
$$A_2 \quad B_2 \quad C_2$$
$$A_3 \quad B_3 \quad C_3$$

A nonlinear line detector algorithm, proposed by Rosenfeld and Thurston (1971), establishes pixel B_2 as part of a dark vertical line if

$$A_i, C_i > B_i$$

by a prespecified threshold. Similar expressions apply for lines of other orientations and for bright lines on dark backgrounds.

Vanderbrug (1976) has proposed what he calls a semilinear detector. For the pixel array above this determines B_2 as part of a dark vertical line if

$$\sum_{i=1}^{3} A_i \quad \text{and} \quad \sum_{i=1}^{3} C_i > \sum_{i=1}^{3} B_i$$

by some prespecified threshold.

Gurney (1980) has noted that the semilinear detector works better than the non linear algorithm although line thickening results and computational cost is high. These disadvantages are obviated by the use of the additional constraint with the semilinear algorithm:

$$A_2 > B_2 \quad \text{and} \quad C_2 > B_2$$

Gurney also discusses means by which the thresholds for the semilinear detector can be effectively established.

5.8 General Convolution Filtering

It is clear that smoothing, edge and line detection represent just particular ways of defining the template entries in (5.1) and that more general spatial filtering operations could be defined by loading the template in different fashions. For example, edge enhancement by subtractive smoothing treated in Sect. 5.6.4 could be implemented by the single template

$$
\begin{bmatrix} -a & -a & -a \\ -a & 2-a & -a \\ -a & -a & -a \end{bmatrix} =
\begin{bmatrix} 0 & 0 & 0 \\ 0 & 1 & 0 \\ 0 & 0 & 0 \end{bmatrix} +
\begin{bmatrix} 0 & 0 & 0 \\ 0 & 1 & 0 \\ 0 & 0 & 0 \end{bmatrix} -
\begin{bmatrix} a & a & a \\ a & a & a \\ a & a & a \end{bmatrix}
$$

where $a = 1/9$. This template implements a high spatial frequency boosting.

By expanding the size of the template it is possible to determine detectors that are sensitive to edges and lines in other than the four common orientations. In addition templates can be used for recognition of large objects in imagery, where the templates are loaded with zeros, except for those locations corresponding to the shape and orientation of an object of interest. In this case the procedure is referred to as template matching and is more akin to correlation than convolution (Rosenfeld, 1978).

5.9 Shape Detection

Recognition of shapes in image data has not been considered in remote sensing as extensively as it has in object recognition exercises, such as robot vision. Presumably this is because the resolution generally available in the past has been insufficient to define shape with any degree of precision. However with ground resolution elements better than 20 to 30 m in imagery, shapes such as those of rectangular fields in agriculture, and circular pivotal irrigation systems are quite well defined.

Shape recognition can be carried out using template techniques, in which the templates are chosen according to the shape of interest (Hord, 1982). The operation required is one of correlation and not the convolution operation of (5.4). Correlation is defined by that same expression but with additions in place of subtractions. A major difficulty with this approach, which as a consequence renders the technique of limited value in practice, is that the template must match not only the shape of interest, but also its size and orientation. Other methods therefore are often employed. These include the adoption of shape factors (Underwood, 1970), moments of area (Pavlidis, 1978) and Fourier transforms of shape boundaries (Pavlidis, 1980). In each of these the shape must first be delineated from the rest of the image. This is achieved by edge and line detection processes.

References for Chapter 5

The template techniques that form the basis of much of the material presented in this chapter are treated also by Duda and Hart (1973), Moik (1980) and Hord (1982). Gonzalez and Woods (1992) present the method in a vector formulation, noting that the convolution operation in (5.1) can be expressed as the scalar product of the vector of template entries and the vector of pixel brightnesses currently covered by the template. Such an approach allows templates to be designed to detect combinations of lines and edges. Since the template entries are expressed in vector form they can be used to define vector sub-spaces into which an image has projections. A large projection into an edge sub-space implies edges in the image of the pixel currently being assessed, and so on. This is assessed in terms of the vector angle between the image pixel vector and the subspace basis vectors (template entries).

Gradient methods are covered also by Moik (1980), Gonzalez and Woods (1992), Hord (1982) and to an extent by Castleman (1996). Gonzalez and Woods also include discussions on the use of thresholds applied to the response of gradient operators. Vanderbrug (1976) and Gurney (1980) consider the properties of nonlinear and semilinear line detecting templates.

Paine and Lodwick (1989) provide a good discussion of the application of edge detection methods, while Brzakovic et al. (1991) consider the use of rule-based methods to assist in edge detection when a number of templates is involved. While the edge detection methods treated here have been applicable only to single bands of data, Cumani (1991) and Drewnick (1994) have proposed operators for use on multispectral data.

K. R. Babu and R. Nevatia, 1980: Linear Feature Extraction and Description. Computer Graphics and Image Processing, 13, 257–269.

E. O. Brigham, 1974: The Fast Fourier Transform, N. J. Prentice-Hall.

E. O. Brigham, 1988: The Fast Fourier Transform and its Applications, N. J. Prentice-Hall.

D. Brzakovic, R. Patton and R. L. Wang, 1991: Rule-based Multitemplate Edge Detector. CVGIP: Graphical Models and Image Processing, 53, 258–268.

K. R. Castleman, 1996: Digital Image Processing, N. J. Prentice-Hall.

A. Cumani, 1991: Edge Detection in Multispectral Images. CVGIP: Computer Models and Image Processing, 53, 40–51.

R. O. Duda and P. E. Hart, 1973: Pattern Classification and Scene Analysis, N. Y., Wiley.

C. Drewnick, 1994: Multi-Spectral Edge Detection. Some Experiments on Data from Landsat TM. Int. J. Remote Sensing, 15, 3743–3765.

E. M. Eliason and A. S. McEwan, 1990: Adaptive Box Filters for Removal of Random Noise from Digital Images. Photogrammetric Engineering and Remote Sensing, 56, 453–458.

R. C. Gonzalez and R. E. Woods, 1992: Digital Image Processing, Mass., Addison-Wesley.

C. M. Gurney, 1980: Threshold Selection for Line Detection Algorithms. IEEE Trans. Geoscience and Remote Sensing, GE-18, 204–211.

R. M. Hord, 1982: Digital Image Processing of Remotely Sensed Data, N. Y. Academic.

C. D. McGillem and G. R. Cooper, 1984: Continuous and Discrete Signal and Systems Analysis, 2e, N. Y., Holt, Reinhard and Winston.

J. G. Moik, 1980: Digital Processing of Remotely Sensed Images, N. Y., Academic.

S. H. Paine and G. D. Lodwick, 1989: Edge Detection and Processing of Remotely Sensed Digital Images. Photogrammetria (PRS), 43, 323–336.

C. Paul and K. S. Shanmugan, 1982: A Fast Thinning Operator. IEEE Trans. Systems, Man & Cybernetics, SMC-12, 567–569.

T. Pavlidis, 1978: A Review of Algorithms for Shape Analysis. Computer Graphics and Image Processing, 7, 243–258.

T. Pavlidis, 1980: Algorithms for Shape Analysis of Contours and Waveforms. IEEE Trans. Pattern Analysis and Machine Intelligence, PAMI-2, 301–312.

A. Rosenfeld, 1978: Image Processing and Recognition, Technical Report 664, Computer Vision Laboratory, University of Maryland.

A. Rosenfeld and M. Thurston, 1971: Edge and Curve Detection for Visual Scene Analysis, IEEE Trans. Computers, C-20, 562–569.

E. E. Underwood, 1970: Quantitative Stereology, Mass., Addison-Wesley.
G. J. Vanderbrug, 1976: Line Detection in Satellite Imagery, IEEE Trans. Geoscience Electronics, GE-14, 37–44.

Problems

5.1 The template entries for line and edge detection sum to zero whereas those for smoothing do not. Why do you think that is so?

5.2 Repeat the example of Fig. 5.10 but by using a [5 × 1] smoothing operation in part (a), rather than [3 × 1] smoothing.

5.3 Repeat the example of Fig. 5.10 but by using a [3 × 1] median filtering operation in part (a) rather than [3 × 1] mean value smoothing.

5.4 An alternative smoothing process to median and mean value filtering using template methods is known as modal filtering. In this approach a pixel at the centre of the template neighbourhood is replaced by the brightness value that occurs most frequently in the neighbourhood. Apply [3 × 1] and [5 × 1] modal filters to the image data of Fig. 5.6. Note differences in the results compared with mean value and median smoothing, particularly around the edges.

5.5 Suppose S is a template operation that implements smoothing and 0 is the template operator that leaves an image unchanged (see Sect. 5.8). Then an edge enhanced image created by the subtractive smoothing approach of Sect. 5.6.4 can be expressed according to

$$\text{New image} = 0 \text{ (old image)} + 0 \text{ (old image)} - S \text{ (old image)}$$

Rewrite this expression to incorporate two user defined parameters α and β that will cause the formula to implement any of smoothing, edge detection or edge enhancement.

5.6 (Requires vector algebra background)
Show that template methods for line and edge detection can be expressed as the scalar product of a vector composed from the template entries and a vector formed from the neighbourhood of pixels currently covered by the template. Show how the angle between the template and pixel vectors can be used to assess the edge or line feature a current pixel most closely corresponds to. (See Gonzalez and Woods (1992)).

5.7 The following kernel is sometimes convolved with image data. What operation will it implement?

0	−1	0
−1	4	−1
0	−1	0

5.8 Consider the middle pixel shown in the figure below and calculate its new value if
 (i) a 3 × 3 median filtering is applied,
 (ii) a 3 × 3 template which performs edge enhancement by subtractive smoothing is applied,
 (iii) a 3 × 1 image smoothing template with a threshold 2 is applied,
 (iv) the Sobel operator is applied for edge detection.

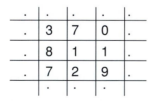

5.9 Image smoothing can be performed by template operators that implement averaging or median filtering. Compare the methods, particularly as they affect edges. Would you expect median filtering to be useful in edge enhancement by the technique of subtracting a smoothed image from the original?

Chapter 6
Multispectral Transformations of Image Data

The multispectral or vector character of most remote sensing image data renders it amenable to spectral transformations that generate new sets of image components or bands. These components then represent an alternative description of the data, in which the new components of a pixel vector are related to its old brightness values in the original set of spectral bands via a linear operation. The transformed image may make evident features not discernable in the original data or alternatively it might be possible to preserve the essential information content of the image (for a given application) with a reduced number of the transformed dimensions. The last point has significance for displaying data in the three dimensions available on a colour monitor or in colour hardcopy, and for transmission and storage of data.

The role of this chapter is to present image transformations of value in the enhancement of remote sensing imagery, although some also find application in preconditioning image data prior to classification by the techniques of Chaps. 8 and 9. The techniques covered, which appeal directly to the vector nature of the image, include the principal components transformation and so-called band arithmetic. The latter includes the creation of ratio images. Some specialised transformations, such as the Kauth-Thomas tasseled cap transform are also treated.

6.1 The Principal Components Transformation

The multispectral or multidimensional nature of remote sensing image data can be accommodated by constructing a vector space with as many axes or dimensions as there are spectral components associated with each pixel. In the case of Landsat Thematic Mapper data it will have seven dimensions while for SPOT HRV data it will be three dimensional. For hyperspectral data there may be several hundred axes. A particular pixel in an image is plotted as a point in such a space with co-ordinates that correspond to the brightness values of the pixels in the appropriate spectral components. For simplicity the treatment to be developed in this topic will be based upon a two dimensional multispectral space (say Landsat Multispectral Scanner bands 5 and 7) since the diagrams are then easily understood and the mathematical detail is readily assimilated. The results derived however are perfectly general and apply to data of any dimensionality.

6.1.1 The Mean Vector and Covariance Matrix

The positions of pixel points in multispectral space can be described by vectors, whose components are the individual spectral responses in each band. Strictly, these are

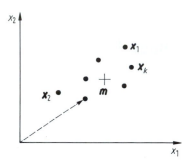

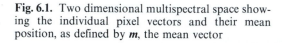

Fig. 6.1. Two dimensional multispectral space showing the individual pixel vectors and their mean position, as defined by **m**, the mean vector

vectors drawn from the origin to the pixel point as seen in Appendix C, but this concept is not used explicitly. Consider a multispectral space with a large number of pixels plotted in it as shown in Fig. 6.1, with each pixel described by its appropriate vector **x**. The mean position of the pixels in the space is defined by the expected value of the pixel vector **x**, according to

$$\boldsymbol{m} = \mathscr{E}\{\boldsymbol{x}\} = \frac{1}{K} \sum_{k=1}^{K} \boldsymbol{x}_k \tag{6.1}$$

where **m** is the mean pixel vector and the $\boldsymbol{x}_k$ are the individual pixel vectors of total number K; $\mathscr{E}$ is the expectation operator.

While the mean vector is useful to define the average or expected position of the pixels in multispectral vector space, it is of value to have available a means by which their scatter or spread is described. This is the role of the covariance matrix which is defined as

$$\Sigma_x = \mathscr{E}\{(\boldsymbol{x} - \boldsymbol{m})(\boldsymbol{x} - \boldsymbol{m})^t\} \tag{6.2a}$$

in which the superscript 't' denotes vector transpose. (See Appendix C).

An unbiased estimate of the covariance matrix is given by

$$\Sigma_x = \frac{1}{K-1} \sum_{k=1}^{K} (\boldsymbol{x}_k - \boldsymbol{m})(\boldsymbol{x}_k - \boldsymbol{m})^t \tag{6.2b}$$

The covariance matrix is one of the most important mathematical concepts in the analysis of multispectral remote sensing data, as a result of which it is of value to consider some sample calculations to enable its properties to be emphasised. In these it will be seen that if there is correlation between the responses in a pair of spectral bands the corresponding off-diagonal element in the covariance matrix will be large by comparison to the diagonal terms. On the other hand, if there is a little correlation, the off-diagonal terms will be close to zero. This behaviour can also be described in terms of the correlation matrix R whose elements are related to those of the covariance matrix by

$$\varrho_{ij} = v_{ij}/\sqrt{v_{ii}v_{jj}} \tag{6.3}$$

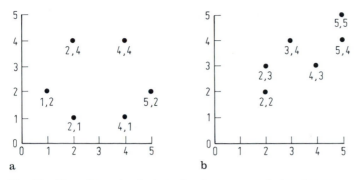

Fig. 6.2. Two dimensional data showing no correlation between components **a** and high correlation between components **b**

where ϱ_{ij} is an element of the correlation matrix and υ_{ij}, etc. are elements of the covariance matrix; υ_{ii} and υ_{jj} are the variances of the ith and jth bands of data. The ϱ_{ij} describe the correlation between band i and band j.

Consider the two, two-dimensional sets of data shown in Fig. 6.2. That in Fig. 6.2a shows little correlation between the two components: in other words, both components are necessary to describe where a pixel lies in the space. The data shown in Fig. 6.2b however exhibits a high degree of correlation between its two components, evident in the elongated spread of the data at an angle to the axes. One dimension on its own is almost sufficient to predict where a pixel lies in the space, and an increase or decrease in either component suggests a corresponding increase or decrease in the other. This is not the case with Fig. 6.2 a. In terms of the individual images corresponding to the bands of multispectral data, highly correlated bands as depicted in Fig. 6.2 b would yield image components very similar in appearance. Where one is dark the other will be dark and so on. The image components corresponding to Fig. 6.2a, however, would display no similar consistently common behaviour.

Table 6.1 shows a sample set of hand calculations undertaken to find the covariance and correlation matrices for Fig. 6.2 a. Normally this would be carried out by computer, particularly for data with higher dimensionality. As noted from the correlation matrix there is no correlation between the individual components of the data, a fact which is evident also in the zero off-diagonal entries in the covariance matrix. The entry of 2.40 in the upper left hand corner of the covariance matrix signifies that the data points have a variance of 2.40 along the horizontal axis, or a standard deviation of 1.55 about the mean. Similarly, the variance and standard deviation vertically are 1.87 and 1.37 respectively.

For the data in Fig. 6.2b, it is shown by a similar set of calculations to those in Table 6.1 that

$$m = \begin{bmatrix} 3.50 \\ 3.50 \end{bmatrix} \quad \Sigma_x = \begin{bmatrix} 1.900 & 1.100 \\ 1.100 & 1.100 \end{bmatrix}$$

and

$$R = \begin{bmatrix} 1.000 & 0.761 \\ 0.761 & 1.000 \end{bmatrix}$$

Table 6.1. Computation of covariance and correlation matrices.

The mean vector is $m = \begin{bmatrix} 3.00 \\ 2.33 \end{bmatrix}$

x	$x - m$	$[x - m][x - m]^t$
$\begin{bmatrix} 1 \\ 2 \end{bmatrix}$	$\begin{bmatrix} -2.00 \\ -0.33 \end{bmatrix}$	$\begin{bmatrix} 4.00 & 0.66 \\ 0.66 & 0.11 \end{bmatrix}$
$\begin{bmatrix} 2 \\ 1 \end{bmatrix}$	$\begin{bmatrix} -1.00 \\ -1.33 \end{bmatrix}$	$\begin{bmatrix} 1.00 & 1.33 \\ 1.33 & 1.77 \end{bmatrix}$
$\begin{bmatrix} 4 \\ 1 \end{bmatrix}$	$\begin{bmatrix} 1.00 \\ -1.33 \end{bmatrix}$	$\begin{bmatrix} 1.00 & -1.33 \\ -1.33 & 1.77 \end{bmatrix}$
$\begin{bmatrix} 5 \\ 2 \end{bmatrix}$	$\begin{bmatrix} 2.00 \\ -0.33 \end{bmatrix}$	$\begin{bmatrix} 4.00 & -0.66 \\ -0.66 & 0.11 \end{bmatrix}$
$\begin{bmatrix} 4 \\ 4 \end{bmatrix}$	$\begin{bmatrix} 1.00 \\ 1.67 \end{bmatrix}$	$\begin{bmatrix} 1.00 & 1.67 \\ 1.67 & 2.79 \end{bmatrix}$
$\begin{bmatrix} 2 \\ 4 \end{bmatrix}$	$\begin{bmatrix} -1.00 \\ 1.67 \end{bmatrix}$	$\begin{bmatrix} 1.00 & -1.67 \\ -1.67 & 2.79 \end{bmatrix}$

whereupon $\Sigma_x = \begin{bmatrix} 2.40 & 0 \\ 0 & 1.87 \end{bmatrix}$

and $R = \begin{bmatrix} 1.00 & 0 \\ 0 & 1.00 \end{bmatrix}$

where R is the correlation matrix.

Thus components 1 and 2 of the data are 76% correlated.

It should be noted that both the covariance and correlation matrices are symmetric and that an image data set, in which there is no correlation between any of its multispectral components, will have a diagonal covariance (and correlation) matrix.

6.1.2 A Zero Correlation, Rotational Transform

It is fundamental to the development of the principal components transformation to ask whether there is a new co-ordinate system in the multispectral vector space in which the data can be represented without correlation; in other words, such that the covariance matrix in the new co-ordinate system is diagonal. For a particular two dimensional vector space such a new co-ordinate system is depicted in Fig. 6.3. If the vectors describing the pixel points are represented as y in the new co-ordinate system then it is desired to find a linear transformation G of the original co-ordinates, such that

$$y = Gx \tag{6.4}$$

subject to the constraint that the covariance matrix of the pixel data in y space is diagonal. In y space the covariance matrix is, by definition,

$$\Sigma_y = \mathscr{E}\{(y - m_y)(y - m_y)^t\}$$

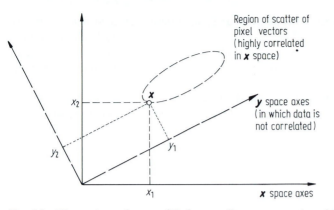

Fig. 6.3. Illustration of a modified co-ordinate system in which the pixel vectors have uncorrelated components

where m_y is the mean vector expressed in terms of the y co-ordinates. It is shown readily that

$$m_y = \mathscr{E}\{y\} = \mathscr{E}\{Gx\} = G\mathscr{E}\{x\} = Gm_x{}^1$$

where m_x is the data mean in x space. Therefore

$$\Sigma_y = \mathscr{E}\{(Gx - Gm_x)(Gx - Gm_x)^t\}$$

which can be written as

$$\Sigma_y = G\mathscr{E}\{(x - m_x)(x - m_x)^t\}\, G^t \quad {}^2$$

i.e. $\quad \Sigma_y = G\Sigma_x G^t \hspace{5cm}$ (6.5)

where Σ_x is the covariance of the pixel data in x space. Since Σ_y must, by demand, be diagonal, G can be recognised as the transposed matrix of eigenvectors of Σ_x, provided G is an orthogonal matrix. This can be seen from the material presented in Appendix C dealing with the diagonalization of a matrix. As a result, Σ_y can then be identified as the diagonal matrix of eigenvalues of Σ_x,

i.e.

$$\Sigma_y = \begin{bmatrix} \lambda_1 & 0 & & \\ 0 & \lambda_2 & & \\ & & \ddots & \\ & & & \lambda_N \end{bmatrix}$$

---

[1] $\mathscr{E}\{Gx\} = \dfrac{1}{K}\sum_{k=1}^{K} Gx_k = G\dfrac{1}{K}\sum_{k=1}^{K} x_k = Gm_x$

i.e. G, being a matrix of constants, can be taken outside an expectation operator.

[2] Since $[Gx]^t = x^t G^t$ etc.

where N is the dimensionality of the data. Since Σ_y is, by definition, a covariance matrix and is diagonal, its elements will be the variances of the pixel data in the respective transformed co-ordinates. It is arranged such that $\lambda_1 > \lambda_2 > \ldots \lambda_N$ so that the data exhibits maximum variance in y_1, the next largest variance in y_2 and so on, with minimum variance in y_N.

The principal components transform defined by (6.4) subject to the diagonal constraint of (6.5) is also known as the Karhunen-Loève or Hotelling transform.

Before proceeding it is of value at this stage to pursue further the examples of Fig. 6.2, to demonstrate the computational aspects of principal components analysis. Recall that the original x space covariance matrix for the highly correlated image data of Fig. 6.2b is

$$\Sigma_x = \begin{bmatrix} 1.90 & 1.10 \\ 1.10 & 1.10 \end{bmatrix}$$

To determine the principal components transformation it is necessary to find the eigenvalues and eigenvectors of this matrix. The eigenvalues are given by the solution to the characteristic equation

$$|\Sigma_x - \lambda I| = 0, \quad I \text{ being the identity matrix.}$$

i.e.
$$\begin{vmatrix} 1.90 - \lambda & 1.10 \\ 1.10 & 1.10 - \lambda \end{vmatrix} = 0$$

or $\lambda^2 - 3.0\lambda + 0.88 = 0$

which yields $\lambda = 2.67$ and 0.33

As a check on the analysis it may be noted that the sum of the eigenvalues is equal to the trace of the covariance matrix, which is the sum of its diagonal elements.

The covariance matrix in the appropriate y co-ordinate system (with principal components as axes) is therefore

$$\Sigma_y = \begin{bmatrix} 2.67 & 0 \\ 0 & 0.33 \end{bmatrix}$$

Note that the first principal component, as it is called, accounts for $2.67/(2.67 + 0.33) \equiv 89\%$ of the total variance of the data in this particular example. It is now of interest to find the actual principal components transformation matrix G. Recall that this is the transposed matrix of eigenvectors of Σ_x. Consider first, the eigenvector corresponding to $\lambda_1 = 2.67$. This is the vector solution to the equation

$$[\Sigma_x - \lambda_1 I] \, g_1 = 0$$

with $g_1 = \begin{bmatrix} g_{11} \\ g_{21} \end{bmatrix}$ for the two dimensional example at hand.

Substituting for Σ_x and λ_1 gives the pair of equations

$$-0.77 g_{11} + 1.10 g_{21} = 0$$
$$1.10 g_{11} - 1.57 g_{21} = 0$$

which are not independent, since the set is homogeneous. It does have a non-trivial solution however because the coefficient matrix has a zero determinant. From either equation it can be seen that

$$g_{11} = 1.43 g_{21} \qquad (6.6)$$

At this stage either g_{11} or g_{21} would normally be chosen arbitrarily, and then a value would be computed for the other. However the resulting matrix G has to be orthogonal so that $G^{-1} \equiv G^t$. This requires the eigenvectors to be normalised, so that

$$g_{11}^2 + g_{21}^2 = 1 \qquad (6.7)$$

This is a second equation that can be solved simultaneously with (6.6) to give

$$g_1 = \begin{bmatrix} 0.82 \\ 0.57 \end{bmatrix}$$

In a similar manner it can be shown that the eigenvector corresponding to $\lambda_2 = 0.33$ is

$$g_2 = \begin{bmatrix} -0.57 \\ 0.82 \end{bmatrix}$$

The required principal components transformation matrix therefore is

$$G = \begin{bmatrix} 0.82 & -0.57 \\ 0.57 & 0.82 \end{bmatrix}^t = \begin{bmatrix} 0.82 & 0.57 \\ -0.57 & 0.82 \end{bmatrix}$$

Now consider how these results can be interpreted. First of all, the individual eigenvectors g_1 and g_2 are vectors which define the principal component axes in terms of the original co-ordinate space. These are shown in Fig. 6.4: it is evident that the data is uncorrelated in the new axes and that the new axes are a rotation of the original set. For this reason (even in more than two dimensions) the principal components transform is classed as a rotational transform.

Secondly consider the application of the transformation matrix G to find the positions (i. e., the brightness values) of the pixels in the new uncorrelated co-ordinate system. Since $y = Gx$, this example gives

$$\begin{bmatrix} y_1 \\ y_2 \end{bmatrix} = \begin{bmatrix} 0.82 & 0.57 \\ -0.57 & 0.82 \end{bmatrix} \begin{bmatrix} x_1 \\ x_2 \end{bmatrix} \qquad (6.8)$$

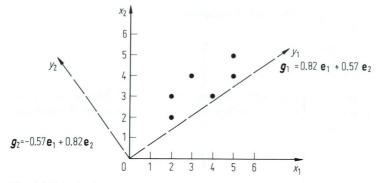

Fig. 6.4. Principal component axes for the data set of Fig. 2b; e_1 and e_2 are horizontal (x_1) and vertical (x_2) direction vectors

which is the actual principal components transformation to be applied to the image data. Thus, for

$$x = \begin{bmatrix} 2 \\ 2 \end{bmatrix}, \begin{bmatrix} 4 \\ 3 \end{bmatrix}, \begin{bmatrix} 5 \\ 4 \end{bmatrix}, \begin{bmatrix} 5 \\ 5 \end{bmatrix}, \begin{bmatrix} 3 \\ 4 \end{bmatrix}, \begin{bmatrix} 2 \\ 3 \end{bmatrix}$$

there is

$$y = \begin{bmatrix} 2.78 \\ 0.50 \end{bmatrix}, \begin{bmatrix} 4.99 \\ 0.18 \end{bmatrix}, \begin{bmatrix} 6.38 \\ 0.43 \end{bmatrix}, \begin{bmatrix} 6.95 \\ 1.25 \end{bmatrix}, \begin{bmatrix} 4.74 \\ 1.57 \end{bmatrix}, \begin{bmatrix} 3.35 \\ 1.32 \end{bmatrix}.$$

The pixels plotted in y space are shown in Fig. 6.5. Several points are noteworthy. First, the data exhibits no discernable correlation between the pair of new axes (i.e., the principal components). Secondly, most of the data spread is in the direction of the first principal component. It could be interpreted that this component contains most of the information in the image. Finally, if the pair of principal component images are produced by using the y_1 and y_2 component brightness values for the pixels, the first principal component image will show a high degree of contrast whereas the second will show only limited use of the available brightness value range. By comparison to the first component, the second will make use of only a few available brightness levels. It will be

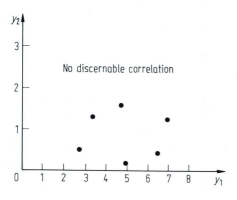

Fig. 6.5. Pixel points located in (uncorrelated) principal components space

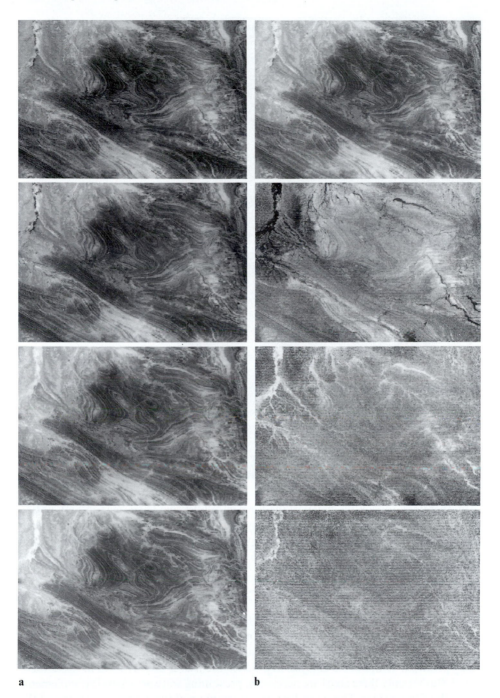

a   b

Fig. 6.6. a Four Landsat multispectral scanner bands for the region of Andamooka in central Australia; **b** The four principal components of the image segment; **c** (overleaf) Comparison of standard false colour composite (band 7 to red, band 5 to green and band 4 to blue) with a principal component composite (first component to red, second to green and third to blue)

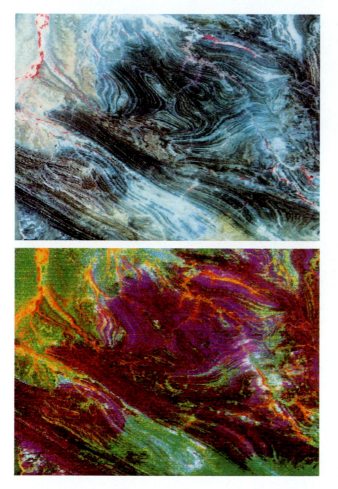

Fig. 6.6 c

seen, therefore, to lack the detail of the former. Whilst this phenomenon may not be particularly evident for a simple two dimensional example, it is especially noticeable in the fourth component of a principal component transformed Landsat multispectral scanner image as can be assessed in Fig. 6.6.

6.1.3 An Example – Some Practical Considerations

The material presented in Sect. 6.1.2 provides the background and rationale for the principal components transform. By working through the numerical example in detail the importance of eigenanalysis of the covariance matrix can be seen. However when using principal components analysis in practice the user is not involved in this level of detail. Rather only three steps are necessary, presuming software exists for implementing each of those steps. These are, first, the assembling of the covariance matrix of the image to be transformed according to (6.2). Normally, in an image analysis system, software will be available for this step, usually in conjunction with the need to generate signatures for classification as described in Chap. 8. The second step necessary is to determine the eigenvalues and eigenvectors of the covariance matrix. Either special purpose software will be available for this or general purpose matrix eigenanalysis routines can be used. The latter are found in packages such as MATLAB, Mathematica and Maple. At this stage the eigenvalues are used simply to assess the distribution of data variance over the respective components. A rapid fall off in the size of the eigenvalues indicates that the original band description of the image data exhibits a high degree of correlation and that significant results will be obtained in the transformation to follow.

The final step is to form the components using the eigenvectors of the covariance matrix as the weighting coefficients. As seen in (6.4) (noting that G is a transposed matrix of eigenvectors) and as demonstrated in (6.8), the components of the eigenvectors act as coefficients in determining the principal component brightness values for a pixel as a weighted sum of its brightnesses in the original spectral bands. The first eigenvector produces the first principal component from the original data, the second eigenvector gives rise to the second component, and so on.

Figure 6.6a shows the four original bands of an image acquired by the Landsat multispectral scanner for a small image segment in central Australia. The covariance matrix for this image is

$$\Sigma_x = \begin{bmatrix} 34.89 & 55.62 & 52.87 & 22.71 \\ 55.62 & 105.95 & 99.58 & 43.33 \\ 52.87 & 99.58 & 104.02 & 45.80 \\ 22.71 & 43.33 & 45.80 & 21.35 \end{bmatrix}$$

and its eigenvalues and eigenvectors are:

eigenvalues	253.44	7.91	3.96	0.89
eigenvector	0.34	−0.61	0.71	−0.06
components	0.64	−0.40	−0.65	−0.06
(vertically)	0.63	0.57	0.22	0.48
	0.28	0.38	0.11	−0.88

The first principal component image will be expected therefore to contain 95% of the data variance. By comparison, the variance in the last component is seen to be negligible. It is to be expected that this component will appear almost totally as noise of low amplitude.

The four principal component images for this example are seen in Fig. 6.6b in which the information redistribution and compression properties of the transformation are illustrated. By association with Fig. 6.5 it would be anticipated that the later components should appear dull and poor in contrast. The high contrasts displayed are a result of a contrast enhancement applied to components for the purpose of display. This serves to highlight the poor signal to noise ratio.

Figure 6.6c shows a comparison of a standard false colour composite formed from the original Landsat bands and a colour composite formed by displaying the first principal component as red, the second as green and the third as blue. Owing to the noise in the second and third components these were smoothed with a 3 × 3 mean value template first.

Notwithstanding the anticipated negligible information content of the last, or last few, image components resulting from a principal components analysis it is important in photointerpretation to examine all components since often local detail may appear in a later component. The covariance matrix used to generate the principal component transformation matrix is a global measure of the variability of the original image segment. Abnormal local detail therefore may not necessarily be mapped into one of the earlier components but could just as easily appear later. This is often the case with geological structure.

6.1.4 The Effect of an Origin Shift

It will be evident that some principal component pixel brightnesses could be negative owing to the fact that the transformation is a simple axis rotation. Clearly a combination of positive and negative brightnesses cannot be displayed. Nor can negative brightness pixels be ignored since their appearance relative to the other pixels in a component serve to define detail. In practice, the problem with negative values is accommodated by shifting the origin of the principal components space to yield all components with positive and thus displayable brightnesses. This has no effect on the properties of the transformation as can be seen by inserting an origin shift term in the definition of the covariance matrix in the principal components axes. Define $y' = y - y_0$ were y_0 is the position of a new origin. In the new y' co-ordinates

$$\Sigma_{y'} = \mathscr{E}\{(y' - m_{y'})(y' - m_{y'})^t\}$$

Now $m_{y'} = m_y - y_0$ so that

$$y' - m_{y'} = y - y_0 - m_y + y_0 = y - m_y.$$

Thus $\Sigma_{y'} = \Sigma_y$ – i.e. the origin shift has no influence on the covariance of the data in the principal components axes, and can be used for convenience in displaying principal component images.

6.1.5 Application of Principal Components in Image Enhancement and Display

In constructing a colour display of remotely sensed data only three dimensions of information can be mapped to the three colour primaries of the display device. In the case of Landsat multispectral scanner data, usually bands 4, 5 and 7 are chosen for this, band 6 being ignored, since it is often strongly correlated with band 7. A less ad hoc means for colour assignment rests upon performing a principal components transform and assigning the first three components to the red, green and blue colour primaries.

Examination of a typical set of principal component images for Landsat data, such as those seen in Fig. 6.6, reveals that there is very little detail in the fourth component so that, in general, it could be ignored without prejudicing the ability to extract meaningful information from the scene. A difficulty with principal components colour display, however, is that there is no longer a one to one mapping between sensor wavelength bands and colours. Rather each colour now represents a linear combination of spectral components, making photointerpretation difficult for many applications. An exception would be in exploration geology where structural differences may be enhanced in principal components imagery, there being little interest (in the case of Landsat MSS data) in the meanings of the actual colours.

6.1.6 The Taylor Method of Contrast Enhancement

It will be demonstrated below that application of the contrast modification techniques of Chap. 4 to each of the individual components of a highly correlated vector image will yield an enhanced image in which certain highly saturated hues are missing. An interesting contrast stretching procedure which can be used to create a modified image with good utilisation of the range of available hues rests upon the use of the principal components transformation. It was developed by Taylor (1973) and has also been presented by Soha and Schwartz (1978). A more recent and general treatment has been given by Campbell (1996).

Consider a two dimensional image with the (two dimensional) histogram shown in Fig. 6.7. As observed the two components are highly correlated as revealed also from an inspection of the covariance matrix for the image which is

$$\Sigma_x = \begin{bmatrix} 0.885 & 0.616 \\ 0.616 & 0.879 \end{bmatrix} \tag{6.9}$$

The range of brightness values occupied in the histogram suggests that there is value in performing a contrast stretch. Suppose a simple linear stretch is decided upon; the conventional means then for implementing such an enhancement with a multicomponent image is to apply it to each component independently. This requires the one dimensional histogram for each component to be constructed. These are obtained by counting the number of pixels with a given brightness value in each component, irrespective of their brightness in the other component – in other words they are marginal distributions of the two dimensional distribution. The single dimensional histograms corresponding to Fig. 6.7 are shown in Fig. 6.8a and the result of applying linear contrast enhancement to each of these is seen in Fig. 6.8b. The two dimensional histogram resulting from the contrast stretches applied to the individual components is shown in Fig. 6.9 wherein it is seen that the correlation between the components is still present and that if component 1 is displayed as red and component 2 as green, no highly saturated reds or greens will be evident in the enhanced image, although brighter yellows will be more obvious than in the original data. It is a direct result of the correlation in the image that the highly saturated colour primaries are not displayed. The situation is even worse for display of three dimensional correlated image data.

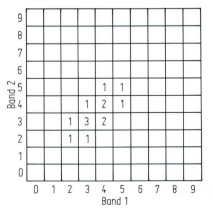

Fig. 6.7. Histogram for a hypothetical two dimensional image showing correlation in its components. The numbers indicated on the bars (out of page) are the counts

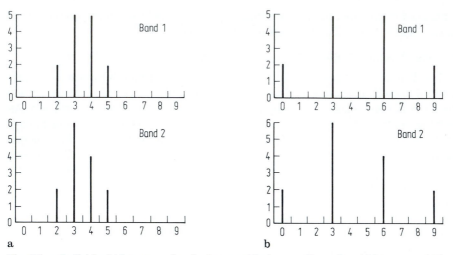

Fig. 6.8. a Individual histograms for the image with the two dimensional histogram of Fig. 6.7; **b** The individual histograms after a simple linear contrast stretch over all available brightness values

Simple contrast enhancement of each component independently will yield an image without highly saturated reds, blues and greens but also without saturated yellows, cyans and magentas. The procedure recommended by Taylor overcomes this, as demonstrated now. This fills the available colour space on the display more fully.

Let x be the vector of brightness values of the pixels in the original image and y be the corresponding vector of intensities after principal components transformation, such that $y = Gx$. G is the principal components transformation matrix, composed of transposed eigenvectors of the original covariance matrix Σ_x. The covariance matrix which describes the scatter of pixel points in the principal components (y) vector space is a diagonal matrix of eigenvalues which, for three dimensional data, is of the form

$$\Sigma_y = \begin{bmatrix} \lambda_1 & 0 & 0 \\ 0 & \lambda_2 & 0 \\ 0 & 0 & \lambda_3 \end{bmatrix}$$

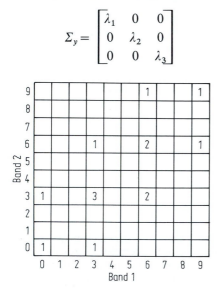

Fig. 6.9. Histogram of a two dimensional image after simple linear contrast stretch of the components individually

Suppose now the individual principal components are enhanced in contrast such that they each cover the corresponding range of brightness values and, in addition, have the same variances; in other words the histograms of the principal components are matched, for example, to a Gaussian histogram that has the same variance in all dimensions. The new covariance matrix will therefore be of the form

$$\Sigma'_y = \begin{bmatrix} \sigma^2 & 0 & 0 \\ 0 & \sigma^2 & 0 \\ 0 & 0 & \sigma^2 \end{bmatrix} = \sigma^2 I$$

where I is the identity matrix. Since the principal components are uncorrelated, enhancement of the components independently yields an image with good utilisation of the available colour space, with all hues possible. The axes in the colour space however are principal components axes and, as noted in the previous section, are not as desirable for photointerpretation as having a colour space based upon the original components of the image. It would be of particular value therefore if the image data could be returned to the original x space to give a one-to-one mapping between the display colours and image components. Let the contrast enhanced principal components be represented by the vector y'. These can be transformed back to the original axes for the image by using the inverse of the principal components transformation matrix G^{-1}. Since G is orthogonal its inverse is simply its transpose, which is readily available. The new covariance matrix of the data back in the original image domain is

$$\Sigma'_x = G^t \mathscr{E} \{(y' - \mathscr{E}(y')) (y' - \mathscr{E}(y'))^t\} G$$

where $x' = G^t y'$, is the modified pixel vector in the original space. Consequently

$$\begin{aligned} \Sigma'_x &= G^t \mathscr{E} \{(y' - \mathscr{E}(y')) (y' - \mathscr{E}(y'))^t\} G \\ &= G^t \Sigma'_y G \\ &= G^t \sigma^2 I G \end{aligned}$$

i.e. $\quad \Sigma'_x = \sigma^2 I$.

Thus the covariance matrix of the enhanced principal components data is preserved on transformation back to the original image space. No correlation is introduced and the data shows good utilisation of the colour space using the original image data components. In practice, one problem encountered with the Taylor procedure is the noise introduced into the final results by the contrast enhanced third principal component. Should all possible brightness values be available in the components this would not occur. However because most image analysis software treats image data in integer format in the range 0 to 255, rounding of intermediate results to integer form produces the noise. One possible remedy is to filter the noisy components before the inverse transform is carried out.

It will be appreciated from the foregoing discussion that colour composite principal component imagery will appear more colourful than a colour composite product formed from original image bands. This is a direct result of the ability to fill the colour space completely by contrast enhancing the uncorrelated components, by comparison to the poor utilization of colour by the original correlated data, as seen in the illustration of Fig. 6.9 and as demonstrated in Fig. 6.6c.

6.1.7 Other Applications of Principal Components Analysis

Owing to the information compression properties of the principal components transformation it lends itself to reduced representation of image data for storage or transmission. In such a situation only the uppermost significant components are retained as a representation of an image, with the information content so lost being indicated by the sum of the eigenvalues corresponding to the components ignored. Thereafter if the original image is to be restored, either on reception through a communications channel or on retrieval from memory, then the inverse of the transformation matrix is used to reconstruct the image from the reduced set of components. Since the matrix is orthogonal its inverse is simply its transpose. This technique is known as bandwidth compression in the field of telecommunications. Until recently it had not found great application in satellite remote sensing image processing, because hitherto image transmission has not been a consideration and available memory has not placed stringent limits on image storage. With increasing use of imaging spectro-metry data however (Sect. 1.4.5), bandwidth compression has become more important, as discussed in Sect. 13.7.

An interesting application of principal components analysis is in the detection of features that change with time between images of the same region. This is described by example in Chap. 11.

6.2 The Kauth-Thomas Tasseled Cap Transformation

The principal components transformation treated in the previous section yields a new co-ordinate description of multispectral remote sensing image data by establishing a diagonal form of the global covariance matrix. The new co-ordinates (components) are linear combinations of the original spectral bands. Other linear transformations are of course possible. One is a procedure referred to as canonical analysis, treated in Chap. 10. Another, to be developed below, is application-specific in that the new axes in which data are described have been devised to maximise information of importance, in this case, to agriculture. Other similar special transformations would also be possible.

The so-called "tasseled cap" transformation (Crist and Kauth, 1986) developed by Kauth and Thomas (1976) is a means for highlighting the most important (spectrally observable) phenomena of crop development in a way that allows discrimination of specific crops, and crops from other vegetative cover, in Landsat multitemporal, multi-spectral imagery. Its basis lies in an observation of crop trajectories in band 6 versus band 5, and band 5 versus band 4 subspaces. Consider the former as shown in Fig. 6.10a.

A first observation that can be made is that the variety of soil types on which specific crops might be planted appear as points along a diagonal in the band 6, band 5 space as shown. This is well-known and can be assessed from an observation of the spectral reflectance characteristics for soils. (See for example Chap. 5 of Swain and Davis, 1978.) Darker soils lie nearer the origin and lighter soils at higher values in both bands. The actual slope of this line of soils will depend upon global external variables such as atmospheric haze and soil moisture effects. If the transformation to be derived is to be

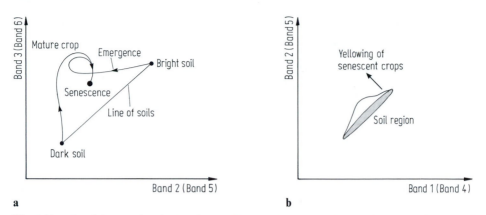

Fig. 6.10. a Band 6 versus band 5 Landsat multispectral scanner subspace showing trajectories of crop development; **b** Band 5 versus band 4 subspace also depicting crop development

used quantitatively these effects need to be modelled and the data calibrated or corrected beforehand.

Consider now the trajectories followed in the band 6 versus band 5 subspace for crop pixels corresponding to growth on different soils – in this case take the extreme light and dark soils as depicted in Fig. 6.10a. For both regions at planting the multispectral response is dominated by soil types, as expected. As the crops emerge the shadows cast over the soil dominate any green matter response. As a result there is considerable darkening of the response of the lighter soil crop field and only a slight darkening of that on dark soil. When both crops reach maturity their trajectories come together implying closure of the crop canopy over the soil. The response is then dominated by the green biomass, being in a high band 6 and low band 5 region, as is well known. When the crops senesce and turn yellow their trajectories remain together and move away from the green biomass point in the manner depicted in the diagram. However whereas the development to maturity takes place almost totally in the same plane, the yellowing development in fact moves out of this plane, as can be assessed by how the trajectories develop in the band 5 versus band 4 subspace during senescence as illustrated in Fig. 6.10b.

Should the crops then be harvested the trajectories beyond senescence move, in principle, back towards their original soil positions.

Having made these observations, the two diagrams of Fig. 6.10 can now be combined into a single three dimensional version in which the stages of the crop trajectories can be described according to the parts of a cap, with tassels, from which the name of the subsequent transformation is derived. This is shown in Fig. 6.11. The first point to note is that the line of soils used in Fig. 6.10a is shown now as a plane of soils. Its maximum spread is along the three dimensional diagonal as indicated; however it has a scatter about this line consistent with the spread in band 5 versus band 4 as shown in Fig. 6.10b. Kauth and Thomas note that this plane of soils forms the brim and base of the cap. As crops develop on any soil type their trajectories converge essentially towards the crown of the cap at maturity whereupon they fold over and continue to yellowing as indicated. Thereafter they break up to return ultimately to various soil positions, forming tassels on the cap as shown.

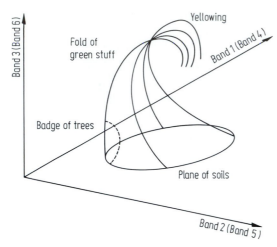

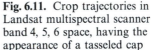

Fig. 6.11. Crop trajectories in Landsat multispectral scanner band 4, 5, 6 space, having the appearance of a tasseled cap

The behaviour observable in Fig. 6.11 led Kauth and Thomas to consider the development of a linear transformation that would be useful in crop discrimination. As with the principal components transform, this transformation will yield four orthogonal axes. However the axis directions are chosen according to the behaviour seen in Fig. 6.11.

Three major orthogonal directions of significance in agriculture can be identified. The first is the principal diagonal along which soils are distributed. This was chosen by Kauth and Thomas as the first axis in the tasseled cap transformation. The development of green biomass as crops move towards maturity appears to occur orthogonal to the soil major axis. This direction was then chosen as the second axis, with the intention of providing a greeness indicator. Crop yellowing takes place in a different plane to maturity. Consequently choosing a third axis orthogonal to the soil line and greeness axis will give a yellowness measure. Finally a fourth axis is required to account for data variance not substantially associated with differences in soil brightness or vegetative greenness or yellowness. Again this needs to be orthogonal to the previous three. It was called "non-such" by Kauth and Thomas in contrast to the names "soil brightness", "green-stuff" and "yellow-stuff" they applied to the previous three.

The transformation that produces the new description of the data may be expressed as

$$u = Rx + c \qquad (6.10)$$

where x is the original Landsat multispectral scanner pixel vector, and u is the vector of transformed brightness values. This has soil brightness as its first component, greenness as its second and yellowness as its third. These can therefore be used as indices, respectively. R is the transformation matrix and c is a constant vector chosen (arbitrarily) to avoid negative values in u.

The transformation matrix R is the transposed matrix of column unit vectors along each of the transformed axes (compare with the principal components transformation matrix). For a particular agricultural region Kauth and Thomas chose the first unit vector as a line of best fit through a set of soil classes. The subsequent unit vectors were generated by using a Gram-Schmidt orthogonalization procedure in the directions required. The transformation matrix generated was

$$R = \begin{bmatrix} 0.433 & 0.632 & 0.586 & 0.264 \\ -0.290 & -0.562 & 0.600 & 0.491 \\ -0.829 & 0.522 & -0.039 & 0.194 \\ 0.223 & 0.012 & -0.543 & 0.810 \end{bmatrix}$$

From this it can be seen, at least for the region investigated by Kauth and Thomas, that the soil brightness is a weighted sum of the original four Landsat bands with approximately equal emphasis. The greeness measure is the difference between the infrared and visible responses. In a sense therefore this is more a biomass index. The yellowness measure can be seen to be substantially the difference between the Landsat visible red and green bands.

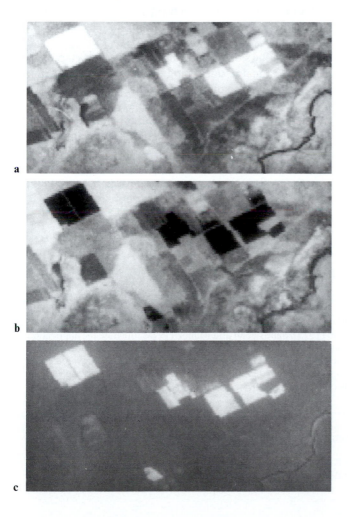

a

b

c

Fig. 6.12. Landsat multi-spectral scanner band 7 a and band 5, b images of an arid region containing irrigated crop fields. The ratio of these two images c shows vegetated regions as bright, soils as mid to dark grey and water as black

Just as new images can be synthesised to correspond to various principal components so can the actual transformed images be created for this approach. By applying (6.10) to every pixel in a Landsat multispectral scanner image, soil brightness, greeness, yellowness and non-such images can be produced. These can then be used to assess stages in crop development. The method can also be applied to other sensors.

6.3 Image Arithmetic, Band Ratios and Vegetation Indices

Addition, subtraction, multiplication and division of the pixel brightnesses from two bands of image data to form a new image are particularly simple transformations to apply and can also be implemented in hardware using look up tables. Multiplication seems not to be as useful as the others, band differences and ratios being most common.

Differences can be used to highlight regions of change between two images of the same area. This requires that the images be registered using the techniques of Chap. 2 beforehand. The resultant difference image must be scaled to remove negative brightness values. Normally this is done so that regions of no change appear mid-grey, with changes shown as brighter or duller than mid-grey according to the sign of the difference.

Ratios of different spectral bands from the same image find use in reducing the effect of topography, as a vegetation index, and for enhancing subtle differences in the spectral reflectance characteristics for rocks and soils. As an illustration of the value of band ratios for providing a single vegetation index image, Fig. 6.12 shows Landsat multispectral scanner band 5 and band 7 images of an agricultural region along with the band 7/band 5 ratio. As seen, healthy vegetated areas are bright, soils are mid to dark grey, and water is black. These shades are readily understood from an examination of the corresponding spectral reflectance curves. Variations on simple arithmetic operations between bands are also sometimes used as indices. Some of these are treated in Sect. 10.4.3. Note that band ratioing is not a linear transformation.

References for Chapter 6

An easily read treatment of the principal components transformation has been given by Jensen and Waltz (1979), although the degree of mathematical detail has been kept to a minimum. Theoretical treatments can be found in many books on pattern recognition, image analysis and data analysis, although often under the alternative titles of Karhunen-Loève and Hotelling transforms. Treatments of this type that could be consulted include Andrews (1972), Gonzalez and Woods (1992) and Ahmed and Rao (1975). Santisteban and Muñoz (1978) illustrate the application of the technique. The transformation has also been looked at as a method for detecting changes between successive images of the same region. This is illustrated in Sect. 11.7 and covered more fully in the papers by Byrne, Crapper and Mayo (1980), Howarth and Boasson (1983), Ingebritsen and Lyon (1985) and Richards (1984).

N. Ahmed and K. R. Rao, 1975: Orthogonal Transforms for Digital Signal Processing, Berlin, Springer-Verlag

H. C. Andrews, 1972: Introduction to Mathematical Techniques in Pattern Recognition, New York, Wiley.

E. F. Byrne, P. F. Crapper and K. K. Mayo, 1980: Monitoring Land-Cover Change by Principal Components Analysis of Multitemporal Landsat Data. Remote Sensing of Environment, 10, 175–184.

N.A. Campbell, 1996: The Decorrelation Stretch Transformation. Int. J. Remote Sensing, 17, 1939–1949.

E. P. Crist and R. T. Kauth, 1986: The Tasseled Cap De-Mystified. Photogrammetric Engineering and Remote Sensing, 52, 81–86.

R.C. Gonzalez and R.E. Woods, 1992: Digital Image Processing, Mass., Addison-Wesley.

P.J. Howarth and E. Boasson, 1983: Landsat Digital Enhancements for Change Detection in Urban Environments. Remote Sensing of Environment, 13, 149–160.

S.E. Ingebritsen and R.J.P. Lyon, 1985: Principal Components Analysis of Multitemporal Image Pairs. Int. J. Remote Sensing, 6, 687–696.

S.K. Jensen and F.A. Waltz, 1979: Principal Components Analysis and Canonical Analysis in Remote Sensing. Proc. American Photogrammetric Soc. 45th Ann. Meeting, 337–348.

R.J. Kauth and G.S. Thomas, 1976: The Tasseled Cap – A Graphic Description of the Spectral-Temporal Development of Agricultural Crops as Seen by Landsat. Proc. LARS 1976 Symp. on Machine Process. Remotely Sensed Data, Purdue University.

J.A. Richards, 1984: Thematic Mapping from Multitemporal Image Data Using the Principal Components Transformation. Remote Sensing of Environment, 16, 35–46.

A. Santisteban and L. Muñoz, 1978: Principal Components of a Multispectral Image: Application to a Geologic Problem. IBM J. Research and Development, 22, 444–454.

J.M. Soha and A.A. Schwartz, 1978: Multispectral Histogram Normalization Contrast Enhancement. Proc. 5th Canadian Symp. on Remote Sensing, 86–93.

P.H. Swain and S.M. Davis (Eds), 1978: Remote Sensing: The Quantitative Approach, New York, McGraw-Hill.

M.M. Taylor, 1973: Principal Components Colour Display of ERTS Imagery. Third Earth Resources Technology Satellite–1 Symposium, NASA SP-351, 1877–1897.

Problems

6.1 (a) At a conference research group A and research group B both presented papers on the value of the principal components transformation (also known as the Karhunen-Loève or Hotelling transform) for reducing the number of features required to represent image data. Group A described very good results that they had obtained with the method whereas Group B indicated that they felt it was of little use. Both groups were using image data with only two spectral components. The covariance matrices for their respective images are:

$$\Sigma_A = \begin{bmatrix} 5.4 & 4.5 \\ 4.5 & 6.1 \end{bmatrix} \qquad \Sigma_B = \begin{bmatrix} 28.0 & 4.2 \\ 4.2 & 16.4 \end{bmatrix}$$

Explain the points of view of both groups.

(b) If information content can be related directly to variance indicate how much information is discarded if only the first principal component is retained by both groups.

6.2 Suppose you have been asked to describe the principal components transformation to a non-specialist. Write a single paragraph summary of its essential features, using diagrams if you wish, but no mathematics.

6.3 (For those mathematically inclined), Demonstrate that the principal components transformation matrix developed in Section 6.1.2 is orthogonal.

6.4 Colour image products formed from principal components generally appear richer in colour than a colour composite product formed by combining the original bands of remote sensing image data. Why do you think that is so?

6.5 (a) The steps involved in computing principal component images may be summarised as:

    calculation of the image covariance matrix
    eigenanalysis of the covariance matrix
    computation of the principal components.

Assessments can be made in the first two steps as to the likely value in proceeding to compute the components. Describe what you would look for in each case.

(b) The covariance matrix need not be computed over the full image to produce a principal components transformation. Discuss the value of using training areas to define the portion of image data to be taken into account in compiling the covariance matrix.

6.6 Imagine you have two images from a sensor which has a single band in the range 0.9 to 1.1 μm. One image was taken before a flood occurred. The second shows the extent of flood inundation. Produce a sketch of what the "two-date" multispectral space would look like if the image from the first date contained rich vegetation, sand and water and that in the second date contains the same cover types but with an expanded region of water. Demonstrate how a two dimensional principal components transform can be used to highlight the extent of flooding.

6.7 Describe the nature of the correlations between the pairs of axis variables (e.g. bands) in each of the cases in Fig. 6.13.

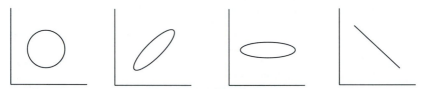

Fig. 6.13. Examples of two dimensional correlations

Chapter 7
Fourier Transformation of Image Data

7.1 Introduction

Many of the geometric enhancement techniques used with remote sensing image data can be carried out using the simple template-based techniques of Chap. 5. More flexibility is offered however if procedures are implemented in the so-called spatial frequency domain by means of the Fourier transformation. As a simple illustration, filters can be designed to extract periodic noise from an image that is unable to be removed by practical templates. As demonstrated in Sect. 5.4 the computational cost of using Fourier transformation for geometric operations is high by comparison to the template methods usually employed. However with the computational capacity of modern workstations, and the flexibility available in Fourier transform processing, this approach is one that should not be ignored.

Development of Fourier transform theory depends upon a knowledge of complex numbers and facility with integral calculus. The reader without that background may wish to pass over this Chapter and may do so without detracting from material in the remainder of the book. It is the purpose of the Chapter to present an overview of the significant aspects of the theory of Fourier transformation of image data. In its entirety the topic is an extensive one and well beyond the scope of this treatment. Instead the material presented in the following will serve to introduce the operational aspects of the topic, with little dependence on proofs and theory. Should the treatment be found to be too brief, particularly in the background material of Sect. 7.2 to 7.5, more details can be found in Brigham (1974, 1988), and McGillem and Cooper (1984).

7.2 Special Functions

A number of mathematical functions are important in both developing and understanding the Fourier transformation. These are reviewed in this section along with some properties that will be of use later on.

Although functions of interest in image processing have position as their independent variable, it will be convenient here to use functions of time. These will be interpreted as functions of position as required.

7.2.1 The Complex Exponential Function

The complex exponential is defined by

$$f(t) = Re^{j\omega t} \tag{7.1a}$$

where $j = \sqrt{-1}$, R is the amplitude of the function and ω is called its radian frequency. The units of ω are radians per second (or radians per unit of spatial variable). Frequently ω is expressed in terms of "natural" frequency

$$f = \omega/2\pi \qquad (7.1\,\text{b})$$

where f has units of hertz (or cycles per spatial variable). The complex exponential is periodic, with period $T = 2\pi/\omega$. This is appreciated by plotting it as a function of the independent variable on the complex (argand) plane. Alternatively, we can express

$$f(t) = Re^{\pm j\omega t} = R\cos\omega t \pm jR\sin\omega t \qquad (7.1\,\text{c})$$

to see its periodic behaviour in terms of sinusoids. For convenience we will now choose $R = 1$. From this last expression we see

$$\cos\omega t = \mathcal{Re}\{e^{j\omega t}\}$$
$$\sin\omega t = \mathcal{Im}\{e^{j\omega t}\}$$

where $\mathcal{Re}$ and $\mathcal{Im}$ are operators that select the real and imaginary parts of a complex number.

Finally, it can be seen from (7.1 c)

$$\cos\omega t = \frac{1}{2}(e^{j\omega t} + e^{-j\omega t}) \qquad (7.2\,\text{a})$$

$$\sin\omega t = \frac{1}{2j}(e^{j\omega t} - e^{-j\omega t}) \qquad (7.2\,\text{b})$$

7.2.2 The Dirac Delta Function

A function of particular importance in determining properties of sampled signals, which include digital image data, is the impulse function, also referred to as the Dirac delta function. This is a spike-like function of infinite amplitude and infinitessimal duration. It cannot be defined explicitly. Instead it is defined by a limiting operation as in the following manner.

Consider the rectangular pulse of duration α and amplitude $1/\alpha$ as seen in Fig. 7.1. Note that the area under the curve is 1. Accordingly the delta function $\delta(t)$ is defined as the pulse in the limit as α goes to zero. As a formal definition, the best that can be done is

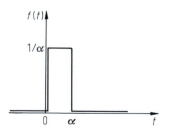

Fig. 7.1. Pulse which approaches an impulse in the limit as $\alpha \to 0$

$$\delta(t) = 0 \quad \text{for} \quad t \neq 0 \tag{7.3a}$$

and

$$\int_{-\infty}^{\infty} \delta(t)\, dt = 1 \tag{7.3b}$$

This turns out to be sufficient for our purposes. Equation (7.3) defines a delta function at the origin; an impulse at time t_0 is defined by

$$\delta(t - t_0) = 0 \quad \text{for} \quad t \neq t_0 \tag{7.4a}$$

and

$$\int_{-\infty}^{\infty} \delta(t - t_0)\, dt = 1 \tag{7.4b}$$

7.2.2.1 Properties of the Delta Function

From the definition of the delta function it can be seen that the product of a delta function with another function is

$$\delta(t - t_0)\, f(t) = \delta(t - t_0)\, f(t_0), \tag{7.5a}$$

from which we can see

$$\int_{-\infty}^{\infty} \delta(t - t_0)\, f(t)\, dt = \int_{-\infty}^{\infty} \delta(t - t_0)\, f(t_0)\, dt$$

$$= f(t_0) \int_{-\infty}^{\infty} \delta(t - t_0)\, dt$$

i.e.

$$\int_{-\infty}^{\infty} \delta(t - t_0)\, f(t)\, dt = f(t_0) \tag{7.5b}$$

This is known as the sifting property of the impulse.

7.2.3 The Heaviside Step Function

Figure 7.2 shows the Heaviside step function defined by

$$u(t - t_0) = 1 \quad \text{for} \quad t \geq t_0 \tag{7.6a}$$
$$= 0 \quad \text{for} \quad t < t_0 \tag{7.6b}$$

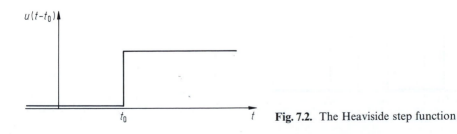

Fig. 7.2. The Heaviside step function

Note that it is 1 when its argument is zero or positive, and is zero for a negative argument. It can be seen that $u(t)$ is related to $\delta(t)$ by

$$\delta(t) = \frac{\mathrm{d}u(t)}{\mathrm{d}t}$$

7.3 Fourier Series

If a function $f(t)$ is periodic with period T — i.e. $f(t) = f(t+T)$ — then it can be expressed as an infinite sum of complex exponentials in the manner

$$f(t) = \sum_{n=-\infty}^{\infty} F_n e^{jn\omega_0 t}, \quad \omega_0 = \frac{2\pi}{T} \tag{7.7a}$$

in which n is an integer and the complex expansion coefficients F_n are given by

$$F_n = \frac{1}{T} \int_{-T/2}^{T/2} f(t)\, e^{-jn\omega_0 t}\, \mathrm{d}t \tag{7.7b}$$

The expressions in (7.7) are referred to as the exponential form of the Fourier series; (7.1c) also allows a trigonometric expression to be derived (McGillem and Cooper, 1984). Although (7.7a) is expressed in exponentials we often colloquially talk of (7.7a) as showing the sinusoidal spectral composition of $f(t)$. Equation (7.2) shows that this is acceptable and quite accurate.

As an illustration consider the need to determine the Fourier series of the square waveform in Fig. 7.3. From (7.7b) it can be seen that

$$F_n = \frac{1}{T} \int_{-T/4}^{T/4} e^{-jn\omega_0 t}\, \mathrm{d}t$$
$$= \frac{1}{n\pi} \sin\frac{n\pi}{2}.$$

This tells the amount of each of the constituent $e^{jn\omega_0 t}$ in (7.7a) required to represent the square waveform – i.e. it describes its sinusoidal composition. Note that when $n=0$, $F_0 = 1/2$ as expected from Fig. 7.3. For $n > 1$ the coefficients decrease in amplitude

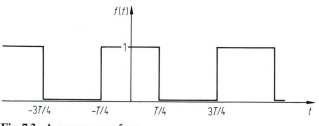

Fig. 7.3. A square waveform

according to $1/n$. In general the F_n are complex and thus can be expressed in the form of an amplitude and phase, referred to respectively as amplitude and phase spectra.

7.4 The Fourier Transform

The Fourier series of the preceding section is a description of a periodic function in terms of a sum of sinusoidal terms (expressed in complex exponentials) at integral multiples of the so-called fundamental frequency ω_0. For functions that are non-periodic, or *aperiodic* as they are sometimes called, decomposition into sinusoidal components requires use of the Fourier transformation. The transform itself, which is equivalent to the Fourier series coefficients of (7.7b), is defined by

$$F(\omega) = \int_{-\infty}^{\infty} f(t) e^{-j\omega t} \, dt \tag{7.8a}$$

In general, an aperiodic function requires a continuum of sinusoidal frequency components for a Fourier description. Indeed if we plot $F(\omega)$, or for that matter its amplitude and phase, as a function of frequency it will be a continuous function of ω. The function $f(t)$ can be reconstructed from the spectrum according to

$$f(t) = \frac{1}{2\pi} \int_{-\infty}^{\infty} F(\omega) e^{j\omega t} \, d\omega \tag{7.8b}$$

A Fourier transform of some importance is that of the unit pulse shown in Fig. 7.4a. From (7.8a) this is seen to be

$$F(\omega) = \int_{-a}^{a} e^{-j\omega t} \, dt = 2a \, \frac{\sin a\omega}{a\omega}$$

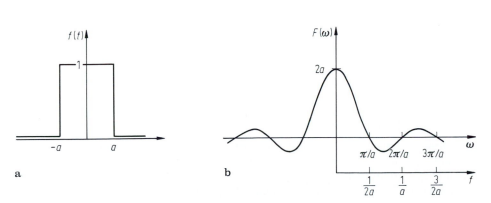

Fig. 7.4. a Unit pulse and **b** its Fourier transform

which is shown plotted in Fig. 7.4 b. Note that the frequency axis accommodates both positive and negative frequencies. The latter have no physical meaning but rather are an outcome of using complex exponentials in (7.8 a) instead of sinusoids.

It is also of interest to note the Fourier transform of an impulse

$$F(\omega) = \int_{-\infty}^{\infty} \delta(t) \, e^{-j\omega t} \, dt = 1$$

from the sifting property of the impulse (7.5 b); the Fourier transform of a constant is

$$F(\omega) = \int_{-\infty}^{\infty} c \, e^{-j\omega t} \, dt = 2\pi c \, \delta(\omega).$$

This result is easily shown by working from the spectrum $F(\omega)$ to the time function and again using the sifting property. In a like manner it can be shown that the Fourier transform of a periodic function is given by

$$F(\omega) = 2\pi \sum_{n=-\infty}^{\infty} F_n \, \delta(\omega - n\omega_0)$$

where F_n is the Fourier *series* coefficient corresponding to the frequency $n\omega_0$.

7.5 Convolution

7.5.1 The Convolution Integral

In Sect. 5.3 the concept of convolution was introduced as a means for determining the response of a linear system. It is also a very useful signal synthesis operation in general and finds particular application in the description of digital data, as will be seen in later sections. Here we express the convolution of two functions $f_1(t)$ and $f_2(t)$ as

$$y(t) = \int_{-\infty}^{\infty} f_1(\tau) f_2(t-\tau) \, d\tau \overset{\Delta}{=} f_1(t) * f_2(t) \tag{7.9}$$

It is a commutative operation, i.e. $f_1(t) * f_2(t) = f_2(t) * f_1(t)$ a fact that can sometimes be exploited in evaluating the integral.

The convolution operation can be illustrated by interpreting the defining integral as representing the following four operations:

(i) folding – form $f_2(-\tau)$ by taking its mirror image about the ordinate axis
(ii) shifting – form $f_2(t-\tau)$ by shifting $f_2(-\tau)$ by the amount t
(iii) multiplication – form $f_1(\tau) f_2(t-\tau)$
(iv) integration – compute the area under the product.

These steps are illustrated in Fig. 7.5.

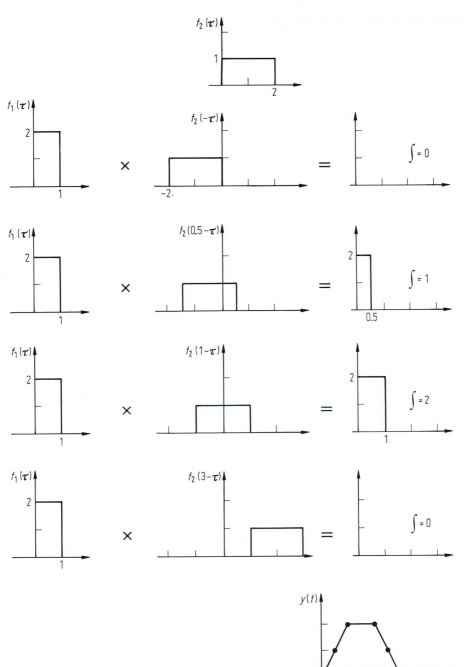

Fig. 7.5. Graphical illustration of the convolution operation

7.5.2 Convolution with an Impulse

Convolution of a function with an impulse is important in sampling. The sifting theorem for the delta function, along with (7.9), shows

$$f(t) * \delta(t-t_0) = \int_{-\infty}^{\infty} f(\tau)\, \delta(t-\tau-t_0)\, d\tau$$
$$= f(t-t_0).$$

Thus the effect is to shift the function $f(t)$ to a new origin.

7.5.3 The Convolution Theorem

This theorem is readily verified using the definition of convolution and the definition of the Fourier transform. It has two forms (Papoulis, 1980). These are:

If

$$y(t) = f_1(t) * f_2(t)$$

then

$$Y(\omega) = F_1(\omega)\, F_2(\omega), \tag{7.10a}$$

and, if

$$Y(\omega) = F_1(\omega) * F_2(\omega)$$

then

$$y(t) = \frac{1}{2\pi} f_1(t)\, f_2(t) \tag{7.10b}$$

7.6 Sampling Theory

The foregoing sections have dealt with functions that are continuous with time (or with position, as the case may be). However our interest principally is in functions, and images, that are discrete with time or position. Discrete time functions and digital images can be considered to be the result of the corresponding continuous functions having been sampled on a regular basis. Again, we will develop the concepts of sampling using functions of a single variable, such as time; the concepts are readily extended to two dimensional image functions.

A periodic sequence of impulses, spaced T apart,

$$\Delta(t) = \sum_{k=-\infty}^{\infty} \delta(t-kT) \tag{7.11}$$

can be considered as a sampling function, i.e. it can be used to extract a uniform set of samples from a function $f(t)$ by forming the product

$$f = f(t)\, \Delta(t). \tag{7.12}$$

According to (7.5a), f is a sequence of samples of value $f(kT)\, \delta(t-kT)$. Despite the undefined magnitude of the delta function we will be content in this treatment to regard that product as a sample of the function $f(t)$. Strictly this should be interpreted in terms of so-called distribution theory; a simple interpretation of (7.12) as a set of uniformly spaced samples of $f(t)$ however will not compromise our subsequent development.

It is important to consider the Fourier transform of the set of samples in (7.12) so that the frequency composition of a sampled function can be appreciated. This can be done using the convolution theorem (7.10 b) provided the Fourier transform of $\varDelta(t)$ can be found.

The Fourier transform of $\varDelta(t)$ can be determined via its Fourier series. From (7.7 b) and (7.5 b) the Fourier series coefficients of $\varDelta(t)$ are given by

$$\varDelta_n = \frac{1}{T} \int_{-T/2}^{T/2} \delta(t)\, e^{-jn\omega_0 t}\, \mathrm{d}t = \frac{1}{T}$$

which, with the expression for the Fourier transform of a periodic function in Sect. 7.4, gives the Fourier transform of $\varDelta(t)$ as

$$\varDelta(\omega) = \frac{2\pi}{T} \sum_{n=-\infty}^{\infty} \delta(\omega - n\omega_s) \tag{7.13}$$

where $\omega_s = 2\pi/T$. Thus the Fourier transform of the periodic sequence of impulses spaced T apart in time is itself a periodic sequence of impulses in the frequency domain, spaced $2\pi/T$ rad $\cdot$ s^{-1} apart (or $1/T$ Hz apart). Thus if $f(t)$ has the spectrum $F(\omega)$ (i.e. Fourier Transform) depicted in Fig. 7.6a then the spectrum of the set of samples in (7.12) is as shown in Fig. 7.6c. This is given by convolving $F(\omega)$ with the sequence of impulses in (7.13), according to (7.10 b). Recall that convolution with an impulse shifts a function to a new origin centred on the impulse.

Figure 7.6c demonstrates that the spectrum of a sampled function is a periodic repetition of the spectrum of the unsampled function, with the repetition period in the frequency domain determined by the rate at which the time function is sampled. If the sampling rate is high then the segments of the spectrum are well separated. If the sampling rate is low then the segments in the spectrum are close together.

In the illustration shown in Fig. 7.6 the spectrum of $f(t)$ is shown to be limited to frequencies below B Hz. ($2\pi B$ rad $\cdot$ s^{-1}); B is referred to as the bandwidth of $f(t)$. Not all real non-periodic functions have a limited bandwidth – the single pulse of Fig. 7.4 is an example of this – however it suits our purpose here to assume there is a limit to the frequency composition of functions of interest to us, defined by the signal bandwidth.

If adjacent segments are to remain separated as depicted in Fig. 7.6c then it is clear that

$$\frac{1}{T} > 2B \tag{7.14}$$

i.e. that the rate at which the function $f(t)$ is sampled must exceed twice the bandwidth of $f(t)$. Should this not be the case then the segments of the spectrum of the sampled function overlap as shown in Fig. 7.6d, causing a form of distortion called *aliasing*.

A sampling rate of $2B$ in (7.14) is referred to as the Nyquist rate; Eq. (7.14) itself is often referred to as the *sampling theorem*.

Time functions   Spectra

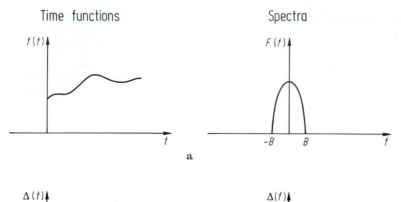

a

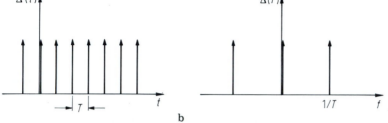

b

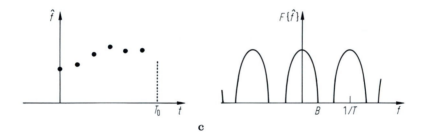

c

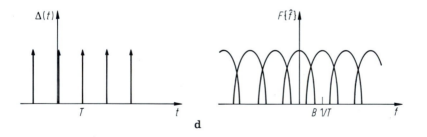

d

Fig. 7.6. Development of the Fourier transform of a sampled function. **a** Unsampled function and its spectrum; **b** Periodic sequence of impulses and its spectrum; **c** Sampled function and its spectrum; **d** Sub-Nyquist rate sampling impulses and spectrum with aliasing. F represents Fourier transformation

7.7 The Discrete Fourier Transform

7.7.1 The Discrete Spectrum

Consider now the problem of finding the spectrum (i.e. of computing the Fourier transform) of a sequence of samples. This is the first stage in our computation of the Fourier transform of an image. Indeed, the sequence of samples to be considered here could be looked at as a single line of pixels in digital image data.

Figure 7.7 a shows that the spectrum of a set of samples is itself a continuous function of frequency. For digital processing clearly it is necessary that the spectrum be also represented by a set of samples, that would, for example, exist in computer memory. Therefore we have to introduce a suitable sampling function also in the frequency domain. For this purpose consider an infinite periodic sequence of impulses in the frequency domain spaced Δf (i.e. $\Delta\omega/2\pi$) apart as shown in Fig. 7.7 b. It can be shown

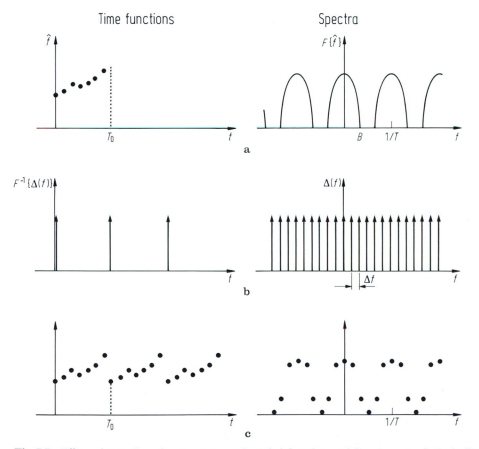

Fig. 7.7. Effect of sampling the spectrum. **a** Sampled function and its spectrum; **b** Periodic sequence of impulses used to sample the spectrum (right) and its time domain equivalent (left); **c** Sampled version of the spectrum (right) and its time domain equivalent (left); the latter is a periodic version of the samples in **a.** In these F^{-1} represents an inverse Fourier transformation

that the inverse transform of this sequence is another sequence of impulses in the time domain, spaced $T_0 = 1/\Delta f$ apart. This can be appreciated readily from (7.11) and (7.13), although here we are going from the frequency domain to the time domain rather than vice versa.

If the (periodic) spectrum $F(\omega)$ in Fig. 7.7a is multiplied by the frequency domain sampling function of Fig. 7.7b then the convolution theorem (7.10a) implies that the samples of $f(t)$ will be formed into a periodic sequence with period T_0 as illustrated in Fig. 7.7c. It is convenient if the number of samples used to represent the spectrum is the same as the actual number of samples taken of $f(t)$. Let this number be K. (There is a distortion introduced by using a finite rather than infinite number of samples. This will be addressed later.) Since the time domain has samples spaced T apart, the duration of sampling is KT seconds. It is pointless sampling the time domain over a period longer than T_0 since no new information is added. Simply other periods are added. Consequently the optimum sampling time is T_0, so that $T_0 = KT$. Thus the sampling increment in the frequency domain is $\Delta f = 1/T_0 = 1/KT$. It is the inverse of the sampling duration. Likewise the total unambiguous bandwidth in the frequency domain is $K \times \Delta f = 1/T$, covering just one segment of the spectrum.

With those parameters established we can now consider how the Fourier transform operation can be modified to handle digital data.

7.7.2 Discrete Fourier Transform Formulae

Let the sequence $\phi(k)$, $k = 0, \ldots K-1$ be the set of K samples taken of $f(t)$ over the sampling period 0 to T_0. The samples correspond to times $t_k = kT$.

Let the sequence $F(r)$, $r = 0, \ldots K-1$ be the set of samples of the frequency spectrum. These can be derived from the $\phi(k)$ by suitably modifying (7.8a). For example, the integral over time can be replaced by the sum over $k = 0$ to $K-1$, with dt replaced by T, the sampling increment. The continuous function $f(t)$ is replaced by the samples $\phi(k)$ and $\omega = 2\pi f$ is replaced by $2\pi r\Delta f$, with $r = 0, 1, \ldots K-1$. Thus $\omega = 2\pi r/T_0$. The time variable t is replaced by $kT = kT_0/K$, $k = 0, \ldots K-1$. With these changes (7.8a) can be written in sampled form as

$$F(r) = T \sum_{k=0}^{K-1} \phi(k) \, W^{rk}, \, r = 0, \ldots K-1 \tag{7.15}$$

with

$$W = e^{-j2\pi/K}. \tag{7.16}$$

Equation (7.15) is known as the *discrete Fourier transform* (DFT). In a similar manner a *discrete inverse Fourier transform* (DIFT) can be derived that allows reconstruction of the time sequence $\phi(k)$ from the frequency samples $F(r)$. This is

$$\phi(k) = \frac{1}{T_0} \sum_{r=0}^{K-1} F(r) \, W^{-rk}, \, k = 0, \ldots K-1 \tag{7.17}$$

Substitution of (7.15) into (7.17) shows that those two expressions form a Fourier transform pair. This is achieved by putting $k = l$ in (7.17) so that

$$\phi(l) = \frac{1}{T_0} \sum_{r=0}^{K-1} F(r)\, W^{-rl}$$

$$= \frac{1}{T_0} \sum_{r=0}^{K-1} T \sum_{k=0}^{K-1} \phi(k)\, W^{r(k-l)}$$

$$= \frac{1}{K} \sum_{k=0}^{K-1} \phi(k) \sum_{r=0}^{K-1} W^{r(k-l)}$$

The second sum in this expression is zero for $k \neq l$; when $k = l$ it is K, so that the right hand side of the equality then becomes $\phi(l)$ as required. An interesting aspect of this development has been that T has cancelled out, leaving $1/K$ as the net constant from the forward and inverse transforms. As a result (7.15) and (7.17) could conveniently be written

$$F(r) = \sum_{k=0}^{K-1} \phi(k)\, W^{rk}, r = 0, \ldots K-1 \tag{7.15'}$$

$$\phi(k) = \frac{1}{K} \sum_{r=0}^{K-1} F(r)\, W^{-rk}, k = 0, \ldots K-1 \tag{7.17'}$$

7.7.3 Properties of the Discrete Fourier Transform

Three properties of the discrete Fourier transform and its inverse are of importance here.

Linearity: Both the DFT and DIFT are linear operations. Thus if $F_1(r)$ is the DFT of $\phi_1(k)$ and $F_2(r)$ is the DFT of $\phi_2(k)$ then for any complex constants a and b, $aF_1(r) + bF_2(r)$ is the DFT of $a\phi_1(k) + b\phi_2(k)$.

Periodicity: From (7.16), $W^K = 1$ and $W^{kK} = 1$ for k integral. Thus for $r' = r + K$

$$F(r') = T \sum_{k=0}^{K-1} \phi(k)\, W^{(r+K)k} = F(r).$$

Therefore in general

$$F(r + mK) = F(r) \tag{7.18a}$$

$$\phi(k + mK) = \phi(k) \tag{7.18b}$$

where m is an integer. Thus both the sequence of time samples and the sequence of frequency samples are periodic with period K. This is consistent with the development of Sect. 7.7.1 and has two important implications. First, to generate the Fourier series components of a periodic function, samples need only be taken over one period. Secondly, sampling converts an aperiodic sequence into a periodic one, the period being determined by the sampling duration.

Symmetry: Let $r' = K - r$ in (7.15), to give $F(r') = T \sum_{k=0}^{K-1} \phi(k)\, W^{-rk}\, W^{kK}$. Since $W^{kK} = 1$ this shows $F(K - r) = F(r)*$ where here * represents complex conjugate. This implies that the amplitude spectrum is symmetric about $K/2$ and the phase spectrum is antisymmetric (i.e. odd).

7.7.4 Computation of the Discrete Fourier Transform

It is convenient to consider the reduced form of (7.15):

$$A(r) = \frac{1}{T} F(r) = \sum_{k=0}^{K-1} \phi(k) \, W^{rk}, r = 0, \dots K-1 \tag{7.19}$$

Computation of the K values of $A(r)$ from the K samples $\phi(k)$ requires K^2 multiplications and K^2 additions, assuming that the required values of W^{rk} would have been calculated beforehand and stored. Since the W^{rk} are complex, the multiplications and additions necessary to evaluate $A(r)$ are complex. Thus, as the number of samples $\phi(k)$ becomes large, the time required to compute the sampled spectrum $A(r)$ increases enormously (as the square of the number of samples). Between 1000 and 10,000 samples may in fact require unacceptably high computing time. A technique is required therefore to reduce substantially the number of arithmetic operations required in computing discrete Fourier transforms.

7.7.5 Development of the Fast Fourier Transform Algorithm

Assume K is even; in fact the algorithm to follow will require K to be expressible as $K = 2^m$ where m is an integer. From $\phi(k)$ form two sequences $Y(k)$ and $Z(k)$ each of $K/2$ samples. The first contains the even numbered samples of $\phi(k)$ and the second the odd numbered samples, viz.

$$Y(k) : \phi(0), \ \phi(2), \dots \phi(K-2)$$

$$Z(k) : \phi(1), \ \phi(3), \dots \phi(K-1)$$

so that

$$Y(k) = \phi(2k)$$
$$Z(k) = \phi(2k+1) \qquad k = 0, \dots \frac{K}{2} - 1.$$

Equation (7.19) can then be written

$$A(r) = \sum_{k=0}^{K/2-1} \{ Y(k) \, W^{2rk} + Z(k) \, W^{r(2k+1)} \}$$

$$= \sum_{k=0}^{K/2-1} Y(k) \, W^{2rk} + W^r \sum_{k=0}^{K/2-1} Z(k) \, W^{2rk}$$

$$= B(r) + W^r C(r)$$

where $B(r)$ and $C(r)$ will be recognised as the discrete Fourier transforms of the sequences $Y(k)$ and $Z(k)$. These are periodic, with period $K/2$, according to (7.18). Since $W^{K/2} = -1$ it can be shown that the first $K/2$ samples of $A(r)$ and the last $K/2$ samples of $A(r)$ can be obtained from the same amount of computation, viz;

$$A(r) = B(r) + W^r C(r)$$
$$A\left(r + \frac{K}{2}\right) = B(r) - W^r C(r) \left.\right] \quad r = 0 \ldots \frac{K}{2} - 1 \qquad (7.20)$$

Furthermore values of W^r only up to $W^{K/2}$ are required.

The procedure of (7.20) can be represented conveniently in flow chart form. This is shown for $K = 8$ in Fig. 7.8.

Equation (7.20) requires the Fourier transforms $B(r)$ and $C(r)$. The same procedure can again be used to advantage for these; $Y(k)$ and $Z(k)$ are each broken up into sequences of odd and even samples, requiring $Y(k)$ and $Z(k)$ to contain an even number each. This in turn means that K had to be divisible at least by 4. Let $S(k)$ contain the even numbered samples of $Y(k)$ and $T(k)$ the odd numbered samples. Also let $U(k)$ contain the even numbered samples of $Z(k)$ and $V(k)$ the odd numbered samples:

$$S(k): Y(0), Y(2), \ldots \quad (\text{i.e. } \phi(0), \phi(4), \ldots)$$
$$T(k): Y(1), Y(3), \ldots \quad (\text{i.e. } \phi(2), \phi(6), \ldots)$$
$$U(k): Z(0), Z(2), \ldots \quad (\text{i.e. } \phi(1), \phi(5), \ldots)$$
$$V(k): Z(1), Z(3), \ldots \quad (\text{i.e. } \phi(3), \phi(7), \ldots)$$

If the discrete Fourier transforms of these are denoted $D(r)$, $E(r)$, $G(r)$ and $H(r)$ respectively, each containing $K/4$ points, then

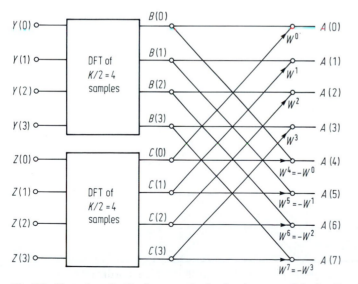

Fig. 7.8. Flow chart for the first stage in the development of the fast Fourier transform algorithm, for the case of $K = 8$

$$B(r) = \sum_{k=0}^{K/2-1} Y(k) \, W^{2rk}$$
$$= D(r) + W^{2r} E(r)$$

which can be written

$$\left.\begin{array}{l} B(r) = D(r) + W^{2r} E(r) \\ B\left(r + \dfrac{K}{4}\right) = D(r) - W^{2r} E(r) \end{array}\right] r = 0, \ldots \frac{K}{4} - 1$$

again showing that the first and second halves of the set of $B(r)$ can be obtained by the same calculations. Similarly

$$\left.\begin{array}{l} C(r) = G(r) + W^{2r} H(r) \\ C\left(r + \dfrac{K}{4}\right) = G(r) - W^{2r} H(r) \end{array}\right] r = 0, \ldots \frac{K}{4} - 1 \; .$$

Figure 7.9 shows how the flow chart of Fig. 7.8 can be modified to take account of this development.

Clearly the procedure followed to this point can be repeated as many times as there are discrete Fourier transforms left to compute. Ultimately transforms will be required on sequences with just two samples each. For example if $K = 8$, the sequences S, T, U and V will each contain only two samples and their discrete Fourier transforms will be of the form

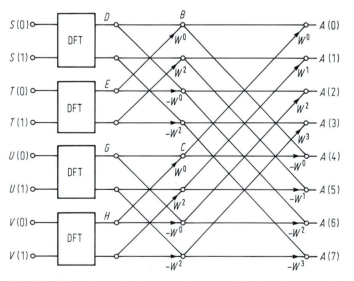

Fig. 7.9. Flow chart for the second stage of the development of the fast Fourier transform algorithm, for the case of $K = 8$

$$D(r) = \sum_{k=0}^{1} S(k) \, W^{4rk}, \qquad r = 0, 1$$

$$= S(0) \, W^0 + S(1) \, W^{4r} \qquad r = 0, 1$$

i.e.

$$D(0) = S(0) + S(1)$$

$$D(1) = S(0) - S(1),$$

showing that the discrete Fourier transform of two samples is obtained by simple addition and subtraction. Doing likewise for the other sequences gives the final flow chart for $K = 8$ as shown in Fig. 7.10.

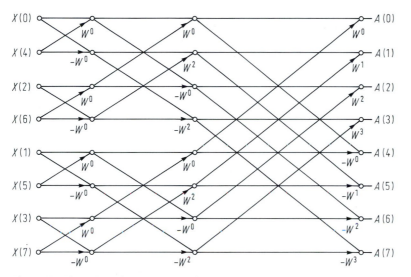

Fig. 7.10. Flow chart for a complete fast Fourier transform evaluation when $K = 8$

7.7.6 Computational Cost of the Fast Fourier Transform

The information contained in Fig. 7.10 can be used to determine the computational cost of the fast Fourier transform algorithm and therefore to find its speed advantage over a direct evaluation of the discrete formula in (7.19). The figure shows that the only multiplications required are by the values of W. While this is strictly not necessary in the first set of calculations on the left of the figure (Since $W^0 = 1$, and $-W^0 = -1$) it is simpler in programming if the multiplications are retained. Thus the left hand set of computations requires $K/2$ complex multiplications and K additions (or subtractions). The next column of operations requires another $K/2$ multiplications as does the last set for the case of $K = 8$. Altogether for this illustration $3/2 \, K$ multiplications and $3 \, K$ additions are required. It is easy to generalize this to:

$$\text{number of complex multiplications} = \frac{1}{2} K \log_2 K$$

$$\text{number of complex additions} = K \log_2 K.$$

On the basis of multiplications alone the fast Fourier transform (FFT) is seen, from the material in Sect. 7.7.4, to be faster than direct evaluation of the discrete Fourier transform (DFT) by a factor of $2K/\log_2 K$. Moreover its cost increases almost linearly with the number of samples, whereas that for the DFT increases quadratically. This is illustrated in Fig. 7.11.

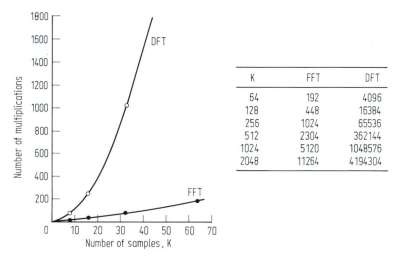

K	FFT	DFT
64	192	4096
128	448	16384
256	1024	65536
512	2304	362144
1024	5120	1048576
2048	11264	4194304

Fig. 7.11. Number of multiplications required in the evaluation of a discrete Fourier transform directly (DFT) and by means of the fast Fourier transform method (FFT)

7.7.7 Bit Shuffling and Storage Considerations

Application of the fast Fourier transform requires K to be continuously divisible by 2 (i.e. $K = 2^m$ as indicated above). Although other versions of the algorithm can also be derived (Brigham, 1974) the case of $K = 2^m$ is most common, and is used here.

Inspection of the flow chart in Fig. 7.10 reveals that the order of the data fed into the algorithm needs to be rearranged before the technique can be employed. This can be achieved very simply by a process known as bit shuffling. To do this the index of the input samples is expressed in binary notation (see Appendix B), the binary digits are reversed, and the new binary number converted back to decimal form, as illustrated in the following for $K = 8$.

$$
\begin{array}{ccccccc}
X(0) & \rightarrow & X(000) & \rightarrow & X(000) & \rightarrow & X(0) \\
X(1) & & X(001) & & X(100) & & X(4) \\
X(2) & & X(010) & & X(010) & & X(2) \\
X(3) & & X(011) & & X(110) & & X(6) \\
X(4) & & X(100) & & X(001) & & X(1) \\
X(5) & & X(101) & & X(101) & & X(5) \\
X(6) & & X(110) & & X(011) & & X(3) \\
X(7) & & X(111) & & X(111) & & X(7)
\end{array}
$$

Apart from the immense savings in time, use of the FFT also leads to a savings in memory. Apart from storing the $K/2$ values of W^r, the entire computation can be carried out using a complex vector of length $K+1$. This is because there exist pairs of elements in each vector or column of the operation whose values are computed from numbers stored in the same pair of locations in the previous column.

7.8 The Discrete Fourier Transform of an Image

7.8.1 Definition

The foregoing sections have treated functions with a single independent variable. That variable could have been time, or even position along a line of an image. We now need to turn our attention to functions with two independent variables, to allow Fourier transforms of images to be determined. Despite this apparent increase in complexity we will find that full advantage can be taken of the material of the previous sections. Let

$$\phi(i, j), \quad i, j = 0, \ldots K - 1 \tag{7.21}$$

be the brightness of a pixel at location i, j in an image of $K \times K$ pixels. The Fourier transform of the image, in discrete form, is described by

$$\Phi(r, s) = \sum_{i=0}^{K-1} \sum_{j=0}^{K-1} \phi(i, j) \exp\left[-j2\pi(ir + js)/K\right]. \tag{7.22}$$

An image can be reconstructed from its transform according to

$$\phi(i, j) = \frac{1}{K^2} \sum_{i=0}^{K-1} \sum_{j=0}^{K-1} \Phi(r, s) \exp\left[+j2\pi(ir + js)/K\right]. \tag{7.23}$$

7.8.2 Evaluation of the Two Dimensional, Discrete Fourier Transform

Equation (7.22) can be rewritten as

$$\Phi(r, s) = \sum_{i=0}^{K-1} W^{ir} \sum_{j=0}^{K-1} \phi(i, j) W^{js} \tag{7.24}$$

with $W = e^{-j2\pi/K}$ as before. The term involving the right hand sum can be recognised as the one dimensional discrete Fourier transform

$$\Phi(i, s) = \sum_{j=0}^{K-1} \phi(i, j) W^{js}, \quad i = 0, \ldots K - 1. \tag{7.25}$$

In fact it is the one dimensional transform of the ith row of pixels in the image. The result of this operation is that the rows of an image are replaced by their Fourier transforms; the transformed pixels are then addressed by the spatial frequency index s across a row rather than by the positional index j. Using (7.25) in (7.24) gives

$$\Phi(r, s) = \sum_{i=0}^{K-1} \Phi(i, s) \; W^{ir} \qquad\qquad (7.26)$$

which is the one dimensional discrete Fourier transform of the sth column of the image, after the row transforms of (7.25) have been performed.

Thus, to compute the two dimensional Fourier transform of an image, it is only necessary to transform each row individually to generate an intermediate image, and then transform this by column to yield the final result. Both the row and column transforms would be carried out using the fast Fourier transform algorithm of Sect. 7.7.5. From the information provided in Sect. 7.7.6 it can be seen therefore that the number of multiplications required to transform an image is $K^2 \log_2 K$.

7.8.3 The Concept of Spatial Frequency

Entries in the Fourier transformed image $\Phi(r, s)$ represent the composition of the original image in terms of spatial frequency components, both vertically and horizontally. Spatial frequency is the image analog of the frequency of a signal in time. A sinusoidal signal with high frequency alternates rapidly, whereas a low frequency signal changes slowly with time. Similarly, an image with high spatial frequency, say in the horizontal direction, exhibits frequent changes of brightness with position horizontally. A picture of a crowd of people would be a particular example. By comparison a head and shoulders view of a person is likely to be characterised mainly by low spatial frequencies. Typically an image is composed of a collection of both horizontal and vertical spatial frequency components of differing strengths and these are what the discrete Fourier transform indicates. The upper left hand pixel in $\Phi(r, s)$ – i. e. $\Phi(0, 0)$ – is the average brightness value of the image. This is the component in the spectrum with zero frequency in both directions. Thereafter pixels of $\Phi(r, s)$ both horizontally and vertically represent components with frequencies that increment by $1/K$ where the original image is of size $K \times K$. Should the scale of the image be known then the spatial frequency increment can be calibrated in terms of metres^{-1}. For example the increment in spatial frequency for a 512×512 pixel image that covers 15.36 km (i. e. Landsat TM) is $65 \times 10^{-6}\text{m}^{-1}$.

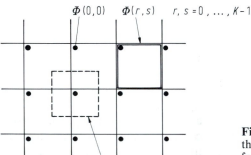

Displayed spectrum, having $\Phi(0,0)$ at the centre

$\Phi(0,0)$ $\Phi(r,s)$ $r, s = 0, \ldots, K-1$

Fig. 7.12. Illustration of the periodic nature of the two dimensional discrete Fourier transform, showing how an array centred on $\Phi(0,0)$ is chosen for symmetrical display purposes

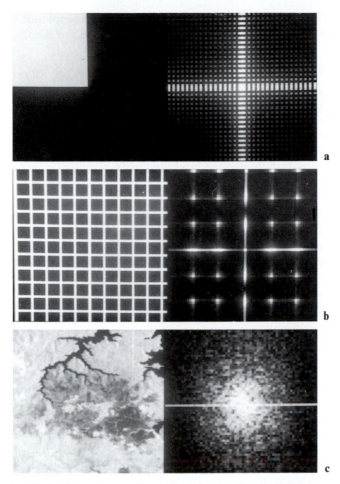

Fig. 7.13. Illustrations of Fourier transforms of images. **a** Single square; **b** Bar pattern; **c** Landsat MSS image. *The material for this figure was produced by Brenton Hamilton*

In Sect. 7.7.3 it is shown that the one dimensional discrete Fourier transform is periodic with period K. The same is true of the discrete two dimensional form. Indeed the $K \times K$ pixels of $\Phi(r, s)$ computed according to (7.22) can be viewed as one period of an infinite periodic two dimensional array in the manner depicted in Fig. 7.12. It is also shown that the amplitude of the discrete Fourier transform is symmetric about $K/2$. Similarly $\Phi(r, s)$ is symmetric about its centre. This can be interpreted by implying that no new amplitude information is shown by displaying pixels horizontally and vertically beyond $K/2$. Rather than ignore them (since their accompanying phase is important) the display is adjusted in the manner shown in Fig. 7.12 to bring $\Phi(0, 0)$ to the centre. In this way the pixel at the centre of the Fourier transform array represents the image average brightness value. Pixels away from the centre represent the proportions of increasing spatial frequency components in the image. This is the usual method of presenting two dimensional image transforms. Examples of spectra displayed in this manner are given in Fig. 7.13. To make visible components with smaller amplitudes, a logarithmic amplitude scaling has been used, according to (Gonzalez and Woods, 1992)

$$D(r, s) = \log\left[1 + |\Phi(r, s)|\right].$$

7.8.4 Image Filtering for Geometric Enhancement

The high spatial frequency content of an image is that associated with frequent changes of brightness with position. Edges, lines and some types of noise are examples of high frequency data. In contrast, gradual changes of brightness with position, such as associated with more general tonal variations, account for the low frequency content in the spectrum. Since ranges of spatial frequency are identified with regions in the spectrum we can envisage how the spectrum of an image could be altered to produce different geometric enhancements of the image itself. For example, if the region near the centre of the spectrum is removed, leaving behind only the high frequencies, and the image is then reconstructed from the modified spectrum, a version containing only edges and line-like features will be produced. On the other hand, if the high frequency components are removed, leaving behind only the region near the centre of the spectrum, the reconstructed image will appear smoothed, since edges, lines and high frequency noise will have been deleted.

Modification of the spectrum in the manner just described can be expressed as a multiplicative operation:

$$Y(r, s) = \Phi(r, s)\, H(r, s) \quad \text{for all} \quad r, s \tag{7.27}$$

where $H(r, s)$ is the filter function and $Y(r, s)$ is the new spectrum. To implement simple sharpening or smoothing as described above $H(r, s)$ would be set to 0 for those frequency components to be removed and 1 for those components to be retained. Often sharpening is called high pass filtering, and smoothing low pass filtering, because of the nature of the modification to the spectrum. Both can be implemented also with the template methods of Chap. 5. However (7.27) allows more complicated filtering operations to be carried out. As an example, a specific band of spatial frequency could be excluded readily. $H(r, s)$ can also be chosen to have values other than 0 and 1 to allow more versatile modification of the spectrum.

The overall process of geometric enhancement via the frequency domain involves three steps. First, the image has to be Fourier transformed to produce its spectrum. Secondly, the spectrum is modified according to (7.27). Finally the image is reconstructed from the modified spectrum using (7.23), which can also be implemented by rows and columns. Together these three operations require $2K^2 \log_2 K + K^2$ multiplications, as used in Sect. 5.4 to compare this approach to that based upon simple templates.

7.8.5 Convolution in Two Dimensions

The convolution theorem for functions (Sect. 7.5.3) has a two dimensional counterpart, again in two forms. These are:

If

$$y(i, j) = \phi(i, j) * h(i, j)$$

then

$$Y(r, s) = \Phi(r, s) \, H(r, s) \tag{7.28a}$$

and, if

$$y(i, j) = \phi(i, j) \, h(i, j)$$

then

$$Y(r, s) = \Phi(r, s) * H(r, s). \tag{7.28b}$$

Unlike (7.10b) there is no $1/2\pi$ scaling factor here since the spatial frequency variables r and s are equivalent to frequency f in Hz and not the radian frequency ω in rad $\cdot$ s^{-1} used in (7.10b).

The convolution operation implied in (7.28) is defined by (5.3). However when digital images are of concern its discrete version is of interest. This is defined, in the image domain, as

$$y(i, j) = \sum_m \sum_n \phi(m, n) \, h(i - m, j - n) \tag{7.29}$$

where m and n are dummy variables. As with one dimensional convolution described in Sect. 7.5.1 evaluation of (7.29) requires that one function, in this case the filter function, be folded about the origin (which in two dimensions amounts to a 180° rotation) to produce $h(-m, -n)$ and then delayed by variable amounts i, j. The delayed folded version is then multiplied pixel by pixel with the image $\phi(m, n)$ and the sum over all spectral pixels taken. This produces one pixel $y(i, j)$ in the modified image.

Equation (7.28) implies that any of the geometric enhancement operations that can be carried out by modifying the spectrum can also be carried out by performing a convolution between the image and the inverse Fourier transform of the filter function $H(r, s)$. Conversely, operations such as simple mean value filtering with an $M \times N$ template as described in Sect. 5.5.1, can also be described in the spatial frequency domain. This requires the Fourier transform of the template to be found. To do this requires the template to be regarded as of the same dimensions as the image but with a value of zero everywhere except for a set of $M \times N$ pixels with the appropriate non-zero value.

7.9 Concluding Remarks

Geometric modification of an image via the frequency domain is a particularly powerful technique owing to the ease with which the filter function $H(r, s)$ may be designed. The material presented in this Chapter has been intended as an introduction to the concepts and operations involved. For the user contemplating using Fourier domain methods, several other issues should be taken into consideration including the use of so-called window functions. This is illustrated most easily by a return to the material on sampling in Sect. 7.6. In that section it was noted that a sampled function could be regarded as the unsampled version multiplied by an infinite periodic sequence of impulses. The spectrum of the infinite set of samples so produced is the spectrum of the original function convolved with the spectrum of the sequence of impulses as shown in Fig. 7.6. However in practice it is not possible to take an infinite number of samples of a function. Instead sampling is commenced at a given time and terminated after some period τ. This finite time sampling window can be considered as a long pulse of unit amplitude and duration τ that multiplies the infinite sequence of samples. The spectrum of the set of samples is, as a consequence, modified by being convolved by the spectrum of the sampling window. Since the window is a long pulse, its Fourier transform is as shown in Fig. 7.4 although compressed to near the origin. If the sampling duration is long enough this approximates an impulse and there is little effect on the spectrum. For shorter sampling times however the sidelobes in Fig. 7.4b cause distortion of the spectrum. To minimise this effect sampling windows different to a long pulse are sometimes used. A good consideration of these is found in Brigham (1974).

In the preceding sections we have referred to the Fourier transform approach as a means for geometric enhancement since it can implement operations such as sharpening and smoothing. In the material of Chapter Five these are referred to explicitly as neighbourhood operations. To appreciate that the Fourier transform is also a neighbourhood operation consider the flow chart for the fast Fourier transform implementation in Fig. 7.10. If we pick one output value – i.e. one point on the spectrum – it can be traced back through the flow chart and be seen to have a contribution from every one of the input samples. In a similar manner the pixels in the Fourier transform of an image have contributions from all of the pixels in the original image.

Other image transforms also exist, perhaps the most notable being the Hadamard transform. Whereas the Fourier transform is based on expressing an image in terms of horizontal and vertical sinusoidal spatial frequency components, the Hadamard transform is based upon Walsh functions. These are discrete functions that take on values of $+1$ or -1 and in some ways are a natural basis for an image transform. Moreover the transform is readily computed (Gonzalez and Woods, 1992 and Billingsley, 1983).

References for Chapter 7

Treatments of digital image processing in the fields of electrical engineering and computer science invariably contain detailed considerations of the use of the Fourier transform and frequency domain techniques for geometric modification of image data. Particular texts that could be consulted include Castleman (1996), Gonzalez and Woods (1992) and Moik (1980). An excellent presentation of the discrete Fourier transform, discrete convolution and the fast Fourier transform algorithm will be found in Brigham (1974, 1988). While Brigham relates to the one dimensional case it will be clear from the material in Sect. 7.8.2 above that it can be used also with images.

F. C. Billingsley, 1983: Data Processing and Reprocessing. In R. N. Colwell (Ed.) Manual of Remote Sensing, 2e. Falls Church, American Society of Photogrammetry.
E. O. Brigham, 1974: The Fast Fourier Transform. N. J. Prentice-Hall.
E. O. Brigham, 1988: The Fast Fourier Transform and its Applications. N. J. Prentice-Hall.
K. R. Castleman, 1996: Digital Image Processing. N. J. Prentice-Hall.
R. C. Gonzalez and R. E. Woods, 1992: Digital Image Processing. Mass. Addison-Wesley.
C. D. McGillem and G. R. Cooper, 1984: Continous and Discrete Signal and System Analysis. N. Y. Holt, Rinehart and Winston.
J. G. Moik, 1980: Digital Processing of Remotely Sensed Images. Washington, NASA.
A. Papoulis, 1980: Circuits and Systems: A Modern Approach. Tokyo, Holt-Saunders.

Problems

7.1 Compute the discrete Fourier transform of the square wave shown in Fig. 7.3 using $K = 2$, 4 and 8 samples per period of the waveform respectively. You can use the flow chart of Fig. 7.10 to help in this.

7.2 Compute the discrete Fourier transform of the unit pulse shown in Fig. 7.4. Use respectively $K = 2$, 4 and 8 samples over a time interval equal to 8 a, where 2 a is the width of the pulse as shown in the Figure. Compare the results with those obtained in problem 7.1.

7.3 (a) A common technique for smoothing an image is to compute averages over square or rectangular windows as discussed in Sect. 5.5. Consider a 3×1 smoothing template used to smooth a single line of image data in the manner of Fig. 5.4. Determine the corresponding filter function in the spatial frequency domain by finding the discrete Fourier transform of the template. You may find the material of Fig. 7.4 to be of value.

(b) Imagine an ideal low pass filter function in the spatial frequency domain that could be used to smooth just the lines of an image. Determine the corresponding function in the image domain by computing the inverse Fourier transform of the ideal filter. Taking into account the discrete pixel nature of the image, approximate the inverse transform by an appropriate one dimensional template.

7.4 Verify the results in Sect. 7.5.2 graphically.

7.5 (a) The periodic sequence of impulses of (7.11) is an idealised sampling function. In practice it is not possible to take infinitessimally short samples of a function; rather the samples will have a finite, albeit small duration. This could be modelled mathematically by replacing $\Delta(t)$ in (7.12) by a periodic pulse waveform. This periodic sequence of pulses can be represented by the convolution of a single pulse with the periodic sequence of impulses in (7.11). With this in mind describe what modifications are needed to Fig. 7.6 to account for samples of finite duration.

(b) Suppose the total period of sampling is equivalent to ten sample intervals. Describe the effect this has on Fig. 7.6.

7.6 In Fig. 7.6 a suppose the function $f(t)$ is a sinewave of frequency B Hz. Its frequency spectrum will consist of two impulses, one at $+B$ Hz and the other at $-B$ Hz. Produce the spectrum of the sampled sinusoid if only three samples are taken every two periods. Suppose the waveform is then reconstructed by feeding the samples through a low pass filter that will pass all frequency components unattenuated, up to $1/2T$ Hz, where T is the sampling interval, and will exclude all components with frequencies in excess of $1/2T$ Hz. Describe the shape of the reconstructed signal; this will give an appreciation of aliasing distortion.

Chapter 8
Supervised Classification Techniques

The purpose of this Chapter is to present the algorithms used regularly for the supervised classification of single sensor remote sensing image data.

When data from a variety of sensors or sources (such as found in the integrated spatial data base of a Geographical Information System) requires analysis, more sophisticated tools may be required. These are the subject of Chapter 12 which deals with the topic of Data Fusion.

8.1 Steps in Supervised Classification

Supervised classification is the procedure most often used for quantitative analysis of remote sensing image data. It rests upon using suitable algorithms to label the pixels in an image as representing particular ground cover types, or classes. A variety of algorithms is available for this, ranging from those based upon probability distribution models for the classes of interest (such as outlined in Chap. 3) to those in which the multispectral space is partitioned into class-specific regions using optimally located surfaces. Irrespective of the particular method chosen, the essential practical steps are:

1. Decide the set of ground cover types into which the image is to be segmented. These are the information classes and could, for example, be water, urban regions, croplands, rangelands, etc.
2. Choose representative or prototype pixels from each of the desired set of classes. These pixels are said to form *training data*. Training sets for each class can be established using site visits, maps, air photographs or even photointerpretation of a colour composite product formed from the image data (either in hardcopy form or on a colour display). Often the training pixels for a given class will lie in a common region enclosed by a border. That region is then often called a *training field*.
3. Use the training data to estimate the parameters of the particular classifier algorithm to be used; these parameters will be the properties of the probability model used or will be equations that define partitions in the multispectral space. The set of parameters for a given class is sometimes called the *signature* of that class.
4. Using the trained classifier, label or classify every pixel in the image into one of the desired ground cover types (information classes). Here the whole image segment of interest is typically classified. Whereas training in Step 2 may have required the user to identify perhaps 1% of the image pixels by other means, the computer will label the rest by classification.

5. Produce tabular summaries or thematic (class) maps which summarise the results of the classification.

It is our objective now to consider the range of algorithms that could be used in 3 and 4. In so doing it will be assumed that the information classes each consists of only one spectral class, so that the two names will be used synonomously. (See Chap. 3 for a discussion of the two class types.) By making this assumption, problems with establishing sub-classes will not distract from the algorithm development to be given. Handling sub-classes is taken care of explicitly in Chaps. 9 and 11.

In the following sections it is assumed that the reader is familiar at least with the sections on quantitative analysis contained in Chap. 3. This relates particularly to definitions and terminology.

8.2 Maximum Likelihood Classification

Maximum likelihood classification is the most common supervised classification method used with remote sensing image data. This is developed in the following in a statistically acceptable manner; it can be derived however in a more general and rigorous manner and this is presented for completeness in Appendix E. The present approach is sufficient though for most remote sensing exercises.

8.2.1 Bayes' Classification

Let the spectral classes for an image be represented by

$$\omega_i, \ i = 1, \ldots M$$

where M is the total number of classes. In trying to determine the class or category to which a pixel at a location x belongs it is strictly the conditional probabilities

$$p(\omega_i | x), \ i = 1, \ldots M$$

that are of interest. The position vector x is a column vector of brightness values for the pixel. It describes the pixel as a point in multispectral space with co-ordinates defined by the brightnesses, as shown in the simple two-dimensional example of Fig. 3.5. The probability $p(\omega_i | x)$ gives the likelihood that the correct class is ω_i for a pixel at position x. Classification is performed according to

$$x \in \omega_i \quad \text{if} \quad p(\omega_i | x) > p(\omega_j | x) \quad \text{for all} \quad j \neq i \tag{8.1}$$

i.e., the pixel at x belongs to class ω_i if $p(\omega_i | x)$ is the largest. This intuitive *decision rule* is a special case of a more general rule in which the decisions can be biased according to different degrees of significance being attached to different incorrect classifications. The general approach is called Bayes' classification and is the subject of the treatment in Appendix E.

8.2.2 The Maximum Likelihood Decision Rule

Despite its simplicity, the $p(\omega_i|x)$ in (8.1) are unknown. Suppose however that sufficient training data is available for each ground cover type. This can be used to estimate a probability distribution for a cover type that describes the chance of finding a pixel from class ω_i, say, at the position x. Later the form of this distribution function will be made more specific. For the moment however it will be retained in general terms and represented by the symbol $p(x|\omega_i)$. There will be as many $p(x|\omega_i)$ as there are ground cover classes. In other words, for a pixel at a position x in multispectral space a set of probabilities can be computed that give the relative likelihoods that the pixel belongs to each available class.

The desired $p(\omega_i|x)$ in (8.1) and the available $p(x|\omega_i)$ – estimated from training data – are related by Bayes' theorem (Freund, 1992):

$$p(\omega_i|x) = p(x|\omega_i) \, p(\omega_i)/p(x) \qquad (8.2)$$

where $p(\omega_i)$ is the probability that class ω_i occurs in the image. If, for example, 15% of the pixels of an image happen to belong to spectral class ω_i then $p(\omega_i) = 0.15$; $p(x)$ in (8.2) is the probability of finding a pixel from *any* class at location x. It is of interest to note in passing that

$$p(x) = \sum_{i=1}^{M} p(x|\omega_i) \, p(\omega_i),$$

although $p(x)$ itself is not important in the following. The $p(\omega_i)$ are called *a priori* or prior probabilities, since they are the probabilities with which class membership of a pixel could be guessed before classification. By comparison the $p(\omega_i|x)$ are posterior probabilities. Using (8.2) it can be seen that the classification rule of (8.1) is:

$$x \in \omega_i \quad \text{if} \quad p(x|\omega_i) \, p(\omega_i) > p(x|\omega_j) \, p(\omega_j) \quad \text{for all} \quad j \neq i \qquad (8.3)$$

where $p(x)$ has been removed as a common factor. The rule of (8.3) is more acceptable than that of (8.1) since the $p(x|\omega_i)$ are known from training data, and it is conceivable that the $p(\omega_i)$ are also known or can be estimated from the analyst's knowledge of the image. Mathematical convenience results if in (8.3) the definition

$$g_i(x) = \ln \{p(x|\omega_i) \, p(\omega_i)\}$$
$$= \ln p(x|\omega_i) + \ln p(\omega_i) \qquad (8.4)$$

is used, where ln is the natural logarithm, so that (8.3) is restated as

$$x \in \omega_i \quad \text{if} \quad g_i(x) > g_j(x) \quad \text{for all} \quad j \neq i \qquad (8.5)$$

This is, with one modification to follow, the decision rule used in maximum likelihood classification; the $g_i(x)$ are referred to as *discriminant functions*.

8.2.3 Multivariate Normal Class Models

At this stage it is assumed that the probability distributions for the classes are of the form of multivariate normal models. This is an assumption, rather than a demonstrable property of natural spectral or information classes; however it leads to mathematical simplifications in the following. Moreover it is one distribution for which properties of the multivariate form are well-known.

In (8.4) therefore, it is now assumed for N bands that (see Appendix D)

$$p(x|\omega_i) = (2\pi)^{-N/2} |\Sigma_i|^{-1/2} \exp \{-\tfrac{1}{2} (x - m_i)^t \Sigma_i^{-1} (x - m_i)\} \tag{8.6}$$

where m_i and Σ_i are the mean vector and covariance matrix of the data in class ω_i. The resulting term $-N/2 \ln (2\pi)$ is common to all $g_i(x)$ and does not aid discrimination. Consequently it is ignored and the final form of the discriminant function for maximum likelihood classification, based upon the assumption of normal statistics, is:

$$g_i(x) = \ln p(\omega_i) - \tfrac{1}{2} \ln |\Sigma_i| - \tfrac{1}{2}(x - m_i)^t \Sigma_i^{-1} (x - m_i) \tag{8.7}$$

Often the analyst has no useful information about the $p(\omega_i)$, in which case a situation of equal prior probabilities is assumed; as a result $\ln p(\omega_i)$ can be removed from (8.7) since it is then the same for all i. In that case the 1/2 common factor can also be removed leaving, as the discriminant function:

$$g_i(x) = - \ln |\Sigma_i| - (x - m_i)^t \Sigma_i^{-1} (x - m_i) \tag{8.8}$$

Implementation of the maximum likelihood decision rule involves using either (8.7) or (8.8) in (8.5). There is a further consideration however concerned with whether any of the available labels or classes is appropriate. This relates to the use of thresholds as discussed in Sect. 8.2.5 following.

8.2.4 Decision Surfaces

As a means for assessing the capabilities of the maximum likelihood decision rule it is of value to determine the essential shapes of the surfaces that separate one class from another in the multispectral domain. These surfaces, albeit implicit, can be devised in the following manner.

Spectral classes are defined by those regions in multispectral space where their discriminant functions are the largest. Clearly these regions are separated by surfaces where the discriminant functions for adjoining spectral classes are equal. The ith and jth spectral classes are separated therefore by the surface

$$g_i(x) - g_j(x) = 0.$$

This is referred to as a *decision surface* since, if all the surfaces separating spectral classes are known, decisions about class membership of an image pixel can be made on the basis of its position relative to the complete set of surfaces.

The construction $(x - m_i)^t \Sigma_i^{-1} (x - m_i)$ in (8.7) and (8.8) is a quadratic function of x. Consequently the decision surfaces implemented by maximum likelihood classification are quadratic and thus take the form of parabolas, circles and ellipses. Some indication of this can be seen in Fig. 3.8.

8.2.5 Thresholds

It is implicit in the foregoing development that pixels at every point in multispectral space will be classified into one of the available classes ω_i, irrespective of how small the actual probabilities of class membership are. This is illustrated for one dimensional data in Fig. 8.1 a. Poor classification can result as indicated. Such situations can arise if spectral classes (between 1 and 2 or beyond 3) have been overlooked or, if knowing

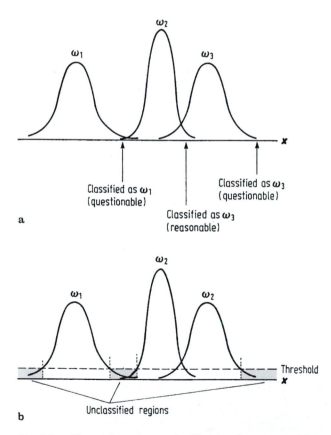

Fig. 8.1. a Illustration of poor classification for patterns lying near the tails of the distribution functions of all spectral classes; **b** Use of a threshold to remove poor classification

other classes existed, enough training data was not available to estimate the parameters of their distributions with any degree of accuracy (see Sect. 8.2.6 following). In situations such as these it is sensible to apply thresholds to the decision process in the manner depicted in Fig. 8.1 b. Patterns which have probabilities for *all* classes below the threshold are not classified.

In practice, thresholds are applied to the discriminant functions and not the probability distributions, since the latter are never actually computed. With the incorporation of a threshold therefore, the decision rule of (8.5) becomes

$$x \in \omega_i \quad \text{if} \quad g_i(x) > g_j(x) \quad \text{for all} \quad j \neq i \tag{8.9 a}$$

$$\text{and} \quad g_i(x) > T_i \tag{8.9 b}$$

where T_i is the threshold seen to be significant for spectral class ω_i. It is now necessary to consider how T_i can be estimated. From (8.7) and (8.9 b) a classification is acceptable if

$$\ln p(\omega_i) - \tfrac{1}{2} \ln |\Sigma_i| - \tfrac{1}{2}(x - m_i)^t \, \Sigma_i^{-1}(x - m_i) > T_i$$

i.e.

$$(x - m_i)^t \, \Sigma_i^{-1}(x - m_i) < -2T_i - \ln |\Sigma_i| + 2 \ln p(\omega_i) \tag{8.10}$$

The left hand side of (8.10) has a χ^2 distribution with N degrees of freedom, if x is (assumed to be) distributed normally (Swain and Davis 1978). N is the dimensionality of the multispectral space. As a result χ^2 tables can be consulted to determine that value of $(x - m_i)^t \, \Sigma_i^{-1}(x - m_i)$ below which a desired percentage of pixels will exist (noting that larger values of that quadratic form correspond to pixels lying further out in the tails of the normal probability distribution). This is depicted in Fig. 8.2.

As an example of how this is used consider the need to choose a threshold for Landsat multispectral scanner data such that 95% of all pixels in a class will be classified (i.e. such that the 5% least likely pixels for each spectral class will be rejected). χ^2 tables show that 95% of all pixels have χ^2 values (in Fig. 8.2) less than 9.488. Thus, from (8.10)

$$T_i = -4.744 - \frac{1}{2} \ln |\Sigma_i| + \ln p(\omega_i)$$

which thus can be calculated from a knowledge of the prior probability and covariance matrix of the ith spectral class.

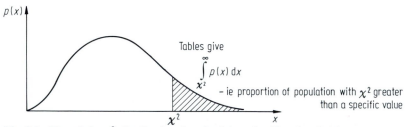

Fig. 8.2. Use of the χ^2 distribution for obtaining classifier thresholds

8.2.6 Number of Training Pixels Required for Each Class

Sufficient training samples for each spectral class must be available to allow reasonable estimates of the elements of the mean vector and the covariance matrix to be determined. For an N dimensional multispectral space at least $N+1$ samples are required to avoid the covariance matrix being singular. Should that happen its inverse in discriminant function expressions cannot be found. Apart from this consideration it is clearly important to have as many training pixels as possible, particularly as the dimensionality of the pixel vector space increases, since in higher dimensional spaces there is an increased chance of having some individual dimensions poorly represented. Swain and Davis (1978) recommend as a practical minimum that $10\,N$ samples per spectral class be obtained for training, with $100\,N$ as being highly desirable if it can be attained.

8.2.7 A Simple Illustration

As an example of the use of maximum likelihood classification, the segment of Landsat multispectral scanner image shown in Fig. 8.3 is chosen. This is a 256×276 pixel array of image data in which four broad ground cover types are evident. These are water, fire burn, vegetation and "developed" land (urban). Suppose one wished to produce a thematic map of these four cover types in order to enable the area and extent of the fire burn to be evaluated.

The first step is to choose training data. For such a broad classification, suitable sets of training pixels for each of the four classes are easily identified visually in the image data. Fig. 8.3 also shows the locations of four training fields used for this purpose. Sometimes, to obtain a good estimate of class statistics it may be necessary to choose several training fields for the one cover type, located in different regions of the image.

The four band signatures for each of the four classes, as obtained from the training fields, are given in Table 8.1. The mean vectors can be seen to agree generally with known spectral reflectance characteristics of the cover types. Also the class variances

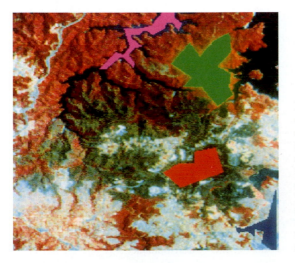

Fig. 8.3. Image segment to be classified, consisting of a mixture of natural vegetation, waterways, urban development and vegetation damaged by fire. Four training regions are identified in solid colour. These are water (violet), vegetation (green), fire burn (red) and urban (dark blue in the bottom right hand corner). Pixels from these were used to generate the signatures in Table 8.1

Table 8.1. Class signatures generated from the training areas in Fig. 8.3. Numbers are on a scale of 0 to 255 (8 bit)

Class	Mean vector	Covariance matrix			
Water	44.27	14.36	9.55	4.49	1.19
	28.82	9.55	10.51	3.71	1.11
	22.77	4.49	3.71	6.95	4.05
	13.89	1.19	1.11	4.05	7.65
Fire burn	42.85	9.38	10.51	12.30	11.00
	35.02	10.51	20.29	22.10	20.62
	35.96	12.30	22.10	32.68	27.78
	29.04	11.00	20.62	27.78	30.23
Vegetation	40.46	5.56	3.91	2.04	1.43
	30.92	3.91	7.46	1.96	0.56
	57.50	2.04	1.96	19.75	19.71
	57.68	1.43	0.56	19.71	29.27
Developed	63.14	43.58	46.42	7.99	−14.86
(urban)	60.44	46.42	60.57	17.38	−9.09
	81.84	7.99	17.38	67.41	67.57
	72.25	−14.86	−9.09	67.57	94.27

Table 8.2. Tabular summary of the thematic map of Fig. 8.4

Class	No. of pixels	Area (ha)
Water	4830	2137
Fireburn	14182	6274
Vegetation	28853	12765
Developed (urban)	22791	10083

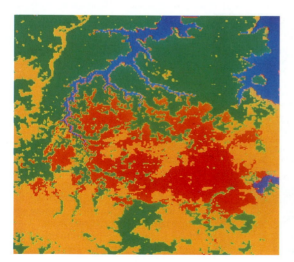

Fig. 8.4. Thematic map produced by maximum likelihood classification. Blue represents water, red is fire damaged vegetation, green is natural vegetation and yellow is urban development

(diagonal elements in the covariance matrices) are small for water as might be expected but on the large side for the developed/urban class, indicative of its heterogeneous nature.

Using these signatures in a maximum likelihood algorithm to classify the four bands of the image in Fig. 8.3, the thematic map shown in Fig. 8.4 is obtained. The four classes, by area, are given in Table 8.2. Note that there are no unclassified pixels, since a threshold was not used in the labelling process. The area estimates are obtained by multiplying the number of pixels per class by the effective area of a pixel. In the case of the Landsat 2 multispectral scanner the pixel size is 0.4424 hectares.

8.3 Minimum Distance Classification

8.3.1 The Case of Limited Training Data

The effectiveness of maximum likelihood classification depends upon reasonably accurate estimation of the mean vector m and the covariance matrix Σ for each spectral class. This in turn is dependent upon having a sufficient number of training pixels for each of those classes. In cases where this is not so, inaccurate estimates of the elements of Σ result, leading to poor classification. When the number of training samples per class is limited it can be more effective to resort to a classifier that does not make use of covariance information but instead depends only upon the mean positions of the spectral classes, noting that for a given number of samples these can be more accurately estimated than covariances. The so-called minimum distance classifier, or more precisely, minimum distance to class means classifier, is such an approach. With this classifier, training data is used only to determine class means; classification is then performed by placing a pixel in the class of the nearest mean.

The minimum distance algorithm is also attractive since it is a faster technique than maximum likelihood classification, as will be seen in Sect. 8.5. However because it does not use covariance data it is not as flexible as the latter. In maximum likelihood classification each class is modelled by a multivariate normal class model that can account for spreads of data in particular spectral directions. Since covariance data is not used in the minimum distance technique class models are symmetric in the spectral domain. Elongated classes therefore will not be well modelled. Instead several spectral classes may need to be used with this algorithm where one might be suitable for maximum likelihood classification. This point is developed further in the case studies of Chap. 11.

8.3.2 The Discriminant Function

The discriminant function for the minimum distance classifier is developed as follows.

Suppose $m_i, i = 1, \ldots M$ are the means of the M classes determined from training data, and x is the position of the pixel to be classified. Compute the set of squared Euclidean distances, defined in vector form as

$$\mathrm{d}(x, m_i)^2 = (x - m_i)^t (x - m_i)$$
$$= (x - m_i) \cdot (x - m_i) \quad i = 1, \ldots M$$

Expanding the product gives

$$d(x, m_i)^2 = x \cdot x - 2\,m_i \cdot x + m_i \cdot m_i.$$

Classification is performed on the basis of

$$x \in \omega_i \quad \text{if} \quad d(x, m_i)^2 < d(x, m_j)^2 \quad \text{for all} \quad j \neq i.$$

Note that $x \cdot x$ is common to all $d(x, m_j)^2$ and thus can be removed. Moreover, rather than classifying according to the smallest of the remaining expressions, the signs can be reversed and classification performed on the basis of

$$x \in \omega_i \quad \text{if} \quad g_i(x) > g_j(x) \quad \text{for all} \quad j \neq i \tag{8.11a}$$

where

$$g_i(x) = 2\,m_i \cdot x - m_i \cdot m_i, \quad \text{etc.} \tag{8.11b}$$

Equation (8.11b) defines the discriminant function for the minimum distance classifier. In contrast to the maximum likelihood approach the decision surfaces for this classifier, separating the distinct spectral class regions in multispectral space, are linear, as seen in Sect. 8.3.4 following. The higher order decision surface possible with maximum likelihood classification renders it more powerful for partitioning multispectral space than the linear surfaces for the minimum distance approach. Nevertheless, as noted earlier, minimum distance classification is of value when the number of training samples is limited and, in such a case, can lead to better accuracies than the maximum likelihood procedure.

Minimum distance classification can be performed also using distance measures other than Euclidean (Wacker and Landgrebe, 1972); notwithstanding this, algorithms based upon Euclidean distance definitions are those generally implemented in software packages for remote sensing image analysis, such as Multispec (http://dynamo.ecn.purdue.edu/~biehl/MultiSpec), ENVI (http://www.rsinc.com) and ERMapper (http://www.ermapper.com).

8.3.3 Degeneration of Maximum Likelihood to Minimum Distance Classification

The major difference between the minimum distance and maximum likelihood classifiers lies in the use, by the latter, of the sample covariance information. Whereas the minimum distance classifier labels a pixel as belonging to a particular class on the basis only of its distance from the relevant mean, irrespective of its direction from that mean, the maximum likelihood classifier modulates its decision with direction, based upon the information in the covariance matrix. Furthermore the entry $-\frac{1}{2} \ln |\Sigma_i|$ in its discriminant function shows explicitly that patterns have to be closer to some means than others to have equivalent likelihoods of class membership. As a result substantially superior performance is expected of the maximum likelihood classifier, in general. The following situation however warrants consideration since then there is no advantage in maximum likelihood procedures. It could occur in practice when class covariance is dominated by systematic noise rather than by natural spectral spreads of the individual spectral classes.

Consider the covariance matrices of all classes to be diagonal and equal and the variances in each component to be identical, so that

$$\Sigma_i = \sigma^2 I \quad \text{for all } i.$$

Under these circumstances the discriminant function for the maximum likelihood classifier, from (8.7) becomes

$$g_i(x) = 1/2 \ln \sigma^{2N} - 1/2 \, \sigma^2 (x - m_i)^t \, (x - m_i) + \ln p(\omega_i)$$

The $\ln \sigma^{2N}$ term is now common to all classes and can be ignored, as can the $x \cdot x$ term that results from the scalar product, leaving

$$g_i(x) = \frac{1}{2\sigma^2} \{2 m_i \cdot x - m_i \cdot m_i\} + \ln p(\omega_i)$$

If the $\ln p(\omega_i)$ are ignored, on the basis of equal prior probabilities, then the $1/2\sigma^2$ factor can be removed giving

$$g_i(x) = 2 m_i \cdot x - m_i \cdot m_i$$

which is the discriminant function for the minimum distance classifier. Thus minimum distance and maximum likelihood classification are equivalent for identical and symmetric spectral class distributions.

8.3.4 Decision Surfaces

The implicit surfaces in multispectral space separating adjacent classes are defined by the respective discriminant functions being equal. Thus the surface between the ith and jth spectral classes is given by

$$g_i(x) - g_j(x) = 0$$

Substituting from (8.11 b) gives

$$2(m_i - m_j) \cdot x - (m_i \cdot m_i - m_j \cdot m_j) = 0$$

This defines a linear surface – often called a hyperplane in more than three dimensions. In contrast therefore to maximum likelihood classification in which the decision surfaces are quadratic and therefore more flexible, the decision surfaces for minimum distance classification are linear and more restricted.

8.3.5 Thresholds

Thresholds can be applied to minimum distance classification by ensuring that not only is a pixel closest to a candidate class but also that it is within a prescribed distance of that class. Such a technique is used regularly. Often the distance threshold is specified according to a number of standard deviations from a class mean.

8.4 Parallelepiped Classification

The parallelepiped classifier is a very simple supervised classifier that is, in principle, trained by inspecting histograms of the individual spectral components of the available training data. Suppose, for example, that the histograms of one particular spectral class for two dimensional data are as shown in Fig. 8.5. Then the upper and lower significant bounds on the histograms are identified and used to describe the brightness value range

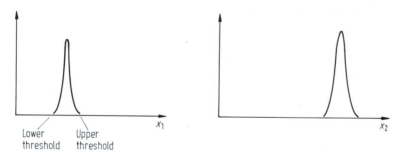

Fig. 8.5. Histograms for the components of a two-dimensional set of training data corresponding to a single spectral class. The upper and lower bounds are identified as the edges of a two-dimensional parallelepiped

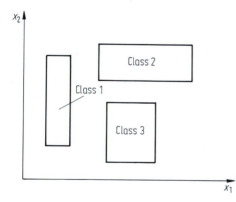

Fig. 8.6. An example of a set of two-dimensional parallelepipeds

for each band for that class. Together, the range in all bands describes a multidimensional box or parallelepiped. If, on classification, pixels are found to lie in such a parallelepiped they are labelled as belonging to that class. A two-dimensional pattern space might therefore be segmented as shown in Fig. 8.6.

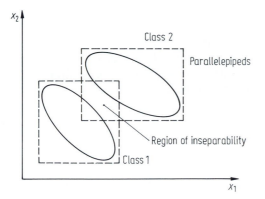

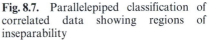

Fig. 8.7. Parallelepiped classification of correlated data showing regions of inseparability

Whilst the parallelepiped method is, in principle, a particularly simple classifier to train and use, it has several drawbacks. One is that there can be considerable gaps between the parallelepipeds; pixels in those regions will not be classified. By comparison the minimum distance and maximum likelihood classifiers will label all pixels in an image, unless thresholding methods are used. Another limitation is that prior probabilities of class membership are not taken account of; nor are they for minimum distance classification. Finally, for correlated data there can be overlap of the parallelepipeds since their sides are parallel to the spectral axes. Consequently there is some data that cannot be separated, as illustrated in Fig. 8.7.

8.5 Classification Time Comparison of the Classifiers

Of the three classifiers commonly used with remote sensing image data the parallelepiped procedure is the fastest in classification since only comparisons of the spectral components of a pixel with the spectral dimensions of the parallelepipeds are required.

For the minimum distance classifier the discriminant function in (8.11 b) requires evaluation for each pixel. In practice $2m_i$ and $m_i \cdot m_i$ would be calculated beforehand, leaving N multiplications and N additions to check the potential membership of a pixel to one class, where N is the number of components in x.

By comparison, evaluation of the discriminant function for maximum likelihood classification in (8.7) requires $N^2 + N$ multiplications and $N^2 + 2N + 1$ additions, to check one pixel against one class, given that

$$-\tfrac{1}{2} \ln |\Sigma_i| + \ln p(\omega_i)$$

would have been calculated beforehand. Ignoring additions by comparison to multiplications, the maximum likelihood classifier takes $N + 1$ times as long as the minimum distance classifier to perform a classification. It is also significant to note that classification time, and thus cost, increases quadratically with number of spectral components for the maximum likelihood classifier but only linearly for minimum distance and parallelepiped classification. This has particular relevance to feature reduction (Chap. 10) and to multitemporal remote sensing image classification.

8.6 The Mahalanobis Classifier

Consider the discriminant function for the maximum likelihood classifier, for the special case of equal prior probabilities, as defined in (8.8). If the sign of this function is reversed it can be considered as a distance squared measure since the quadratic entry has those dimensions and the other term is a constant. Thus we can define

$$d(x, m_i)^2 = \ln|\Sigma_i| + (x - m_i)^t \Sigma_i^{-1} (x - m_i) \tag{8.12}$$

and classify on the basis of the smallest $d(x, m_i)$ as for the Euclidean minimum distance classifier. Thus the maximum likelihood classifier can be regarded as a minimum distance-like classifier but with a distance measure that is direction sensitive and modified according to class.

Consider the case now where all class covariances are equal – i.e. $\Sigma_i = \Sigma$ for all i. Clearly the $\ln|\Sigma_i|$ term is now not discriminating and can be ignored. The distance measure then reduces to

$$d(x, m_i)^2 = (x - m_i)^t \Sigma^{-1} (x - m_i) \tag{8.13}$$

Such a classifier is referred to as a *Mahalanobis distance* classifier, although sometimes the term is applied to the more general measure of (8.12). Mahalanobis distance is understood as the square root of (8.13). Under the additional constraint that $\Sigma = \sigma^2 I$ the Mahalanobis classifier reduces, as before, to the minimum Euclidean distance classifier.

The advantage of the Mahalanobis classifier over the maximum likelihood procedure is that it is faster and yet retains a degree of direction sensitivity via the covariance matrix Σ, which could be a class average or a pooled variance.

8.7 Table Look Up Classification

Since the set of brightness values that can be taken by a pixel in each spectral band is limited, there are a finite, although large, number of pixel vectors in any particular image. For a given class within that image the number of distinct pixel vectors may not be very extensive. Consequently a viable classification scheme is to note the set of pixel vectors corresponding to a given class, using representative training data, and then use these to classify the image by comparing unknown image pixels with each pixel in the training data until a match is found. No arithmetic operations are required and, notwithstanding the number of comparisons that might be necessary to determine a match, it is a fast classifier. It is referred to as a look up table approach since the pixel brightnesses are stored either in software or hardware tables that point to the corresponding classes.

An obvious drawback with this approach is that chosen training data must contain one of every possible pixel vector for each class. Should some be missed then the corresponding pixels in the image will be left unclassified. This is in contrast to the procedures treated above.

8.8 Context Classification

8.8.1 The Concept of Spatial Context

The classifiers treated so far are often referred to as point or pixel-specific classifiers in that they label a pixel on the basis of its spectral properties alone, with no account taken of how any neighbouring pixels are labelled. Yet, in any real image, adjacent pixels are related or correlated, both because imaging sensors acquire significant portions of energy from adjacent pixels[1] and because ground cover types generally occur over a region that is large compared with the size of a pixel. In an agricultural area, for example, if a particular image pixel represents wheat it is highly likely that its neighbouring pixels will also be wheat. This knowledge of neighbourhood relationships is a rich source of information that is not exploited in simple, traditional classifiers. In this section we consider the importance of spatial context and see the benefit of taking it into account when making classification decisions. Not only is the inclusion of context important because it exploits spatial information, as such, but, in addition, sensitivity to the correct context for a pixel can improve a thematic map by helping to remove individual pixel labelling errors that might result from noisy data, or unusual classifier performance (see Prob. 8.6).

Classification methods that take into account the labelling of neighbours when seeking to determine the most appropriate class for a pixel are said to be context sensitive, or simply context classifiers. They attempt to develop a thematic map that is consistent both spectrally and spatially.

The degree to which adjacent pixels are strongly correlated will depend on the spatial resolution of the sensor and the scale of natural and cultural regions on the earth's surface. Adjacent pixels over an agricultural region will be strongly correlated, whereas for the same sensor, adjacent pixels over a busier, urban region would not show strong correlation. Likewise, for a given area, neighbouring Landsat MSS pixels, being larger, may not demonstrate as much correlation as adjacent SPOT HRV pixels. In general terms, context classification techniques usually warrant consideration when processing high resolution imagery.

8.8.2 Context Classification by Image Pre-Processing

Perhaps the simplest method for exploiting spatial context is to process the image data before classification in order to modify or enhance its spatial properties. A median filter (Sect. 5.5.2), for example, will help in reducing salt and pepper noise that would lead to inconsistent class labels. Moreover, the application of simple averaging filters (possibly with edge preserving thresholds) can be used to impose a degree of homogeneity among the brightness values of adjacent pixels thereby increasing the chance that neighbouring pixels may be given the same label.

---

[1] This is referred to as the point spread function effect, which is discussed in Forster (1982).

An alternative is to generate a separate channel of data that associates spatial properties with pixels. For example, a texture channel could be added and classification carried out (using a suitable algorithm such as the minimum distance rule) on the combined multispectral and texture channels. Along this line, Gong and Howarth (1990) have set up a "structural information" channel to bias a classification according to the density of high spatial frequency data in order to improve the classification of image data containing urban segments. The reasoning behind the approach is that urban regions are characterised by high spatial frequency detail whereas, conversely, the high frequency detail present in non-urban regions is low. The additional channel reflects this understanding and accordingly influences the classification which would otherwise be carried out on the basis of spectral data alone.

One of the more useful spatial pre-processing techniques is that used in the ECHO classification methodology. In ECHO (Extraction and Classification of Homogeneous Objects) regions of similar spectral properties are "grown" before classification is performed. Several region growing techniques are available, possibility the simplest of which is to aggregate pixels into small regions by comparing their brightnesses in each channel and then aggregate the small regions into bigger regions in a similar manner. When this is done, ECHO classifies the regions as single objects and only resorts to standard maximum likelihood classification when it has to treat individual pixels that could not be put into regions. Details of ECHO will be found in Kettig and Landgrebe (1976).

8.8.3 Post Classification Filtering

Once a thematic map has been generated using a simple point classifier some degree of spatial context can be developed by logically filtering the map. For example, if the map is examined in 3×3 windows, a label at the centre of the window might be changed to the label most represented in the window. Clearly this must be done carefully, with the user having some control over the minimum size region of a given cover type that is acceptable in the filtered image product (Harris, 1985). Post classification filtering by this approach has been treated by Townsend (1986).

8.8.4 Probabilistic Label Relaxation

Spatial consistency in a classified image product can also be developed using the process of label relaxation. While it has little theoretical foundation, and is more complex than the methods outlined in the previous sections, it does allow the spatial properties of a region to be carried into the classification process in a logically consistent way.

8.8.4.1 The Basic Algorithm
The process commences by assuming that a classification, based on spectral data alone, has already been carried out. There is available therefore, for each pixel, a set of probabilities that describe the chance that the pixel belongs to each of the possible ground cover classes under consideration. This set of probabilities could be computed from (8.6) and (8.7) if maximum likelihood classification had been used first. If another classification method had been employed, then some other assignment process will be required. It

could even be as simple as allocating a high probability to the most favoured class label and lower probabilities to the rest. Let the set of probabilities for a pixel (m) currently of interest be represented by

$$p_m(\omega_i) \qquad i = 1, \dots K \tag{8.14}$$

where K is the total number of classes; $p_m(\omega_i)$ should be read as "the probability that

ω_i is the correct class for pixel m." Note that the full set of $p_m(\omega_i)$ must sum to unity for a given pixel – viz.

$$\sum_i p_m(\omega_i) = 1.$$

Suppose now that a neighbourhood is defined surrounding pixel m. This can be of any size and, in principle, should be large enough to ensure that all the pixels considered to have any spatial correlation with m are included. For high resolution imagery this is not practical and simple neighbourhoods such as that shown in Fig. 8.8 are often adopted.

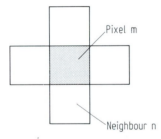

Pixel m

Neighbour n

Fig. 8.8. Definition of a simple neighbourhood about pixel m

Now assume that a *neighbourhood function* $Q_m(\omega_i)$ can be found (by means to be described below) which allows the pixels in the prescribed neighbourhood to influence the possible classification of pixel m. This influence is exerted by multiplying the label probabilities in (8.14) by the $Q_m(\omega_i)$. However, so that the new set of label probabilities sum to one, these new values are divided by their sum:

$$p'_m(\omega_i) = \frac{p_m(\omega_i)\, Q_m(\omega_i)}{\sum_i p_m(\omega_i) Q_m(\omega_i)} \tag{8.15}$$

Such a modification is made to the set of label probabilities for all pixels by moving over the image from its top left hand to bottom right hand corners. In the following it will be seen that the neighbourhood function $Q_m(\omega_i)$ depends on the label probabilities of the neighbouring pixels, so that if all the pixel probabilities are modified in the manner just described then the neighbours for any given pixel have also been altered. Consequently, (8.15) should be applied again to give newer estimates still of the label probabilities.

Indeed, (8.15) is applied as many times as necessary to ensure that the $p'_m(\omega_i)$ have stabilised – i.e. that they do not change with further iteration. It is assumed that the $p'_m(\omega_i)$ then represent the correct set of label probabilities for the pixel, having taken account both of spectral data (in the initial determination of label probabilities) and spatial context (via the neighbourhood functions). Since the process is iterative, (8.15) is usually written as an explicit iteration formula:

$$p^{k+1}_m(\omega_i) = \frac{p^k_m(\omega_i)\, Q^k_m(\omega_i)}{\sum_i p^k_m(\omega_i) Q^k_m(\omega_i)}$$

(8.16)

where k is the iteration counter. Depending on the size of the image and its spatial complexity, the number of iterations required to stabilise the label probabilities may be quite large. However, most change in the label probabilities occurs in the first few iterations and there is good reason to believe that proceeding beyond say 5 to 10 iterations may not be necessary in most cases (see Sect. 8.8.4.4).

8.8.4.2 The Neighbourhood Function

Consider just one of the neighbours of pixel m in Fig. 8.8 – call it pixel n. Suppose there is available a measure of compatibility of the current labelling of pixel m and its neighbouring pixel n. For example let $r_{mn}(\omega_i, \omega_j)$ describe numerically how compatible it is to have pixel m classified as ω_i and neighbouring pixel n classified as ω_j. It would be expected, for example, that this measure will be high if the adjoining pixels are both labelled wheat in an agricultural region, but low if one of the neighbours was classified as snow. There are several ways these *compatibility coefficients*, as they are called, can be defined. An intuitively appealing definition is based on conditional probabilities. Thus, the compatibility measure $p_{mn}(\omega_i|\omega_j)$ is the probability that ω_i is the correct label for pixel m if ω_j is the correct label on pixel n. A small piece of evidence in favour of ω_i being correct for pixel m is $p_{mn}(\omega_i|\omega_j)p_n(\omega_j)$ – i.e. the probability that ω_i is correct for pixel m if ω_j is correct for pixel n multiplied by the probability that ω_j is correct for pixel n [1]. Since probabilities for all possible labels on pixel n are available (even though some might be very small) the total evidence from pixel n in favour of ω_i being the correct class for pixel m will be the sum of the contributions from all pixel n's labelling possibilities, viz.

$$\sum_j p_{mn}(\omega_i|\omega_j)p_n(\omega_j).$$

Consider now the full neighbourhood of the pixel m. In a like manner all the neighbours contribute evidence in favour of labelling pixel m as coming from class ω_i. All these contributions are simply added[2], via the use of *neighbour weights* d_n that recognise that some neighbours may be more influential than others (as for example, pixels along a scan line in MSS data compared with those running down an image, owing to the

---

[1] This is the probability of the joint event that pixel m is labelled ω_i and pixel n is labelled ω_j.

[2] An alternative way of handling the full neighbourhood is to take the geometric mean of the neighbourhood contributions.

oversampling that occurs along rows – see Fig. 1.6). Thus, at the kth iteration, the total neighbourhood support for pixel m being classified as ω_i is:

$$Q^k{}_m(\omega_i) = \sum_n d_n \sum_j p_{mn}(\omega_i|\omega_j)p^k{}_n(\omega_j) \qquad (8.17)$$

This is the definition of the neighbourhood function. In (8.16) and (8.17) it is common to include pixel m in its own neighbourhood so that the modification process is not entirely dominated by the neigbours, particularly if the number of iterations is so large as to take the process quite a long way from its starting point.

Unless there is good reason to do otherwise the neighbour weights are generally chosen all to be the same.

8.8.4.3 Determining the Compatibility Coefficients

Several methods are possible for determining values for the compatibility coefficients $p_{mn}(\omega_i|\omega_j)$. One is to have available a spatial model for the region under consideration, derived from some other data source. In an agricultural region, for example, some general idea of field sizes along with a knowledge of the pixel size of the sensor being used should make it possible to estimate how often one particular class occurs following a given class on an adjacent pixel. Another approach is to compute values for the compatibility coefficients from ground truth pixels, although the ground truth needs to be in the form of training regions that contain heterogeneous and spatially representative cover types.

8.8.4.4 The Final Step – Stopping the Process

While the relaxation process operates on label probabilities, the user is interested in the actual labels themselves. At the completion of relaxation, or at any intervening stage, each of the pixels can be classified according to the highest label probability. Thought has to be given as to how and when the iterations should be terminated. As suggested earlier, the process can be allowed to go to a natural completion at which further iteration leads to no changes in the label probabilities for all pixels. This however presents two difficulties. First, up to several hundred iterations may be involved leading to a costly post classification step. Secondly, it is observed in practice that the relaxation process improves the classification results in the first few iterations, by the embedding of spatial information, often to deteriorate later in the process (Richards, Landgrebe and Swain, 1981). Indeed, if the process is not terminated, the thematic map, after a large number of iterations of relaxation, can be worse than before the technique was applied.

To avoid these difficulties, a stopping rule or other controlling mechanism is needed. As seen in the example of the following section, stopping after just a few iterations may allow most of the benefit to be drawn from the process. Alternatively, the labelling errors remaining at each iteration can be checked against ground truth, if available, and the iterations terminated when the labelling error is seen to be minimised (Gong and Howarth, 1989).

Another approach is to control the propagation of contextual information as iteration proceeds (Lee, 1984). A little thought will reveal that, in the first iteration, only the immediate neighbours of a pixel have an influence on its labelling. In the second iteration the neighbours two away will now have an influence via the intermediary of the intervening pixels. Similarly, as iterations proceed, information from neighbours further away is propagated into the pixel of interest to modify its label probabilities. If the user has a view of the separation between neighbours at which the spatial correlation has dropped to negligible levels, then the appropriate number of iterations should be able to be identified at which to terminate the process without unduly sacrificing any further improvement in labelling accuracy. Noting also that the nearest neighbours should be most influential, with those further out being less important, a useful variation is to reduce the values of the neighbour weights d_n as iteration proceeds so that after say 5 to 10 iterations they have been brought to zero. Further iterations will then have no effect, and degradation in labelling accuracy cannot occur (Lee and Richards, 1989).

8.8.4.5 Examples

Fig. 8.9 illustrates a simple application of relaxation labelling, in which a hypothetical image of 100 pixels has been classified into just two classes – grey and white. The ground truth for the region is shown, along with the thematic map (initial labelling) assumed to have been generated from a point classifier such as the maximum likelihood rule. Also shown are the compatibility coefficients, expressed as conditional probabilities, computed from the ground truth map. Label probabilities were assumed to be 0.9 for the favoured label in the initial labelling and 0.1 for the less likely label. The initial labelling, by comparison with the ground truth, can be seen to have an accuracy of 82 % (there are 12 pixels in error). The labelling (selected on the basis of the largest current label probability) at significant stages during iteration is shown, illustrating the reduction in

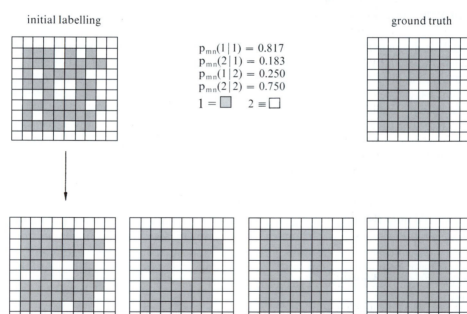

Fig. 8.9. Simple demonstration of pixel relaxation labelling

classification error resulting from the incorporation of spatial information into the process. After 15 interations all initial labelling errors have been removed, leading to a thematic map 100 % in agreement with the ground truth. In this case the relaxation process was allowed to proceed to completion and there have been no ill effects. As pointed out in the previous section, however, this is the exception and stopping rules may have to be applied in most cases. Other simple examples where this is the case will be found in Richards et al, (1981).

As a second example, the leftmost 82×100 pixels of the agricultural image shown in Fig. 3.1 have been chosen. Fig. 8.10 a shows the ground truth for the image segment and Fig. 8.10 b shows the result of a maximum likelihood classification. The initial classification accuracy is 65.6 %. The relaxation process was initialised using actual probability estimates from the maximum likelihood rule. Conditional probabilities as such were not used as compatibility coefficients. Instead, a slightly different set of compatibilities as proposed by Peleg and Rosenfeld (1980) was adopted. Also, to control the propagation of context information and thereby obviate any deleterious effect of allowing the relaxation process to proceed unconstrained, the neighbourhood weights were diminished with iteration count as described in previous section. Fig. 8.10 c shows the final labelling, which has an accuracy of 72.4 %. Full details of this example are available in Lee and Richards (1989).

8.9 Classification Using Neural Networks

Statistical classification algorithms are the most commonly encountered labelling techniques used in remote sensing and, for this reason, have been the principal methods treated in this chapter. One of the valuable aspects of a statistical approach is that a *set* of relative likelihoods is produced. Even though, in the majority of cases, the *maximum* of the likelihoods is chosen to indicate the most possible label for a pixel, there remains nevertheless information in the remaining likelihoods that could be made use of in some circumstances, either to initiate a process such as relaxation labelling (Sect. 8.8.4) or simply to provide the user with some feeling for the other likely classes. Those situations are however not common and, in most applications, the maximum selection is made. That being so, the material in Sects. 8.2.4 and 8.3.4 shows that the decision process has a geometric counterpart in that a comparison of statistically derived discriminant functions leads equivalently to a decision rule that allows a pixel to be classified on the basis of its position in multispectral space compared with the location of a decision surface. This leads us to question whether a geometric interpretation can be adopted in general, without needing first to use statistical models.

8.9.1 Linear Discrimination

8.9.1.1 Concept of a Weight Vector

Consider the simple two class multispectral space shown in Fig. 8.11, which has been constructed intentionally so that a simple straight line can be drawn between the pixels as shown. This straight line, which will be a multidimensional linear surface in general and

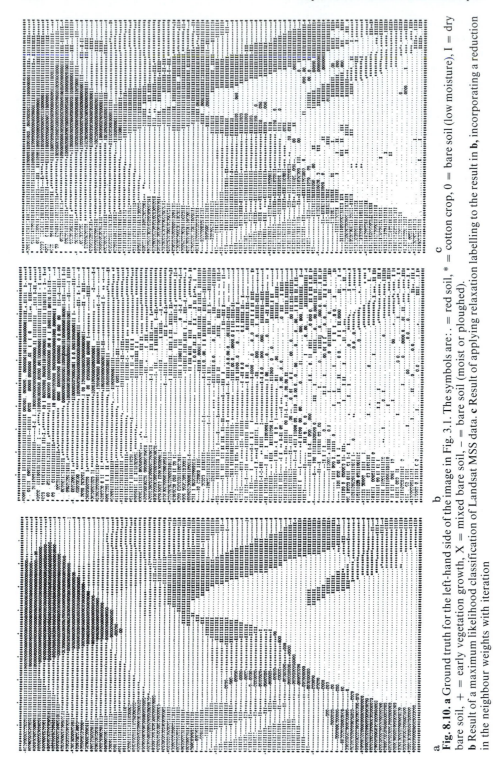

a

Fig. 8.10. a Ground truth for the left-hand side of the image in Fig. 3.1. The symbols are: . = red soil, * = cotton crop, 0 = bare soil (low moisture), I = dry bare soil, + = early vegetation growth, X = mixed bare soil, – = bare soil (moist or ploughed). **b** Result of a maximum likelihood classification of Landsat MSS data. **c** Result of applying relaxation labelling to the result in **b**, incorporating a reduction in the neighbour weights with iteration

which is often called a hyperplane, can function as a decision surface for classification. In the two dimensions shown, the equation of the line can be expressed

$$w_1x_1 + w_2x_2 + w_3 = 0 \tag{8.18}$$

where the x_i are the brightness value co-ordinates of the multispectral space and the w_i are a set of constants, usually called weights. There will be as many weights as the number of channels in the data, plus one. In general, if the number of channels or bands is N, the equation of a linear surface is

$$w_1x_1 + w_2x_2 + ... + w_Nx_N + w_{N+1} = 0 \tag{8.19}$$

which can be written as

$$w^tx + w_{N+1} = 0 \tag{8.20}$$

where x is the co-ordinate vector and w is called the weight vector. The transpose operation has the effect of turning the column vector into a row vector so that the product gives the correct expanded form of the previous equation.

In a real exercise the position of the separating surface would be unknown initially. Training a linear classifier amounts to determining an appropriate set of the weights that places the decision surface between the two sets of training samples. There is not necessarily a unique solution – any of an infinite number of (marginally different) decision hyperplanes will suffice to separate the two classes.

For a given data set, an explicit equation for the separating surface can be obtained using the minimum distance rule, as discussed in Sect. 8.3, which entails finding the mean vectors of the two class distributions. An alternative method is outlined in the following, based on selecting an arbitrary surface and then iterating it into an acceptable position. Even though not often used anymore, this method is useful to consider since it establishes some of the concepts used in neural networks.

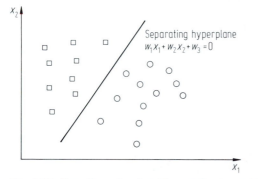

Fig. 8.11. Two dimensional multispectral space, with two classes of pixel that can be separated by a linear surface

8.9.1.2 Testing Class Membership
The calculation in (8.20) will be exactly zero only for values of x lying on the decision surface. If we substitute into that equation values of x corresponding to the pixel points indicated in Fig. 8.11 the left hand side will be non-zero. For pixels in one class a positive

result will be given, while pixels on the other side will give a negative result. Thus, once the decision surface has been identified (i. e. trained), then a decision rule is

$$x \in \text{class 1 if } w'x + w_{N+1} > 0$$

$$x \in \text{class 2 if } w'x + w_{N+1} < 0 \tag{8.21}$$

8.9.1.3 Training

A full discussion of linear classifier training is given in Nilsson (1965); only those aspects helpful to the neural network development following are treated here.

It is expedient to define a new, augmented pixel vector according to

$$y = [x', 1]'$$

If, in (8.20), we also take the term w_{N+1} into the definition of the weight vector, viz.

$$w = [w', w_{N+1}]' \tag{8.22}$$

then the equation of the decision surface, can be expressed more compactly as

$$w'y = 0 \tag{8.23}$$

so that the decision rule of (8.21) can be restated

$$x \in \text{class 1 if } w'y > 0$$

$$x \in \text{class 2 if } w'y < 0 \tag{8.24}$$

We usually think of $w'y = 0$ as defining a linear surface in the x (or now y) multispectral space, in which the coefficients of the variables (y_1, y_2, etc.) are the weights w_1, w_2, etc. However it is also possible to think of the equation as describing a linear surface in which the y's are the coefficients and the w's are the variables. This interpretation will see these surfaces plotted in a co-ordinate system which has axes w_1, w_2, etc. A two-dimensional version of this *weight space*, as it is called, is shown in Fig. 8.12, in which have been

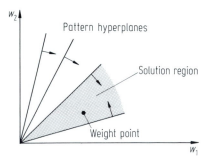

Fig. 8.12. Representation of pixels as hyperplanes and weight vectors as points in so-called weight space. The arrows indicate the side of each pixel plane on which the weight point must lie for correct classification

plotted a number of *pattern hyperplanes*; these are specific linear surfaces in the new co-ordinates that pass through the origin and have, as their coefficients, the components of the (augmented) pixel vectors. Thus, while the pixels plot as points in multispectral space, they plot as linear surfaces in weight space. Likewise, a set of weight coefficients will define a surface in multispectral space, but will plot as a point in weight space. Although this is an abstract concept it will serve to facilitate an understanding of how a linear classifier can be trained.

In weight space the decision rule of (8.24) still applies – however now it tests that the *weight point* is on the appropriate side of the *pattern hyperplane*. For example, Fig. 8.12 shows a single weight point which lies on the correct side of each pixel and thus defines a suitable decision surface in multispectral space. In the diagram, small arrows are attached to each pixel hyperplane to indicate the side on which the weight point must lie in order that the test of (8.24) succeeds for all pixels. The purpose of training the linear classifier is to ensure that the weight point is located somewhere within the *solution region*. If, through some initial guess, the weight point is located somewhere else in weight space then it has to be moved to the solution region.

Suppose an initial guess is made for the weight vector *w*, but that this places the weight point on the wrong side of a particular pixel hyperplane as illustrated in Fig. 8.13.

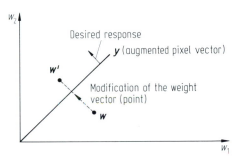

Fig. 8.13. Modification of the weight point to give the correct response

Clearly, the weight point has to be shifted to the other side to give a correct response in (8.24). The most direct manner in which the weight point can be modified is to move it straight across the pixel hyperplane. This can be achieved by adding a scaled amount of the pixel vector to the weight vector[1]. The new position of the weight point is then

$$w' = w + cy \qquad (8.25)$$

where *c* is called the correction increment, the size of which determines by how much the original weight point is moved orthogonal to the pixel hyperplane. If it is large enough the weight point will be shifted right across the pixel plane, as required. Having so

---

[1] The hyperplane in *w* coordinates is given by (8.23); a vector normal to that hyperplane is the vector of the coefficients of *w*. This can be checked for a simple two dimensional example. A line through the origin with unity slope is $-w_1 + w_2 = 0$. A vector normal to the line joins the origin to $(-1,1)$, i. e. $y = [-1,1]^t$.

modified the weight vector, the product in (8.24) then becomes

$$w''y = w'y + cy'y$$
$$= w'y + c|y|^2$$

Clearly, if the initial $w'y$ was erroneously negative a suitable positive value of c will give a positive value of $w''y$; otherwise a negative value of c will correct an erroneous initial positive value of the product.

Using the class membership test in (8.24) and the correction formula of (8.25) the following iterative nonparametric training procedure, referred to as *error correction feedback*, is adopted.

First, an initial position for the weight point is chosen arbitrarily. Then, pixel vectors from training sets are presented one at a time. If the current weight point position classifies a pixel correctly then no action need be taken; otherwise the weight vector is modified as in (8.25) with respect to that particular pixel vector. This procedure is repeated for each pixel in the training set, and the set is scanned as many times as necessary to move the weight point into the solution region. If the classes are linearly separable then such a solution will be found.

8.9.1.4 Setting the Correction Increment

Several approaches can be adopted for choosing the value of the correction increment, c. The simplest is to set c equal to a positive or negative constant (according to the change required in the $w'y$ product). A common choice is to make $c = \pm 1$ so that application of (8.25) amounts simply to adding the augmented pixel vector to or subtracting it from the weight vector, thereby obviating multiplications and giving fast training.

Another rule is to choose the correction increment proportional to the difference between the desired and actual response of the classifier:

$$c = \eta(t - w'y)$$

so that (8.25) can be written

$$w' = w + \Delta w$$

with $\Delta w = \eta(t - w'y)y$ (8.26)

where t is the desired response to the training pattern y and $w'y$ is the actual response; η is a factor which controls the degree of correction applied. Usually t would be chosen as $+1$ for one class and -1 for the other.

8.9.1.5 Classification – The Threshold Logic Unit

After the linear, two category classifier has been trained, so that the final version of the weight vector w is available, it is ready to be presented with pixels it has not seen before in order to attach ground cover class labels to those pixels. This is achieved through application of the decision rule in (8.24). It is useful, in anticipation of neural networks, to picture the classification rule in diagrammatic form as depicted in Fig. 8.14 a. Simply, this consists of weighting elements, a summing device and an output element which, in this case, performs the maximum selection. Together these are referred to as a *threshold logic unit* (TLU). It bears substantial similarity to the concept of a *processing element* used in neural networks for which the output thresholding unit is replaced by a more general function and the pathway for the unity input in the augmented pattern vector is actually incorporated into the output function. The latter can be done for a simple TLU as shown in Fig. 8.14 b, in which the simple thresholding element has been replaced by a functional block which performs the addition of the final weighting coefficient to the weighted sum of the input pixel components, and then performs a thresholding (or more general nonlinear) operation.

8.9.1.6 Multicategory Classification

The foregoing work on linear classification has been based on an approach that can perform separation of pixel vectors into just two categories. Were it to be considered for remote sensing, it needs to be extended to be able to cope with a multiclass problem.

Multicategory classification can be carried out in one of two ways. First a decision tree of linear classifiers (TLUs) can be constructed, as seen in Fig. 8.15, at each decision node of which a decision of the type (water or not water) is made. At a subsequent node the (not water) category might be differentiated as (soil or not soil) etc. It should be noted that the decision process at each node has to be trained separately.

Alternatively, a multicategory version of the simple binary linear classifier can be derived. This reverts, for its derivation, to the concept of a discriminant function and, specifically, defines the linear classifier discriminant function for class *i* as

$$g_i(x) = w_i{}^t y \qquad\qquad i = 1, ... M$$

Class membership is then decided on the basis of the usual decision rule expressed in (8.11 a), i. e. according to the largest of the $g_i(x)$ for the given pixel vector *x*. For training, an initial arbitrary set of weight vectors and thus discriminant functions is chosen. Then

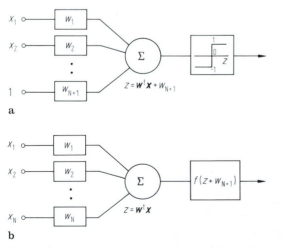

a

b

Fig. 8.14. a Diagrammatic representation of (8.24). **b** More useful representation of a processing element in which the thresholding function is generalised

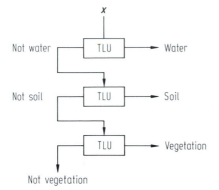

Fig. 8.15. Binary decision tree of TLUs used for multicategory classification

each of the training pixels is checked in turn. Suppose, for a particular pixel the jth discriminant function is erroneously largest, when in fact the pixel belongs the ith category. A correction is carried out by adjusting the weight vectors for these two discriminant functions, to increase that for the correct class for the pixel and to decrease that for the incorrect class, according to

$$w_i' = w_i + cy$$
$$w_j' = w_j - cy \tag{8.27}$$

where c is the correction increment. Again this correction procedure is iterated over the training set of pixels as many times as necessary to obtain a solution. Nilsson (1965) shows that a solution is possible by this approach.

8.9.2 Networks of Classifiers – Solutions of Nonlinear Problems

The decision tree structure shown in Fig. 8.15 is a classifier network in that a collection of simple classifiers (in that case TLUs) is brought together to solve a complex problem. Nilsson (1965) has proposed a general network structure under the name of *layered classifiers* consisting entirely of interconnected TLUs, as shown in Fig. 8.16. The benefit of forming a classifier network is that data sets that are inherently not separable with a simple linear decision surface should, in principle, be able to be handled since the layered classifier is known to be capable of implementing nonlinear surfaces. The drawback however, is that training procedures for layered classifiers, *consisting of TLUs,* are difficult to determine.

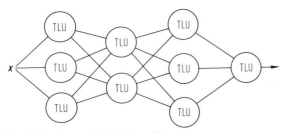

Fig. 8.16. Layered TLU classifier

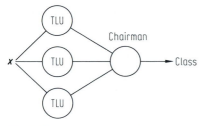

Fig. 8.17. Committee classifier

One specific manifestation of a layered classifier, known as a committee machine, is depicted in Fig. 8.17. Here the first layer consists simply of a set of TLUs, to each of which a pixel vector under test is submitted to see which of two classes is recommended. The second layer is a single element which has the responsibility of judging the recommendations of each of the nodes in the first layer. It is therefore of the nature of a chairman or vote taker. It can make its decision on the basis of several sets of logic. First, it can decide class membership on the basis of the majority vote of the first layer recommendations. Secondly, it can decide on the basis of veto, in which all first layer classifiers have to agree before the vote taker will recommend a class. Thirdly, it could use a form of seniority logic in which the chairman rank orders the decisions of the first layer nodes. It always refers to one first. If that node has a solution then the vote taker accepts it and goes no further. Otherwise it consults the next most senior of the first layer nodes, etc. A committee classifier based on seniority logic has been developed for remote sensing applications by Lee and Richards (1985).

8.9.3 The Neural Network Approach

For the purposes of this treatment a neural network is taken to be of the nature of a layered classifier such as depicted in Fig. 8.16, but with the very important difference that the nodes are not TLUs, although resembling them closely. The node structure in Fig. 8.14 b can be made much more powerful, and coincidentally lead to a training theorem for multicategory nonlinear classification, if the output processing element does not apply a thresholding operation to the weighted input but rather applies a softer, and mathematically differentiable, operation.

8.9.3.1 The Processing Element

The essential processing node in the neural network to be considered here (sometimes called a neuron by analogy to biological data processing from which the term neural network derives) is an element as shown in Fig. 8.14 b with many inputs and with a single output, depicted simply in Fig. 8.18 a. Its operation is described by

$$o = f(w^t x + \theta) \tag{8.28}$$

where θ is a threshold (sometimes set to zero), w is a vector of weighting coefficients and x is the vector of inputs. For the special case when the inputs are the band values of a particular multispectral pixel vector it could be envisaged that the threshold θ takes the place of the weighting coefficient w_{N+1} in (8.20). If the function f is a thresholding operation this processing element would behave as a TLU. In general, the number of inputs to a node will be defined by network topology as well as data dimensionality, as will become evident.

The major difference between the layered classifier of TLUs shown in Fig. 8.16 and the neural network, known as the multilayer perceptron, is in the choice of the function f, called the *activation function*. Its specification is simply that it emulate thresholding in a soft or asymptotic sense and be differentiable. The most commonly encountered expression is

$$f(z) = \frac{1}{1 + e^{-z/\theta_0}} \tag{8.29}$$

where the argument z is $w^t x + \theta$ as seen in (8.28) and θ_0 is a constant. This approaches 1 for z large and positive and 0 for z large and negative and is thus asymptotically thresholding. It is important to recognise that the outcome of the product $w^t x$ is a simple scalar; when plotted with $\theta = 0$, (8.29) appears as shown in Fig. 8.18 b. For θ_0 very small the activation function approaches a thresholding operation. Usually $\theta_0 = 1$.

A neural network for use in remote sensing image analysis will appear as shown in Fig. 8.19, being a layered classifier composed of processing elements of the type shown in Fig. 8.19 a. It is conventionally drawn with an input layer of nodes (which has the function of distributing the inputs to the processing elements of the next layer, and scaling them if necessary) and an output layer from which the class labelling information is provided. In between there may be one or more so-called hidden or other processing layers of nodes. Usually one hidden layer will be sufficient, although the number of nodes to use in the hidden layer is often not readily determined. We return to this issue in Sect. 8.9.3.3 below.

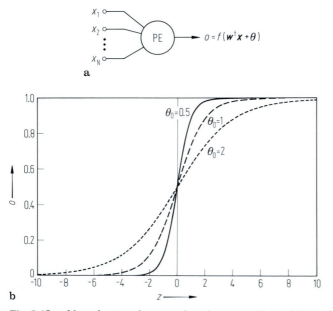

Fig. 8.18. a Neural network processing element. **b** Plots of (8.29) for various θ_0

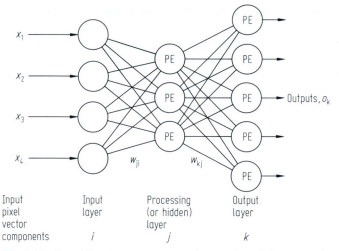

Input
pixel
vector
components

Input
layer

i

Processing
(or hidden)
layer

j

Output
layer

k

Fig. 8.19. A multilayer perceptron neural network, and the nomenclature used in the derivation of the backpropagation training algorithm

8.9.3.2 Training the Neural Network – Backpropagation

Before it can perform a classification, the network of Fig. 8.19 must be trained. This amounts to using labelled training data to help determine the weight vector w and the threshold θ in (8.28) for each processing element connected into the network. Note that the constant θ_0 in (8.29), which governs the gradient of the activation function as seen in Fig. 8.18 b, is generally pre-specified and does not need to be estimated from the use of training data.

Part of the complexity in understanding the training process for a neural net is caused by the need to keep careful track of the parameters and variables over all layers and processing elements, how they vary with the presentation of training pixels and (as it turns out) with iteration count. This can be achieved with a detailed subscript convention, or by the use of a simpler generalised notation. We will adopt the latter approach, following essentially the development given by Pao (1989). The derivation will be focussed on a 3 layer neural net, since this architecture has been found sufficient for many applications. However the results generalise to more layers.

Fig. 8.19 incorporates the nomenclature used. The three layers are lettered as i, j, k with k being the output. The set of weights linking layer i PEs with those in layer j are represented generally by w_{ji}, while those linking layers j and k are represented by w_{kj}. There will be a very large number of these weights, but in deriving the training algorithm it is not necessary to refer to them all individually. Similarly the general activation function arguments z_i and outputs o_i, can be used to represent all the arguments and outputs in the corresponding layer.

For j and k layer PEs (8.28) is

$$o_j = f(z_j) \quad \text{with} \qquad z_j = \sum_j w_{ji} o_i + \theta_j \qquad (8.30a)$$

$$o_k = f(z_k) \quad \text{with} \qquad z_k = \sum_k w_{kj} o_j + \theta_k \qquad (8.30b)$$

The sums in (8.30) are shown with respect to the indices j and k. This should be read as meaning the sums are taken over all inputs of particular layer j and layer k PEs respectively. Note also that the sums are expressed in terms of the outputs of the previous layer since these outputs form the inputs to the PEs in question.

An untrained or poorly trained network will give erroneous outputs. Therefore, as a measure of how well a network is functioning during training, we can assess the outputs at the last layer (k). A suitable measure along these lines is to use the sum of the squared output error. The error made by the network when presented with *a single training pixel* can thus be expressed

$$E = \tfrac{1}{2} \sum_k (t_k - o_k)^2 \tag{8.31}$$

where the t_k represent the desired or target outputs[1] and o_k represents the actual outputs from the output layer PEs in response to the training pixel. The factor of $\tfrac{1}{2}$ is included for arithmetic convenience in the following. The sum is over all output layer PEs.

A useful training strategy is to adjust the weights in the processing elements until the error has been minimised, at which stage the actual outputs are as close as possible to the desired outputs.

A common approach for adjusting weights to reduce (and thus minimise) the value of a function of which they are arguments, is to modify their values proportional to the negative of the partial derivative of the function. This is called a gradient descent technique[2]. Thus for the weights linking the j and k layers let

$$w_{kj}' = w_{kj} + \Delta w_{kj}$$

with $$\Delta w_{kj} = -\eta \frac{\partial E}{\partial w_{kj}}$$

where η is a positive constant that controls the amount of adjustment. This requires an expression for the partial derivative, which can be determined using the chain rule

$$\frac{\partial E}{\partial w_{kj}} = \frac{\partial E}{\partial o_k} \frac{\partial o_k}{\partial z_k} \frac{\partial z_k}{\partial w_{kj}} \tag{8.32}$$

each term of which must now be evaluated.

From (8.30b) and (8.29) we see (for $\theta_0 = 1$)

$$\frac{\partial o_k}{\partial z_k} = f'(z_k) = (1 - o_k)o_k \tag{8.33a}$$

and $$\frac{\partial z_k}{\partial w_{kj}} = o_j \tag{8.33b}$$

---

[1] These will be specified from the training data labelling. The actual value taken by t_k however will depend on how the outputs themselves are used to represent classes. Each output could be a specific class indicator (e.g. 1 for class 1 and 0 class 2); alternatively some more complex coding of the outputs could be adopted. This is considered in Sect. 8.9.3.3.

[2] Another optimisation procedure used successfully for neural network training in remote sensing is the conjugate gradient method (Benediktsson et al., 1993).

Now from (8.31)

$$\frac{\partial E}{\partial o_k} = -(t_k - o_k) \tag{8.33c}$$

Thus the correction to be applied to the weights is

$$\Delta w_{kj} = \eta\,(t_k - o_k)(1 - o_k)o_k o_j \tag{8.34}$$

For a given trial, all of the terms in this expression are known so that a beneficial adjustment can be made to the weights which link the hidden layer to the output layer.

Now consider the weights that link the i and j layers. The weight adjustments are

$$\Delta w_{ji} = -\eta\frac{\partial E}{\partial w_{ji}} = -\eta\frac{\partial E}{\partial o_j}\frac{\partial o_j}{\partial z_j}\frac{\partial z_j}{\partial w_{ji}}$$

In a similar manner to the above development we have

$$\Delta w_{ji} = -\eta\frac{\partial E_j}{\partial o_j}(1 - o_j)o_j o_i$$

Unlike the case with the output layer, however, we cannot obtain an expression for the remaining partial derivative from the error formula, since the o_j are not the outputs at the final layer, but rather those from the hidden layer. Instead we express the derivative in terms of a chain rule involving the output PEs. Specifically

$$\frac{\partial E}{\partial o_j} = \sum_k \frac{\partial E}{\partial z_k}\frac{\partial z_k}{\partial o_j}$$

$$= \sum_k \frac{\partial E}{\partial z_k} w_{kj}$$

The remaining partial derivative can be obtained from (8.33a) and (8.33c) as

$$\frac{\partial E}{\partial z_k} = -(t_k - o_k)(1 - o_k)o_k$$

so that $$\Delta w_{ji} = \eta(1 - o_j)o_j o_i \sum_k (t_k - o_k)(1 - o_k)o_k w_{kj} \tag{8.35}$$

Having determined the w_{kj} from (8.34), it is now possible to find values for the w_{ji} since all other entries in (8.35) are known or can be calculated readily.

For convenience we now define

$$\delta_k = (t_k - o_k)(1 - o_k)o_k \tag{8.36a}$$

and $$\delta_j = (1 - o_j)o_j \sum_k (t_k - o_k)(1 - o_k)o_k w_{kj}$$

$$= (1 - o_j)o_j \sum_k \delta_k w_{kj} \tag{8.36b}$$

so that we have

$$\Delta w_{kj} = \eta\delta_k o_j \tag{8.37a}$$

and $$\Delta w_{ji} = \eta\delta_j o_i \tag{8.37b}$$

both of which should be compared with (8.26) to see the effect of a differentiable activation function.

The thresholds θ_j and θ_k in (8.30) are found in exactly the same manner as for the weights in that (8.37) is used, but with the corresponding inputs chosen to be unity.

Now that we have the mathematics in place it is possible to describe how training is carried out. The network is initialised with an arbitrary set of weights in order that it can function to provide an output. The training pixels are then presented one at a time to the network. For a given pixel the output of the network is computed using the network equations. Almost certainly the output will be incorrect to start with – i. e. the o_k will not match the desired class t_k for the pixel, as specified by its labelling in the training data. Correction to the output PE weights, described in (8.37a), is then carried out, using the definition of δ_k in (8.36a). With these new values of δ_k and thus w_{kj}, (8.36b) and (8.37b) can be applied to find the new weight values in the earlier layers. In this way the effect of the output being in error is propagated back through the network in order to correct the weights. The technique is thus often referred to as *backpropagation*.

Pao (1989) recommends that the weights not be corrected on each presentation of a single training pixel, but rather that the corrections for all pixels in the training set be aggregated into a single adjustment. Thus for p training patterns the bulk adjustments are[1]

$$\Delta w_{kj}' = \sum_p \Delta w_{kj} \text{ and } \Delta w_{ji}' = \sum_p \Delta w_{ji}$$

After the weights have been so adjusted the training pixels are presented to the network again and the outputs re-calculated to see if they correspond better to the desired classes. Usually they will still be in error and the process of weight adjustment is repeated. Indeed the process is iterated as many times as necessary in order that the network respond with the correct class for each of the training pixels or until the number of errors in classifying the training pixels is reduced to an acceptable level.

8.9.3.3 Choosing the Network Parameters

When considering the use of the neural network approach to classification it is necessary to make several key decisions beforehand. First, the number of layers to use must be chosen. Generally, a three layer network is sufficient, with the purpose of the first layer being simply to distribute (or fan out) the components of the input pixel vector to each of the processing elements in the second layer. Thus the first layer does no processing as such, apart perhaps from scaling the input data, if required.

The next choice relates to the number of elements in each layer. The input layer will generally be given as many nodes as there are components (features) in the pixel vectors. The number to use in the output node will depend on how the outputs are used to represent the classes. The simplest method is to let each separate output signify a different class, in which case the number of output processing elements will be the same as the number of training classes. Alternatively, a single PE could be used to represent all classes, in which case a different value or level of the output variable will be attributed to

---

[1] This is tantamount to deriving the algorithm with the error being calculated over all pixels p in the training set, viz $E_p = \sum_p E$, where E is the error expressed for a single pixel in (8.31).

each class. A further possibility is to use the outputs as a binary code, so that two output PEs can represent four classes, three can represent 8 classes and so on.

As a general guide the number of PEs to choose for the hidden or processing layers should be the same as or larger than the number of nodes in the input layer (Lippmann, 1987).

8.9.3.4 Examples
It is instructive to consider a simple example to see how a neural network is able to develop the solution to a classification problem. Figure 8.20 shows two classes of data, with three points in each, arranged so that they cannot be separated linearly. The network shown in Fig. 8.21 will be used to discriminate the data. The two PEs in the first processing layer are described by activation functions with no thresholds – i. e. $\theta = 0$ in (8.28), while the single output PE has a non-zero threshold in its activation function.

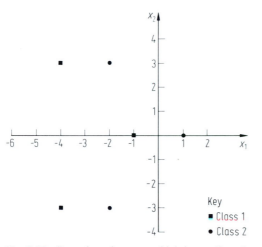

Fig. 8.20. Two-class data set, which is not linearly separable

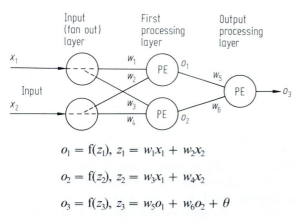

$$o_1 = f(z_1), \; z_1 = w_1 x_1 + w_2 x_2$$

$$o_2 = f(z_2), \; z_2 = w_3 x_1 + w_4 x_2$$

$$o_3 = f(z_3), \; z_3 = w_5 o_1 + w_6 o_2 + \theta$$

Fig. 8.21. Two processing layer neural network to be applied to the data of Fig. 8.20

Table 8.3. Training the network of Fig. 8.21

Iteration	w_1	w_2	w_3	w_4	w_5	w_6	θ	Error
0*	0.050	0.100	0.300	0.150	1.000	0.500	−0.500	0.461
1	0.375	0.051	0.418	0.121	0.951	0.520	−0.621	0.424
2	0.450	0.038	0.455	0.118	1.053	0.625	−0.518	0.408
3	0.528	0.025	0.504	0.113	1.119	0.690	−0.522	0.410
4	0.575	0.016	0.541	0.113	1.182	0.752	−0.528	0.395
5	0.606	0.007	0.570	0.117	1.240	0.909	−0.541	0.391
10	0.642	−0.072	0.641	0.196	1.464	1.034	−0.632	0.378
20	0.940	−0.811	0.950	0.882	1.841	1.500	−0.965	0.279
30	1.603	−1.572	1.571	1.576	2.413	2.235	−1.339	0.135
50	2.224	−2.215	2.213	2.216	3.302	3.259	−1.771	0.040
100	2.670	−2.676	2.670	2.677	4.198	4.192	−2.192	0.010
150	2.810	−2.834	2.810	2.835	4.529	4.527	−2.352	0.007
200	2.872	−2.919	2.872	2.920	4.693	4.692	−2.438	0.006
250	2.901	−2.976	2.902	2.977	4.785	4.784	−2.493	0.005

* arbitrary initial set of weights and θ

Table 8.3 shows the results of training the network with the backpropagation method of the previous sections, along with the error measure of (8.31) at each step. It can be seen that the network approaches a solution quickly (approximately 50 iterations) but takes more iterations (approximately 250) to converge to a final result.

Having trained the network it is now possible to understand how it implements a solution to the nonlinear pattern recognition problem. The arguments of the activation functions of the PEs in the first processing layer each define a straight line (hyperplane in general) in the pattern space. Using the result at 250 iterations, these are:

$$2.901x_1 - 2.976x_2 = 0$$

$$2.902x_1 + 2.977x_2 = 0$$

which are shown plotted in Fig. 8.22. An individual line goes some way towards separating the data but cannot accomplish the task fully. It is now important to consider how the output PE operates on the outputs of the first layer PEs to complete the discrimination of the two classes. For pattern points lying exactly on one of the above lines, the output of the respective PE will be 0.5, given that the activation function of (8.29) has been used. However, for patterns a little distance away from those lines the output of the first layer PEs will be close to 0 or 1 depending on which side of the hyperplane they lie. We can therefore regard the pattern space as being divided into two regions – 0 and 1 – by a particular hyperplane. Using these extreme values, Table 8.4 shows the possible responses of the output layer PE for patterns lying somewhere in the pattern space. As seen, for this example the output PE functions in the nature of a logical OR operation; patterns that lie on the 1 side of EITHER input PE hyperplane are labelled as belonging to one class, while those that lie on the 0 side of both hyperplanes will be labelled as belonging to the other class. Thus patterns which lie in the shaded region shown in Fig. 8.22 will generate a 0 at the output of the network and thus will be labelled as belonging to class 1, while patterns in the unshaded regions will generate a 1 response

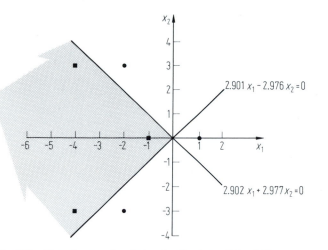

Fig. 8.22. Neural network solution for the data of Fig. 8.20

and thus will be labelled as belonging to class 2. Although this exercise is based only on two classes of data, similar functionality of the various PEs in a network can, in principle, be identified. The input PEs will always set up hyperplane divisions of the data and the later PEs will operate on the results of those simple discriminations.

Table 8.4. Response of the output layer PE

o_1	o_2	o_3
0	0	$0.076 \approx 0$
0	1	$0.908 \approx 1$
1	0	$0.908 \approx 1$
1	1	$0.999 \approx 1$

An alternative way of considering how the network determines a solution is to regard the first processing layer PEs as transforming the data in such a way that later PEs (in this example only one) can exercise linear discrimination. Fig. 8.23 shows a plot of the outputs of the first layer PEs when fed with the training data of Fig. 8.20. As observed, after transformation, the data is linearly separable. The hyperplane shown is that generated by the argument of the activation function of the output layer PE.

To illustrate how the network of Fig. 8.21 functions on unseen (i. e. testing set) data, Table 8.5 shows its response to the testing patterns indicated in Fig. 8.24. The class decision for a pattern is made by rounding the output PE response to 0 or 1 as appropriate. As noted, for this simple example, all patterns are correctly classified.

Benediktsson, Swain and Esroy (1990) have demonstrated the application of a neural network approach to classification in remote sensing, obtaining classification accuracies as high as 95 % on training data although only as high as 52 % when the network was applied to a test data set. It is suggested that the training data may not have been fully representative of the image. This is an important issue with neural nets, more so than with

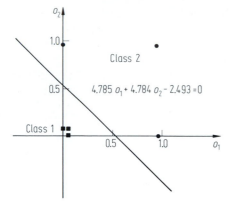

Fig. 8.23. Illustration of how the first processing layer PEs transform the input data into a linearly separable set, which is then discriminated by the output layer hyperplane

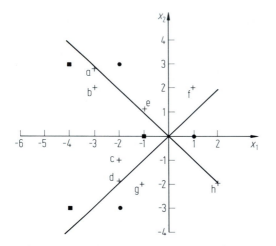

Fig. 8.24. Location of test data, indicated by the lettered crosses

Table 8.5. Performance of the network on the test data

Pattern	x_1	x_2	z_1	o_1	z_2	o_2	z_3	o_3	Class
a	−3.0	2.8	−17.036	0.000	− 0.370	0.408	−0.539	0.368	1
b	−3.0	2.0	−14.655	0.000	− 2.752	0.056	−2.206	0.099	1
c	−2.0	−1.0	− 2.826	0.056	− 8.781	0.000	−2.224	0.098	1
d	−2.0	−1.94	− 0.029	0.493	−11.579	0.000	−0.135	0.466	1
e	−1.0	1.1	− 6.175	0.002	0.373	0.592	0.350	0.587	2
f	1.0	2.0	− 3.051	0.045	8.856	1.000	2.506	0.925	2
g	−1.0	−2.0	3.051	0.955	− 8.856	0.000	2.077	0.889	2
h	2.0	−2.0	11.754	1.000	− 0.150	0.463	4.505	0.989	2

statistical classification methods such as maximum likelihood, since the parameters in the statistical approach are *estimates* of statistics, and are not strongly affected by outlying training samples. The work of Benediktsson also illustrates, to an extent, the dependence of performance on the network architecture chosen.

Hepner (1990) has also used a neural network to perform a classification; in addition to the spectral properties of a pixel, however, he included the spectral measurements of the 3×3 neighbourhood in order to allow spatial context to influence the labelling. Although quantitative accuracies are not given, Hepner is of the view that the results are better than when using a maximum likelihood classifier trained on spectral data only. Lippmann (1987) and Pao (1989) are good general references to consult for a wider treatment of neural network theory than has been given here, including other training methods. Both demonstrate also how neural networks can be used in unsupervised as well as supervised classification. Paola and Schowengerdt (1995a, b) provide a comprehensive review of the use of the multilayer Perception in remove sensing.

A range of neural network tools is available in MATLAB (1992–1998).

References for Chapter 8

The classification techniques used in remote sensing image analysis come from the field of mathematical pattern recognition and, as a consequence, are covered extensively in the standard treatments in that discipline. Particular references that could be consulted to obtain more mathematical detail than has been provided above, and to see other variations on the algorithms often used, include Nilsson (1965), Duda and Hart (1973), and Tou and Gonzalez (1974). For each of these a strong mathematical background is required. The development of classification techniques in Swain and Davis (1978) is far less mathematical and includes introductory conceptual material on the probabilistic aspects of pattern recognition.

The classifiers treated in this chapter have all been single stage in that only one decision is made about a pixel as a result of which it is labelled as belonging to one of the available classes, or is left unclassified. Decision tree procedures are also possible. In these a series of decisions is taken in order to determine the correct label for a pixel. As an illustration, the first decision might allow a distinction to be made between water, shadow and fire burnt pixels on the one hand, and vegetation, urban and cleared regions on the other. Subsequent decisions then allow finer subdivisions leading ultimately to a single label for the pixel. Advantages of this approach include the fact that different sets of features can be used at each decision stage and indeed even different algorithms could be employed. Its use in remote sensing problems is described by Swain and Hauska (1977). The decision tree is one special example of the set of layered classification techniques (Nilsson, 1965). Section 11.8 treats decision trees.

Atkinson et al (1985) discuss the value of image filtering prior to classification using both mean and median value methods. Demonstrations of probabilistic relaxation for pixel classification, and its comparison with simple maximum likelihood classification, will be found in Harris (1985), Gong and Howarth (1989) and Zenzo et al (1987a, 1987b); the latter authors also introduce the concept of fuzzy relaxation. Development of the basic relaxation algorithm will be found in Rosenfeld, Hummel and Zucker (1976). Statistical methods for context classification have been developed by Swain et al (1981), Kittler and Pairman (1985) and Khazenie and Crawford (1990). A selection of context classification methods are compared by Mohn et al (1987).

For a first reading on neural networks, beyond the material presented here, the paper by Lippmann (1987) is recommended.

P. Atkinson, J. L. Cushnie, J. R. Townshend and A. Wilson, 1985: Improving Thematic Mapper Land Cover Classification Using Filtered Data. International Journal of Remote Sensing, 6, 955–961.

J. A. Benediktsson, P. H. Swain and O. K. Esroy, 1990: Neural Network Approaches Versus Statistical Methods in Classification of Multisource Remote Sensing Data. IEEE Trans Geoscience and Remote Sensing, 28, 540–552.

J. A. Benediktsson, P. H. Swain and O. K. Esroy, 1993: Conjugate – Gradient Neural Networks in Classification of Multisource and Very-High-Dimensional Remote Sensing Data. Int. J. Remote Sensing, 14, 2883–2903.

R. O. Duda and P. E. Hart, 1973: Pattern Classification and Scene Analysis, N. Y., Wiley.

B. C. Forster, 1982: The Derivation of Approximate Equations to Correct for the Landsat MSS Point Spread Function. Proc. Commission 1 (Primary Data Acquisition) Int. Soc. for Photogrammetry and Remote Sensing, Canberra, April, 6–10.

J. E. Freund, 1992: Mathematical Statistics, 5e, New Jersey, Prentice Hall.

P. Gong and P. J. Howarth, 1989: Performance Analyses of Probabilistic Relaxation Methods for Land-Cover Classification. Remote Sensing of Environment, 30, 33–42.

P. Gong and P. J. Howarth, 1990: The Use of Structural Information for Improving Land-Cover Classification Accuracies at the Rural-Urban Fringe. Photogrammetric Engineering and Remote Sensing, 56, 67–73.

R. Harris, 1985: Contextual Classification Post-Processing of Landsat Data Using a Probabilistic Relaxation Model. Int. J. Remote Sensing, 6, 847–866.

G. F. Hepner, 1990: Artificial Neural Network Classification Using a Minimal Training Set: Comparison to Conventional Supervised Classification. Photogrammetric Engineering and Remote Sensing, 56, 469–473.

R. L. Kettig and D. A. Landgrebe, 1976: Classification of Multispectral Image Data by Extraction and Classification of Homogeneous Objects. IEEE Trans. Geoscience Electronics, GE-14, 19–26.

N. Khazenie and M. M. Crawford, 1990: Spatial-Temporal Autocorrelation Model for Contextual Classification. IEEE Trans Geoscience and Remote Sensing, 28, 529–539.

J. Kittler and D. Pairman, 1985: Contextual Pattern Recognition Applied to Cloud Detection and Identification. IEEE Trans Geoscience and Remote Sensing, GE-23, 855–863.

T. Lee, 1984: Multisource Context Classification Methods in Remote Sensing. PhD Thesis, The University of New South Wales, Kensington, Australia.

T. Lee and J. A. Richards, 1985: A low Cost Classifier for Multitemporal Applications. Int. J. Remote Sensing, 6, 1405–1417.

T. Lee and J. A. Richards, 1989: Pixel Relaxation Labelling Using a Diminishing Neighbourhood Effect. Proc. IGARSS'89. Vancouver, 634–637.

R. P. Lippmann, 1987: An Introduction to Computing with Neural Nets. IEEE ASSP Magazine, April, 4–22.

MATLAB, 1992–1998: Neural Network Toolbox. The Math Works, Inc, MA.

E. Mohn, N. L. Hjort and G. O. Storvik, 1987: A Simulation Study of Some Contextual Classification Methods for Remotely Sensed Data. IEEE Trans Geoscience and Remote Sensing, 25, 796–804.

N. J. Nilsson, 1965: Learning Machines. N. Y., McGraw Hill.

Y. H. Pao, 1989: Adaptive Pattern Recognition and Neural Networks. Reading, Addison-Wesley.

J. D. Paola and R. A. Schowengerdt, 1995 a. A Review and Analysis of Backpropagation Neural Networks for Classification of Remotely-Sensed Multi-Spectral Imagery. Int. J. Remote Sensing, 16, 3033–3058.

J. D. Paola and R. A. Schowengerdt, 1995 b. A Detailed Comparison of Backpropagation. Neural Network and Maximum-Likelihood Classifiers for Urban Land use Classification. IEEE Trans. Geoscience and Remote Sensing, 33, 981–996.

S. Peleg and A. Rosenfeld, 1980: A New Probabilistic Relaxation Procedure. IEEE Trans. Pattern Analysis and Machine Intelligence, PAMI-2, 362–369.

J. A. Richards, D. A. Landgrebe and P. H. Swain, 1981: On the Accuracy of Pixel Relaxation Labelling. IEEE Trans. Systems, Man and Cybernetics, SMC-11, 303–309.

A. Rosenfeld, R. Hummel and S. Zucker, 1976: Scene Labeling by Relaxation Algorithms. IEEE Trans Systems, Man and Cybernetics, SMC-6, 420–433.

P. H. Swain and S. M. Davis (Eds.), 1978: Remote Sensing: The Quantitative Approach, N. Y., McGraw-Hill.

P. H. Swain and H. Hauska, 1977: The Decision Tree Classifier: Design and Potential. IEEE Trans. Geoscience Electronics, GE-15, 142–147.

P. H. Swain, S. B. Varderman and J. C. Tilton, 1981: Contextual Classification of Multispectral Image Data. Pattern Recognition, 13, 429–441.

J. T. Tou and R. C. Gonzalez, 1974: Pattern Recognition Principles, Mass., Addison Wesley.

F. E. Townsend, 1986: The Enhancement of Computer Classifications by Logical Smoothing. Photogrammetric Engineering and Remote Sensing, 52, 213–221.

A. G. Wacker and D. A. Landgrebe, 1972: Minimum Distance Classification in Remote Sensing. First Canadian Symp. on Remote Sensing, Ottawa.

S. D. Zenzo, R. Bernstein, S. D. Degloria and H. G. Kolsky, 1987: Gaussian Maximum Likelihood and Contextual Classification Algorithms for Multicrop Classification. IEEE Trans Geoscience and Remote Sensing, 25, 805–814.

S. D. Zenzo, S. D. Degloria, R. Bernstein and H. G. Kolsky, 1987: Gaussian Maximum Likelihood and Contextual Classification Algorithms for Multicrop Classification Experiments Using Thematic Mapper and Multispectral Scanner Sensor Data. IEEE Trans Geoscience and Remote Sensing, 25, 815–824.

Problems

8.1 Suppose you have the following training data for three spectral classes, in which each pixel is characterised by only two spectral components λ_1 and λ_2.

Class 1		Class 2		Class 3	
λ_1	λ_2	λ_1	λ_2	λ_1	λ_2
16	13	8	8	19	6
18	13	9	7	19	3
20	13	6	7	17	8
11	12	8	6	17	1
17	12	5	5	16	4
8	11	7	5	14	5
14	11	4	4	13	8
10	10	6	3	13	1
4	9	4	2	11	6
7	9	3	2	11	3

Develop the discriminant functions for a maximum likelihood classifier and use them to classify the patterns

$$x_1 = \begin{bmatrix} 5 \\ 9 \end{bmatrix} \quad x_2 = \begin{bmatrix} 9 \\ 8 \end{bmatrix} \quad x_3 = \begin{bmatrix} 15 \\ 9 \end{bmatrix}$$

under the assumption of equal prior probabilities.

8.2 Repeat question 1 but with the prior probabilities
$$p(1) = 0.048$$
$$p(2) = 0.042$$
$$p(3) = 0.910$$

8.3 Using the data of question 1 develop the discriminant functions for a minimum distance classifier and use them to classify the patterns x_1, x_2 and x_3.

8.4 Develop a parallelepiped classifier from the training data given in question 1 and compare its classifications with those of the maximum likelihood classifier for the patterns x_1, x_2 and x_3, and the new pattern

$$x_4 = \begin{bmatrix} 3 \\ 7 \end{bmatrix}$$

At the conclusion of the tests in questions 1, 3 and 4, it would be worthwhile sketching a multispectral (pattern) space and then locating in it the positions of the training data. Use this to form a subjective impression of the performance of each classifier in questions 1, 3 and 4.

8.5 The following training data represents a subset of that in question 1 for just two of the classes. Develop discriminant functions for both maximum likelihood and minimum distance classifiers and use them to classify the patterns

$$x_5 = \begin{bmatrix} 14 \\ 7 \end{bmatrix} \quad x_6 = \begin{bmatrix} 20 \\ 13 \end{bmatrix}$$

Classify these patterns also using the minimum distance and maximum likelihood classifiers developed on the full training sets of question 1 and compare the results.

Class 1		Class 2	
λ_1	λ_2	λ_1	λ_2
11	12	17	8
10	10	16	4
14	11	14	5
		13	1

8.6 Suppose a particular scene consists of just water and soil, and that a classification into these cover types is to be carried out on the basis Landsat MSS band 7 data using the maximum likelihood rule. When the thematic map is produced it is noticed that some water pixels are erroneously labelled as soil. How can this happen, and what steps could be taken to avoid it? Hint: Sketch some typical one dimensional normal distributions to represent the soil and water in band 7 data, noting that soil would have a very large variance while that for water would be small. Remember the mathematical distribution functions extend to infinity.

Chapter 9

Clustering and Unsupervised Classification

9.1 Delineation of Spectral Classes

The successful application of maximum likelihood classification is dependent upon having delineated correctly the spectral classes in the image data of interest. This is necessary since each class is to be modelled by a normal probability distribution, as discussed in Chap. 8. If a class happens to be multimodal, and this is not resolved, then clearly the modelling cannot be very effective.

Users of remotely sensed data can only specify the information classes. Occasionally it might be possible to guess the number of spectral classes in a particular information class but, in general, the user would have little idea of the number of distinct unimodal groups that the data falls into in multispectral space. Clustering procedures can be used for that purpose; these are methods that have been applied in many data analysis fields to enable inherent data structures to be determined.

Clustering can also be used for unsupervised classification. In this technique an image is segmented into unknown classes. It is the task of the user to label those classes afterwards.

There are a great number of clustering methods. In this chapter those commonly employed with remote sensing data are treated, along with their variations as implemented in image analysis software packages.

9.2 Similarity Metrics and Clustering Criteria

Clustering implies a grouping of pixels in multispectral space. Pixels belonging to a particular cluster are therefore spectrally similar. In order to quantify this relationship it is necessary to devise a similarity measure. Many similarity metrics have been proposed but those used commonly in clustering procedures are usually simple distance measures in multispectral space. The most frequently encountered are Euclidean distance and $L1$ (or interpoint) distance. If x_1 and x_2 are two pixels whose similarity is to be checked then the Euclidean distance between them is

$$\begin{aligned} d(x_1, x_2) &= \|x_1 - x_2\| \\ &= \{(x_1 - x_2)^t (x_1 - x_2)\}^{\frac{1}{2}} \\ &= \left\{\sum_{i=1}^{N} (x_{1_i} - x_{2_i})^2\right\}^{\frac{1}{2}} \end{aligned} \qquad (9.1)$$

where N is the number of spectral components. The $L1$ distance between the pixels is

$$d(\mathbf{x}_1, \mathbf{x}_2) = \sum_{i=1}^{N} |x_{1_i} - x_{2_i}| .$$

$$(9.2)$$

Clearly the latter is computationally faster to determine. However it can be seen as less accurate than the Euclidean distance measure.

By using a distance measure it should be possible to determine clusters in data. Often however there could be several acceptable clusters assignments of the data, as depicted in Fig. 9.1, so that once a candidate clustering has been found it is desirable to have a means by which the "quality" of clustering can be measured. The availability of such a measure should allow one cluster assignment of the data to be chosen over all others.

A common clustering criterion or quality indicator is the sum of squared error (SSE) measure, defined as

$$SSE = \sum_{C_i} \sum_{x \in C_i} (\mathbf{x} - \mathbf{m}_i)^t (\mathbf{x} - \mathbf{m}_i)$$
$$= \sum_{C_i} \sum_{x \in C_i} \|\mathbf{x} - \mathbf{m}_i\|^2$$

$$(9.3)$$

where $\mathbf{m}_i$ is the mean of the ith cluster and $\mathbf{x} \in C_i$ is a pattern assigned to that cluster. The outer sum is over all the clusters. This measure computes the cumulative distance of each pattern from its cluster centre for each cluster individually, and then sums those measures over all the clusters. If it is small the distances from patterns to cluster means are all small and the clustering would be regarded favourably.

Other quality of clustering measures exist. One popular one is to derive a "within cluster scatter measure" by determining the average covariance matrix of the clusters, and a "between cluster scatter measure" by looking at the means of the clusters compared with the global mean of the data. These two measures are combined into a single figure of merit as discussed in Duda and Hart (1973) and Coleman and Andrews (1979). It can be shown that figures of merit such as these are essentially the same as the sum of squared error criterion.

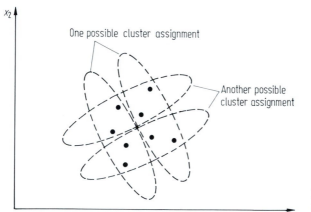

Fig. 9.1. Two apparently acceptable clusterings of a set of two dimensional data

It is of interest to note that SSE has a theoretical minimum of zero, which corresponds to all clusters containing only a single data point. As a result, if an iterative method is used to seek the natural clusters or spectral classes in a set of data then it has a guaranteed termination point, at least in principle. In practice it may be too expensive to allow natural termination. Instead, iterative procedures are often stopped when an acceptable degree of clustering has been achieved.

It is possible now to consider the implementation of an actual clustering algorithm. While it should depend upon a progressive minimisation (and thus calculation) of SSE this is impracticable since it requires an enormous number of values of SSE for the evaluation of all candidate clusterings. For example, there are approximately $C^P/C!$ ways of placing P patterns into C clusters (Duda and Hart, 1973). This number of SSE values would require computation at each stage of clustering to allow a minimum to be chosen. Rather than embark upon such a rigorous and computationally expensive approach the heuristic procedure of the following section is usually adopted in practice.

9.3 The Iterative Optimization (Migrating Means) Clustering Algorithm

The iterative optimization clustering procedure, also called the migrating means technique, is essentially the isodata algorithm presented by Ball and Hall (1965). It is based upon estimating some reasonable assignment of the pixel vectors into candidate clusters and then moving them from one cluster to another in such a way that the SSE measure of the preceding section is reduced.

9.3.1 The Basic Algorithm

The iterative optimization algorithm is implemented by the following set of basic steps:

1. The procedure is initialised by selecting C points in multispectral space to serve as candidate cluster centres. Let these be called

   $$\hat{m}_i, \ i = 1, \ldots C.$$

   The selection of the $\hat{m}_i$ at this stage is arbitrary with the exception that no two may be the same. To avoid anomolous cluster generation with unusual data sets it is generally wise to space the initial cluster means uniformly over the data. This can also serve to enhance convergence.

   Besides choosing the $\hat{m}_i$, the number of clusters C, must be specified beforehand by the user.

2. The location x of each pixel in the segment of the image to be clustered is examined and the pixel is assigned to the nearest candidate cluster. This assignment would be made on the basis of the Euclidean or even $L1$ distance measure.

3. The new set of means that result from the grouping produced in step (2) are computed. Let these be denoted

   $$m_i, \ i = 1, \ldots C.$$

4. If $m_i = \hat{m}_i$ for all i, the procedure is terminated. Otherwise $\hat{m}_i$ is redefined as the current value of m_i and the procedure returns to step (2).

The iterative optimization procedure is illustrated for a simple set of two dimensional patterns in Fig. 9.2.

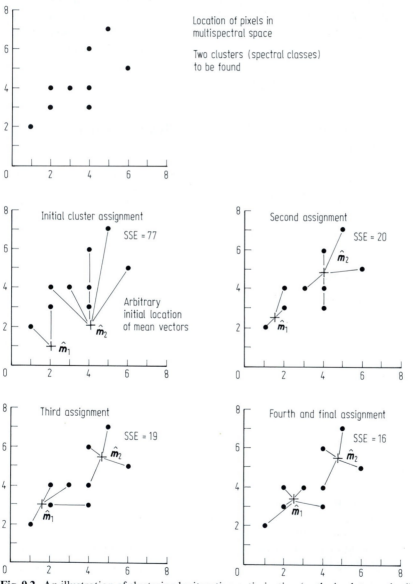

Fig. 9.2. An illustration of clustering by iterative optimization (or the isodata method). As noted, the method leads to a progressive reduction in SSE

9.3.2 Mergings and Deletions

Once clustering is completed, or at any suitable intervening stage, the clusters can be examined to see whether

(i) any clusters contain so few points as to be meaningless (e. g. that they would not give acceptable statistics estimates if used in training a maximum likelihood classifier), or

(ii) some clusters are so close together that they represent an unnecessary or indeed an injudicious division of the data, and thus they should be merged.

In view of the material of Sect. 8.2.6 a guideline exists for (i), viz that a cluster would be of little value for training a maximum likelihood classifier if it did not contain about $10 N$ points where N is the number of spectral components. In Chap. 10, which deals with separability and divergence, means for deciding whether clusters should be merged can also be devised.

9.3.3 Splitting Elongated Clusters

Another stage that can be inserted into the isodata algorithm is to separate elongated clusters into two new clusters. Usually this is done by prespecifying a standard deviation in each spectral band beyond which a cluster should be halved. Again this can be done after a set number of iterations, also specified by the user.

9.3.4 Choice of Initial Cluster Centres

Initialisation of the iterative optimization procedure requires specification of the number of clusters expected, along with their starting positions. In practice the actual or optimum number of clusters to choose will not be known. Therefore it is often chosen conservatively high, having in mind that resulting inseparable clusters can be consolidated after the process is completed, or at intervening iterations, if a merging operation is available.

The choice of the initial locations of the cluster centres is not critical although evidently it will have an influence on the time it takes to reach a final, acceptable clustering. Since no guidance is available in general, the following is a logical procedure (Phillips 1973). The initial cluster centres are chosen uniformly spaced along the multi-dimensional diagonal of the multispectral pixel space. This is a line from the origin to the point corresponding to the maximum brightness value in each spectral component (corresponding to 255 for 8 bit data, etc.). This choice can be refined if the user has some idea of the actual range of brightness values in each spectral component, say by having previously computed histograms. In that case the cluster centres would be initialised along a diagonal through the actual multidimensional extremities of the data.

Choice of the initial locations of clusters in the manner described is a reasonable and effective one since they are then well spread over the multispectral space in a region in which many spectral classes occur, especially for correlated data such as that corresponding to soils, rocks, concretes, etc.

9.3.5 Clustering Cost

Obviously the major limitation of the isodata technique is the need to prespecify the number of cluster centres. If this specification is too high then *a posteriori* merging can be used; however this is an expensive strategy. On the other hand, if too few are chosen initially then some multimodal spectral classes will result which, in turn, will prejudice ultimate classification accuracy.

Irrespective of whether too many or too few clusters are used, the isodata approach is computationally expensive since, at each iteration, every pixel must be checked against all cluster centres. Thus for C clusters and P pixels, PC distances have to be computed at each iteration and the smallest found. For N band data, each Euclidean distance calculation will require N multiplications and N additions, ignoring the square root operation in (9.1) since that need not be carried out. Thus for 20 classes and 10,000 pixels, 100 iterations isodata clustering requires 20 million multiplications per band of data.

9.4 Unsupervised Classification and Cluster Maps

At the completion of clustering, pixels within a given group are usually given a symbol to indicate that they belong to the same cluster or spectral class. Using these symbols a cluster map can be produced; this is a map corresponding to the image which has been clustered, but in which the pixels are represented by their symbol rather than by the original multispectral data. The availability of a cluster map allows a classification to be made. If some pixels with a given label can be identified with a particular ground cover type (by means of maps, site visits or other forms of reference data) then all pixels with the same label can be associated with that class. This method of image classification, depending as it does on *a posteriori* recognition of the classes, is called unsupervised classification since the analyst plays no part until the computational aspects are complete. Often unsupervised classification is used on its own, particularly when reliable training data for supervised classification cannot be obtained or is too expensive to acquire. However, it is also of value, as noted earlier, to determine the spectral classes that should be considered in a subsequent supervised approach. This is pursued in detail in Chap. 11.

9.5 A Clustering Example

To illustrate the nature of the results produced by the iterative optimization algorithm a simple example with Landsat multispectral scanner data is presented. Fig. 9.3 a shows a small image segment (band 7 only for illustration) which consists of regions of crops and background soils. Figure 9.3 b shows a scatter diagram for the image. In this, band 7 versus band 5 brightnesses of the pixels have been plotted. This is a subspace of the full four dimensional multispectral space of the image and gives an illustration of how the data points are distributed.

The data was clustered using the iterative optimization procedure (Kelly, 1983). Only five iterations were used and the algorithm was asked to determine five clusters. Merging and splitting options were employed at the end of each iteration

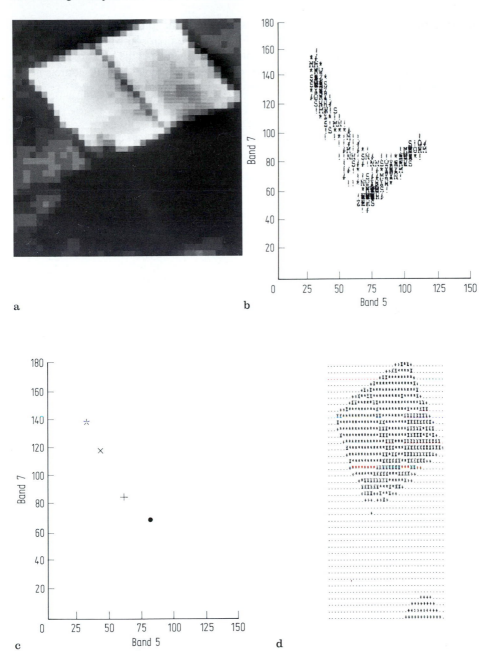

Fig. 9.3. a Image segment used in the clustering illustration; **b** Band 7 versus band 5 scatter diagram for the image; **c** Cluster centres on a band 7 versus band 5 diagram; **d** Cluster map produced by the isodata algorithm.

leading ultimately to the four clusters shown on the plot of cluster means in
Fig. 9.3 c and to the cluster map shown in Fig. 9.3 d. Comparison with Fig. 9.3 a shows
that the vegetation classes have been segmented more finely than the background soils in
this case. Nevertheless the cluster map displays acceptable spatial homogeneity.
Numerical details of the clusters established are given in Table 9.1.

Table 9.1. Cluster means and standard deviations for Fig. 9.3. generated by the iterative
optimization procedure

Cluster	Symbol	Band	Mean	St. Dev.
1	•	4	74.4	9.6
		5	85.5	13.7
		6	89.9	14.2
		7	69.8	12.1
2	*	4	45.0	2.0
		5	32.4	2.2
		6	127.4	6.3
		7	136.8	5.9
3	+	4	60.0	3.2
		5	59.5	4.0
		6	94.4	6.9
		7	83.7	7.5
4	×	4	48.9	3.8
		5	39.1	6.5
		6	114.0	5.9
		7	116.3	8.4

It is important to realise that the results generated in this example are not unique but
depend upon the clustering parameters chosen. In practice the user may need to apply
the algorithm several times with different parameter values to generate the desired
segmentation.

9.6 A Single Pass Clustering Technique

In order to reduce the cost of clustering image data, alternatives to iterative
optimization have been proposed and are widely implemented in software packages for
remote sensing image analysis. Often what they gain in speed they may lose in
accuracy; however if the user is aware of their characteristics they can usually be
employed effectively. One fast clustering procedure which requires only a single pass
through the data is described in the following sub-section.

9.6.1 Single Pass Algorithm

Not all of the region to be clustered must be used in developing cluster centres but
rather, for cost reduction, a randomly selected sample may be chosen and the samples
arranged into a two dimensional array. The first row of samples is then used to obtain a

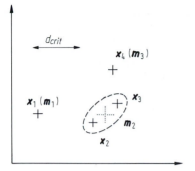

Fig. 9.4. Illustration of generation of cluster centres using the first row of samples

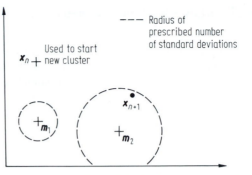

Fig. 9.5. Means by which pixels in the second and subsequent rows of samples are handled in the single pass clustering algorithm

starting set of cluster centres. This is initiated by adopting the first sample as the centre of the first cluster. If the second sample in the first row is further away from the first than a user specified *critical distance* then it is used to form another cluster centre. Otherwise the two samples are said to belong to the same cluster and their mean is computed as the new cluster centre. This procedure, which is illustrated in Fig. 9.4, is applied to all samples in the first row. Once this row has been exhausted the multidimensional standard deviations of the clusters are computed. Each sample in the second and subsequent rows is checked to see which cluster it is closest to. It is assigned to that cluster, and the cluster statistics recomputed, if it lies within a user-prescribed *number of standard deviations*. Otherwise it is used to form a new cluster centre (which is assigned a nominal standard deviation). This is depicted in Fig. 9.5. In this manner all of the samples are clustered and clusters with less than a prescribed number of pixels are deleted. Should a cluster map be required then the original segment of image data is scanned pixel by pixel and each pixel labelled according to the class it is closest to (on the basis usually of Euclidean distance). Should it be an outlying pixel in terms of the available cluster centres it is not labelled.

9.6.2 Advantages and Limitations

Apart from speed, a major advantage of this approach over the isodata procedure is its ability to create cluster centres as it proceeds. It is therefore not necessary for the user to specify beforehand the required number of clusters. However the method has two limitations. First, the user has to have a feel for the parameters required by the algorithm. In particular the user has to specify the critical distance parameter sensibly to enable the initial cluster centres to be established in a reasonable manner. Also he or she has to know how many standard deviations should be used in assigning pixels in the second and subsequent lines of samples to existing clusters. Clearly, with experience, these parameters can be estimated reasonably.

The second limitation is that the method is dependent upon the first line of samples to initiate the clustering. Since it is only a one pass algorithm and has no feedback checking mechanism by way of iteration, its ultimate set of cluster centres can depend significantly on the character of the first line of samples.

9.6.3 Strip Generation Parameter

Adjacent pixels along a line frequently belong to the same cluster, as is to be expected, particularly for images of cultivated regions. A method therefore for enhancing the speed of clustering is to compare a pixel with its predecessor and assign it to the same cluster immediately if it is similar. The similarity check often used is quite straightforward, consisting of a check of the brightness difference in each spectral band. The difference allowable for two pixels to be considered part of the same cluster is called the strip generation parameter.

9.6.4 Variations on the Single Pass Algorithm

The technique outlined in the preceding section has a number of variations. For example, the initial cluster centres can be specified by the user or alternatively can be created from the data using a critical distance parameter as illustrated in Fig. 9.4. Moreover rather than use a multiplier of standard deviation for assigning pixels from the second and subsequent rows of samples, some algorithms proceed exactly as for the first row, with standard deviation information not used at all. Some algorithms use the $L1$ metric of (9.2), rather than Euclidean distance, and some check inter-cluster distances and merge if this is indicated; periodically small clusters can also be eliminated.

The package known as MultiSpec, also uses just critical distance parameters over the full range, although the user can specify a different critical distance for the second and later rows of samples (Landgrebe and Biehl, 1995).

9.6.5 An Example

As an illustration, the single pass procedure has been applied to the data of Fig. 9.3 a. An initial critical distance of 15.0 was used, along with a standard deviation multiplier

Table 9.2. Cluster means and standard deviations for Fig. 9.6. generated by the single pass algorithm

Cluster	Symbol	Band	Mean	St. Dev.
1	•	4	86.0	5.1
		5	102.6	6.1
		6	107.0	5.9
		7	84.2	4.6
2	+	4	68.6	5.6
		5	76.7	8.2
		6	82.6	9.1
		7	64.1	8.8
3	*	4	45.8	2.4
		5	33.7	3.1
		6	123.8	7.5
		7	131.3	9.3
4	×	4	53.4	4.8
		5	48.2	7.6
		6	105.5	6.2
		7	102.5	9.4

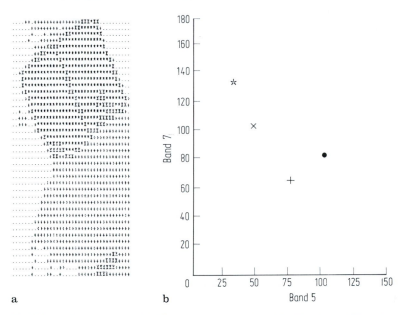

Fig. 9.6. **a** Cluster map and **b** Cluster centres produced for the data of Fig. 9.3 a, using the single pass clustering procedure

of 20.0 and a strip generation parameter of 1.0. The results produced are shown in Table 9.2 and Fig. 9.6. Two points are to be noted. First, different clusters have been found compared with those of the iterative optimization algorithm in Sect. 9.5. In this case there are two soil and two vegetation classes. Secondly, the essential spatial character of the classes has been produced with this algorithm even though the cluster centres generated are also at different locations in the multispectral space. Again, the procedure may need to be used interactively in practice to achieve a desired segmentation.

9.7 Agglomerative Hierarchical Clustering

Another clustering technique that does not require the user to specify the number of classes beforehand is hierarchical clustering. In fact this method produces an output that allows the user to decide the set of natural groupings into which the data falls. The procedure commences by assuming all pixels are individual clusters, it then systematically merges neighbouring clusters by checking distances between means. This is continued until all pixels appear in a single, larger cluster. An important aspect of the approach is that the history of mergings, or fusions as they are usually called in this method, is displayed on a *dendrogram*. This is a diagram that shows at what distances between centres particular clusters are merged. An example of hierarchical clustering, along with its fusion dendrogram is shown in Fig. 9.7. This uses the same two dimensional data set as Fig. 9.2, but note that the ultimate cluster compositions are slightly different. This demonstrates again that different algorithms can and do produce different clusterings.

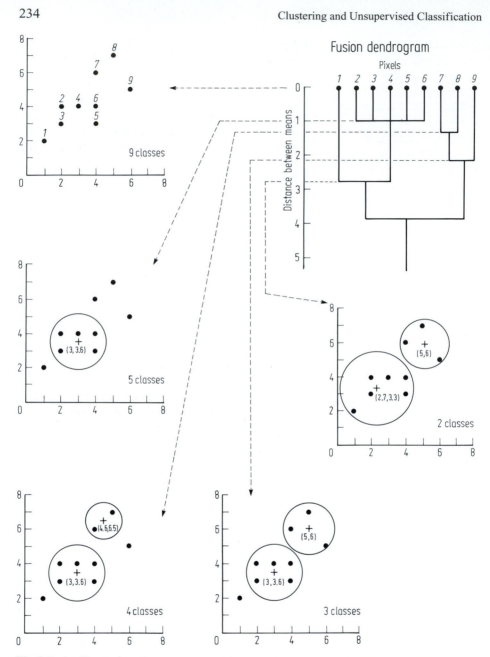

Fig. 9.7. An illustration of agglomerative hierarchical clustering, using Euclidean distance as a similarity measure

The fusion dendrogram of a particular hierarchical clustering exercise can be inspected in an endeavour to determine the intrinsic number of clusters or spectral classes in the data. Long vertical sections in the dendrogram between fusions indicate regions of "stability" which reflect natural data groupings. In Fig. 9.7 the longest region on the distance scale between fusions corresponds to two clusters in the data. One could conclude therefore that this data falls most naturally into two groups.

In the example presented, similarity between clusters was judged on the basis of Euclidean distance. Other similarity measures exist and are sometimes used, including divergence metrics as covered in Chap. 10.

The method given above is called *agglomerative* in view of its starting with a large number of clusters which it fuses progressively into a single cluster. *Divisive* hierarchical clustering procedures also exist in which the data is initialised as a single cluster which is progressively subdivided; these are more expensive computationally and are rarely used. Indeed hierarchical clustering generally does not find a lot of application in remote sensing image analysis since usually a large number of pixels is involved. Nevertheless it is a useful technique for small image data segments particularly since it can reveal data structure.

9.8 Clustering by Histogram Peak Selection

A multidimensional histogram of a segment of image data may exhibit peaks at the locations of spectral classes or clusters. Consequently, a further clustering technique adopted with remote sensing data is to construct such a histogram and then search it to find the location of its peaks. Pixels are then associated with the nearest peak to produce the clusters. This method has been described by Letts (1978).

In using histogram peak selection as a clustering technique it is important to keep in mind that the data and the histogram are discrete in nature and not continuous, as shown in Fig. 9.8. To see the implications of this, consider the following calculation. A 100 pixel by 100 pixel image segment consists of 10,000 pixels. Suppose this corresponds to data with four spectral components each quantised into 256 levels of brightness. Then the corresponding four dimensional histogram will have $(256)^4 = 4295$ million bins or locations into which counts (pixels) will be accumulated. If the bins were filled uniformly than a very sparse histogram would result. Indeed, on the average, there would be only one pixel per half a million bins. Each pixel therefore would appear as a local peak, which clearly would not be a true cluster. The bins of course would not be filled uniformly but nevertheless with bins only one brightness value wide in each spectral component, many artificial peaks will result from some isolated bins occupied by a single pixel and surrounded by empty bins. To circumvent this problem the histogram is accumulated with bins which are several brightness values wide in each dimension. In addition the dynamic range of the data in each dimension is ascertained beforehand from an inspection of the individual histograms in those dimensions. As an illustration, if the individual spectral component histograms for the four bands covered the ranges $(35,95), (25,105), (20,80)$ and $(5,65)$ and bin sizes of 10 brightness values were chosen for each dimension then the total number of four dimensional bins is now $6 \times 8 \times 6 \times 6 = 1728$. With a 100×100 pixel image segment therefore, there are, on the average, 6 pixels per bin which is probably acceptable (although low) to guarantee that peaks determined represent the location of real clusters in the data and not artifacts. Clearly resolution is sacrificed but this is necessary to yield an acceptable clustering by this approach.

The maximum detection algorithm used in this clustering procedure cannot be too sophisticated otherwise the method becomes too expensive to implement. Usually it consists of locating bins in which the count is higher than in the neighbouring bins along the same row and down the same column. For correlated data this can sometimes lead to false indications of peaks, as depicted in Fig. 9.9, in the vicinity of true peaks. This will be so particularly for smaller bin sizes. A better maximum detection procedure is to check diagonal neighbours as well but of course this doubles the search time.

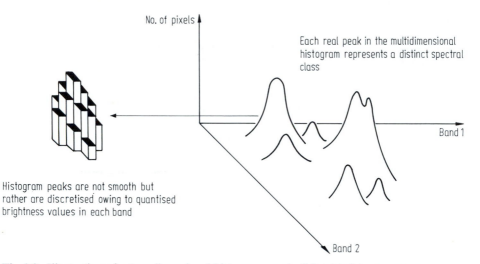

Fig. 9.8. Illustration of a two dimensional histogram emphasising its discrete nature

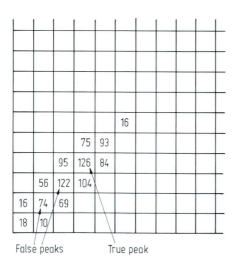

Fig. 9.9. False indication of peaks in a two dimensional histogram, when the peak detection algorithm only searches parallel to the bins

References for Chapter 9

Cluster analysis is a common tool in many applications that involve large amounts of data. Consequently source material on clustering algorithms will be found spread over many disciplines including numerical taxonomy, the social sciences and the physical sciences. However, because of the immense volumes of data to be clustered in remote sensing, the range of techniques that can be used is limited largely to those methods presented in this Chapter and to their variations. Some more general treatments however that may be of value include Anderberg (1973), Hartigan (1975), Tryon and Bailey (1970) and Ryzin (1977).

M. R. Anderberg, 1973: Cluster Analysis for Applications. N. Y. Academic.

G. H. Ball and D. J. Hall, 1965: A Novel Method of Data Analysis and Pattern Classification. Stanford Research Institute, Menlo Park, California.

G. R. Coleman and H. C. Andrews, 1979: Image Segmentation by Clustering. Proc. IEEE, 67, 773–785.

R. O. Duda and P. E. Hart, 1973: Pattern Classification and Scene Analysis. N Y., Wiley.

J. A. Hartigan, 1975: Clustering Algorithms. N. Y., Wiley.

D. J. Kelly, 1983: The Concept of a Spectral Class — A Comparison of Clustering Algorithms. M. Eng. Sc. Thesis. The University of New South Wales, Australia.

D. A. Landgrebe and L. Biehl, 1995: An Introduction to MultiSpec. West Lafayette, Purdue Research Foundation (http: //dynamo.ecn.purdue.edu/~biehl/MultiSpec)

P. A. Letts, 1978: Unsupervised Classification in The Aries Image Analysis System. Proc. 5th Canadian Symp. on Remote Sensing, 61–71.

T. L. Phillips (Ed), 1973: LARSYS Version 3 Users Manual. Laboratory for Applications of Remote Sensing, Purdue University, West Lafayette.

J. van Ryzin, 1977: Classification and Clustering. N. Y., Academic.

R. C. Tryon and D. E. Bailey, 1970: Cluster Analysis, N. Y., McGraw-Hill.

Problems

9.1 Repeat the exercise of Fig. 9.2 but with
 (i) two initial cluster centres at (2,3) and (5,6),
 (ii) three initial cluster centres at (1,1), (3,3) and (5,5), and
(iii) three initial cluster centres at (2,1), (4,2) and (15,15).

9.2 From a knowledge of how a particular clustering algorithm works it is sometimes possible to infer the multidimensional spectral shapes of the clusters generated. For example, methods that depend entirely upon Euclidean distance as a similarity metric would tend to produce hyperspheroidal clusters. Comment on the cluster shapes you would expect to be generated by the migrating means technique based upon Euclidean distance and the single pass procedure, also based upon Euclidean distance.

9.3 Suppose two different techniques have given two different clusterings of a particular set of data and you wish to assess which of the two segmentations is the better. One approach might be to evaluate the sum of square errors measure treated in Sect. 9.2. Another could be based upon covariance matrices. For example it is possible to define an "among clusters" covariance matrix that describes how the clusters themselves are scattered about the data space, and an average "within class" covariance matrix that describes the average shape and size of the clusters. Let these be called Σ_A and Σ_W respectively. How could they be used together to assess the quality of the two clustering results? (See Coleman and Andrews, 1979) Here you may wish to use measures of the "size" of a matrix, such as its trace or determinant (see Appendic C).

9.4 Different clustering methods often produce quite different segmentations of the same set of data, as illustrated in the examples of Fig. 9.3 and 9.6. Yet the results generated for remote sensing applications are generally usable. Why do you think that is the case? (Hint; Is it related to the number of clusters generated?)

9.5 The Mahalanobis distance of (8.13) can be used as the similarity metric for a clustering algorithm. Invent a possible clustering technique based upon (8.13) and comment on the nature of the clusters generated.

9.6 Do you see value in having a two stage clustering process say in which a single pass procedure is used to generate initial clusters and then an iterative technique is used to refine them?

9.7 Recompute the agglomerative hierarchical clustering example of Fig. 9.7 but use the $L1$ distance measure in (9.2) as a similarity metric.

9.8 The histogram peak selection clustering technique of Section 9.8 has some shortcomings. One is related to the need to have large spectral bins in the histogram in order to have a sensible histogram produced when the data dimensionality is high. A consequence is that fine spectral resolution is sacrificed leading to loss of discrimination of spectral classes that are very close. Do you think good spectral discrimination could be regained by applying the technique several times over, on each subsequent occasion clustering just within one of the clusters found previously? Discuss the details of this approach.

Chapter 10
Feature Reduction

10.1 Feature Reduction and Separability

Classification cost increases with the number of features used to describe pixel vectors in multispectral space – i.e. with the number of spectral bands associated with a pixel. For classifiers such as the parallelepiped and minimum distance procedures this is a linear increase with features; however for maximum likelihood classification, the procedure most often preferred, the cost increase with features is quadratic. Therefore it is sensible economically to ensure that no more features than necessary are utilised when performing a classification. Features which do not aid discrimination, by contributing little to the separability of spectral classes, should be discarded since they will represent a cost burden. Removal of least effective features is referred to as feature selection, this being one form of feature reduction. The other is to transform the pixel vector into a new set of co-ordinates in which the features that can be removed are made more evident. Both procedures are considered in some detail in this Chapter.

Feature selection cannot be performed indiscriminantly. Methods must be devised that allow the relative worths of features to be assessed in a quantitative and rigorous way. A procedure commonly used is to determine the mathematical *separability* of classes; in particular, feature reduction is performed by checking how separable various spectral classes remain when reduced sets of features are used. Provided separability is not lowered unduly by the removal of features then those features can be considered of little value in aiding discrimination and therefore to be not cost-effective in the classification process.

10.2 Separability Measures for Multivariate Normal Spectral Class Models
(Adapted in part from Swain and Davis, 1978)

10.2.1 Distribution Overlaps

Consider a two dimensional multispectral space with two spectral classes as depicted in Fig. 10.1. Suppose we wish to see whether the classes could be separated using only one feature – either x_1 or x_2. Of course it is not known which feature offers the best prospects *a priori*. This is what has to be determined by a measure of separability. Consider an assessment of x_1. The spectral classes in the x_1 'subset' or subspace are shown in the figure whereupon some overlap of the single dimensional distributions is indicated. If the distributions are well separated in the x_1 dimension then clearly the

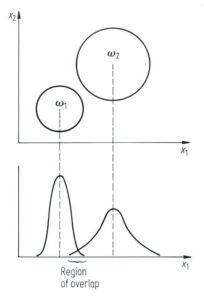

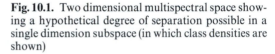

Fig. 10.1. Two dimensional multispectral space showing a hypothetical degree of separation possible in a single dimension subspace (in which class densities are shown)

overlap will be small and it would be unlikely that a classifier would make an error in discriminating between them on the basis of that feature alone. On the other hand for a large degree of overlap substantial classifier error would be expected. The usefulness of the x_1 feature subset therefore can be assessed in terms of the overlap of the distributions in that domain, or more generally, in terms of the similarity of the distributions as a function of x_1 alone.

Consider now an attempt to quantify the separation between a pair of probability distributions (as models of spectral classes) as an indication of the degree of overlap. Clearly distance between means is insufficient since overlap will also be influenced by the standard deviations of the distributions. Instead, a combination of both the distance between means and a measure of standard deviation is required. Moreover this must be a vector-based measure in order to be applicable to multidimensional subspaces. Several such measures are available; only those commonly encountered in connection with remote sensing data are treated in this Chapter. Others may be found in books on statistics that treat similarities in probability distributions. These measures are all referred to as measures of separability which implies the ease with which patterns can be correctly associated with their classes using statistical pattern classification.

10.2.2 Divergence

10.2.2.1 A General Expression

Divergence is a measure of the separability of a pair of probability distributions that has its basis in their degree of overlap. It is defined in terms of the likelihood ratio

$$L_{ij}(x) = p(x|\omega_i)/p(x|\omega_j)$$

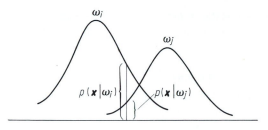

Fig. 10.2. Definition of the probabilities used in the likelihood ratio

where $p(x|\omega_i)$ and $p(x|\omega_j)$ are the values of the ith and jth spectral class probability distributions at the position x. These are shown in an overlap region in Fig. 10.2 whereupon it is evident that $L_{ij}(x)$ is a measure of 'instantaneous' overlap. Clearly for very separable spectral classes $L_{ij}(x) = 0$ or ∞ for all x.

It is of value to choose the logarithm of the likelihood ratio, viz

$$L'_{ij}(x) = \ln p(x|\omega_i) - \ln p(x|\omega_j),$$

by means of which the *divergence* of the pair of class distribution is defined as

$$d_{ij} = \mathscr{E}\{L'_{ij}(x)|\omega_i\} + \mathscr{E}\{L'_{ji}(x)|\omega_j\} \tag{10.1}$$

where $\mathscr{E}\{\}$ is the expectation operator defined as

$$\mathscr{E}\{L'_{ij}(x)|\omega_i\} = \int_x L'_{ij}(x)\, p(x|\omega_i)\, \mathrm{d}x.$$

This is the average or expected value of the likelihood ratio with respect to all patterns in the ith spectral class. Similarly for $\mathscr{E}\{L'_{ji}(x)|\omega_j\}$. From (10.1) it can be seen that

$$d_{ij} = \int_x \{p(x|\omega_i) - p(x|\omega_j)\} \ln \frac{p(x|\omega_i)}{p(x|\omega_j)}\, \mathrm{d}x$$

from which a number of properties of divergence can be established. For example it is always positive and also $d_{ji} = d_{ij}$, as should be the case – i.e., it is symmetric. Moreover, if $p(x|\omega_i) = p(x|\omega_j)$ for all x then $d_{ij} = d_{ji} = 0$ – in other words there is no divergence (or difference) between a distribution and itself.

For statistically independent features (i.e., spectral components) $x_1, x_2 \cdots x_N$ then

$$p(x|\omega_i) = \prod_{n=1}^{N} p(x_n|\omega_i)$$

which leads to

$$d_{ij}(x) = \sum_{n=1}^{N} d_{ij}(x_n).$$

Since divergence is never negative it follows therefore that

$$d_{ij}(x_1, \cdots x_n, x_{n+1}) > d_{ij}(x_1, \cdots x_n).$$

In other words, divergence never decreases as the number of features is increased.

The material to this point has been general, applying to any multivariate spectral class model.

10.2.2.2 Divergence of a Pair of Normal Distributions

Since spectral classes in remote sensing image data are modelled by multidimensional normal distributions it is of particular interest to have available the specific form of (10.1) when $p(x|\omega_i)$ and $p(x|\omega_j)$ are normal distributions with means and covariances of m_i, Σ_i and m_j, Σ_j respectively.

By substitution of the full expressions for the normal distributions it can be shown that

$$
\begin{aligned}
d_{ij} &= \tfrac{1}{2} T_r\{(\Sigma_i - \Sigma_j)(\Sigma_j^{-1} - \Sigma_i^{-1})\} \\
&+ \tfrac{1}{2} T_r\{(\Sigma_i^{-1} + \Sigma_j^{-1})(m_i - m_j)(m_i - m_j)^t\} \\
&= \text{Term 1} + \text{Term 2}.
\end{aligned}
\tag{10.2}
$$

where $T_r\{\}$ is the trace of the subject matrix. Note that Term 1 involves only covariances whereas Term 2 is the square of a normalised (by covariance) distance between the means of the distributions.

Equation (10.2) gives the divergence between a *pair* of spectral classes that are normally distributed. Should there be more than two spectral classes, as is generally the case, all pairwise divergences need to be checked to see whether a particular feature subset gives sufficiently separable data. An average indication of separability is then given by computing the *average divergence*

$$
d_{ave} = \sum_{i=1}^{M} \sum_{j=i+1}^{M} p(\omega_i)\, p(\omega_j)\, d_{ij}
\tag{10.3}
$$

where M is the number of spectral classes and $p(\omega_i)$, $p(\omega_j)$ are the class prior probabilities.

10.2.2.3 Use of Divergence for Feature Selection

Consider the need to select the best three discriminating channels for Landsat multispectral scanner data, for an image in which only three spectral classes exist. The pairwise divergence between each pair of spectral classes would therefore be determined for all combinations of three out of four channels or bands. The feature subset chosen would be that which gives the highest overall indication of divergence – presumably this would be the highest average divergence. Table 10.1 illustrates the number of divergence calculations required for such an example.

In general, for M spectral classes, N total features, and a need to select the best n feature subset, the following set of pairwise divergence calculations are necessary, leaving aside the need finally to compute the average divergence for each subset.

First there are ^NC_n possible combinations of n features from the total N, and for each combination there are $^MC_2$ pairwise divergence measures to be computed. For a complete evaluation therefore

$$
^NC_n \cdot {}^MC_2
$$

Table 10.1. Divergence calculation table

For channel subsets	d_{12}	d_{13}	d_{23}	d_{ave}
1, 2, 3	*	*	*	*
1, 2, 4	*	*	*	*
1, 3, 4	*	*	*	*
2, 3, 4	*	*	*	*

* Entries to be calculated.

measures of pairwise divergence have to be calculated. To assess the best 4 of 7 Landsat Thematic Mapper bands for an image involving 10 spectral classes then

$$^7C_4 \cdot {^{10}C_2} = 1575$$

divergence values have to be computed. Inspection of (10.2) shows each divergence calculation to be considerable. This, together with the large number required in a typical problem, makes the use of divergence to check separability and indeed separability analysis in general, an expensive process computationally.

10.2.2.4 A Problem with Divergence

As spectral classes become further removed from each other in multispectral space, the probability of being able to classify a pattern at a particular location moves asymptotically to 1.0 as depicted in Fig. 10.3a. If divergence is similarly plotted it will be seen from its definition that it increases quadratically with separation between spectral

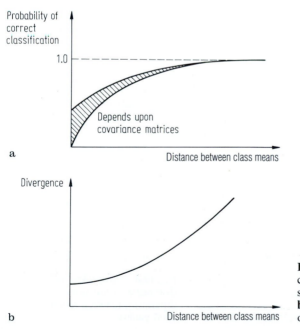

Fig. 10.3. a Probability of correct classification as a function of spectral class separation; **b** Divergence as a function of spectral class separation

class means as depicted in Fig. 10.3b. This behaviour unfortunately is quite misleading if divergence is to be used as an indication of how successfully patterns in the corresponding spectral classes could be mutually discriminated or classified. It implies, for example, that at large separations, further small increases will lead to vastly better classification accuracy whereas in practice this is not the case as observed from the very slight increase in probability of correct classification implied by Fig. 10.3a. Moreover, outlying, easily separable classes will weight average divergence upwards in a misleading fashion to the extent that sub-optimal reduced feature subsets might be indicated as best, as illustrated in Swain and Davis (1978). This problem renders divergence, as it is presently defined, to be unsuitable and indeed unsatifactory. The Jeffries-Matusita distance in the next section does not suffer this drawback.

10.2.3 The Jeffries-Matusita (JM) Distance

10.2.3.1 Definition

The JM distance between a pair of probability distributions (spectral classes) is defined as

$$J_{ij} = \int_x \{ \sqrt{p(x|\omega_i)} - \sqrt{p(x|\omega_j)} \}^2 \, dx \tag{10.4}$$

which is seen to be a measure of the average distance between the two class density functions (Wacker, 1971). For normally distributed classes this becomes

$$J_{ij} = 2(1 - e^{-B}) \tag{10.5}$$

in which

$$B = 1/8(m_i - m_j)^t \left\{ \frac{\Sigma_i + \Sigma_j}{2} \right\}^{-1} (m_i - m_j) + 1/2 \ln \left\{ \frac{|(\Sigma_i + \Sigma_j)/2|}{|\Sigma_i|^{1/2}|\Sigma_j|^{1/2}} \right\} \tag{10.6}$$

which is referred to as the Bhattacharyya distance (Kailath, 1967).

It is of interest to note that the first term in B is akin to the square of the normalised distance between the class means. The presence of the exponential factor in (10.5) gives an exponentially decreasing weight to increasing separations between spectral classes. If plotted as a function of distance between class means it shows a saturating behaviour not unlike that expected for the probability of correct classification, as shown in Fig. 10.4.

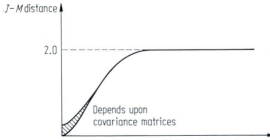

Fig. 10.4. Jeffries-Matusita distance as a function of separation between spectral class means

It is asymptotic to 2.0 so that a JM distance of 2.0 between spectral classes would imply classification of pixel data into those classes, (assuming they were the only two) with 100% accuracy. This saturating behaviour is highly desirable since it does not suffer the difficulty experienced with divergence.

As with divergence, an average pairwise JM distance can be defined according to

$$d_{ave} = \sum_{i=1}^{M} \sum_{j=i+1}^{M} p(\omega_i) p(\omega_j) J_{ij} \tag{10.7}$$

where M is the number of spectral classes and $p(\omega_i)$, $p(\omega_j)$ are the class prior probabilities.

10.2.3.2 Comparison of Divergence and JM Distance

JM distance performs better as a feature selection criterion for multivariate normal classes than divergence for the reasons given above; however it is computationally more complex and thus expensive to use as can be assessed from comparison of (10.2) and (10.6). Suppose a particular problem involves M spectral classes. Consider the cost then of computing all pairwise divergences and all pairwise JM distances. These costs can be assessed largely on the basis of having to compute matrix inverses and determinants, assuming reasonably that they involve similar computational demands using numerical procedures. In the case of divergence it is necessary to compute only M matrix inverses to allow all the pairwise divergences to be found. However for JM distance it is necessary to compute $^{M}C_2 + M$ equivalent matrix inverses since the individual class covariances appear as pairs which have to be added and then inverted. It may be noted that $^{M}C_2 + M = \frac{1}{2} M(M+1)$ so that divergence is a factor of $\frac{1}{2}(M+1)$ more economical to use. When it is recalled how many feature subsets may need to be checked in a feature selection exercise this is clearly an important consideration. However the unbound nature of divergence as discussed in Sect. 10.2.2.4 throws doubt on its usefulness.

10.2.4 Transformed Divergence

10.2.4.1 Definition

A useful modification of divergence becomes apparent by noting the algebraic similarity of divergence to the parameter B in JM distance, as defined in (10.6). Since both involve terms which are functions of the covariance alone, and terms which appear as normalised distances between class means, it should be possible to make use of a heuristic *transformed divergence* measure of the form (Swain and Davis 1978)

$$d_{ij}^{T} = 2(1 - e^{-d_{ij}/8}). \tag{10.8}$$

Because of its exponential character it will have a saturating behaviour with increasing class separation, as does JM distance, and yet it is computationally more economical. This saturating measure is used in the Macintosh package called MultiSpec; it has been demonstrated to be almost as effective as JM distance in feature selection, and considerably better than simple divergence or simple Bhattacharyya distance (Swain et al., 1971, Mausel et al., 1990).

10.2.4.2 Relation between Transformed Divergence and Probability of Correct Classification

It can be shown that the probability of making a classification error in placing a pattern into one of two (equal prior probability) classes with a pairwise divergence d_{ij} is bound by (Kailath, 1967)

$$p_E > \frac{1}{8} e^{-d_{ij}/2},$$

so that the probability of correct classification is bound by

$$p_C < 1 - \frac{1}{8} e^{-d_{ij}/2}.$$

Since $d_{ij} = -8 \ln\left(1 - \frac{1}{2} d_{ij}^T\right)$, from (10.8)

then $P_C < 1 - \frac{1}{8}\left(1 - \frac{1}{2} d_{ij}^T\right)^4.$ (10.9)

This bound on classification accuracy is shown in Fig. 10.5 along with an empirical relationship between transformed divergence and probability of correct (pairwise) classification derived by Swain and King (1973). This figure has considerable value in establishing *a priori* the upper bound achievable on classification accuracy for an existing set of spectral classes.

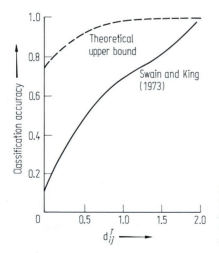

Fig. 10.5. Probability of correct classification as a function of pairwise transformed divergence. The empirical measure, taken from Swain and King (1973), was determined using 2790 sets of multidimensional, normally distributed data, in two classes

10.2.4.3 Use of Transformed Divergence in Clustering

One of the last stages in a practical clustering algorithm is to evaluate the size and relative locations of the clusters produced, as noted in Chap. 9. If clusters are too close to each other they should be merged. The availability of the information in Fig. 10.5 allows merging to be effected based upon a pre-specified transformed divergence, since both cluster mean and covariance data is normally available. By establishing a desired accuracy level (in fact upper bound) for the subsequent classification and then determining the corresponding value of transformed divergence, clusters with separabilities less than this value must be merged.

10.3 Separability Measures for Minimum Distance Classification

The separability measures of Sect. 10.2 relate to spectral classes modelled by multivariate normal distributions, in preparation for maximum likelihood classification. Should another classifier be used this procedure is unduly complex and largely without meaning. For example, if supervised classification is to be carried out using the minimum distance to class means technique there is no advantage in using distribution-based separability measures, since probability distribution class models are not employed. Instead it is better to use a simple measure consistent with the nature of the classification algorithm. For minimum distance calculation this would be a distance measure, computed according to the particular distance metric in use. Commonly this is Euclidean distance. Consequently, when a set of spectral classes have been determined, ready for the classification step, the complete set of pairwise Euclidean distances will provide an indication of class similarities. Unfortunately this cannot be related to an error probability (for misclassification) but finds major application as an *indicator* of what pairs of classes could be merged, if so desired.

10.4 Feature Reduction by Data Transformation

The emphasis of the preceding sections has been *feature selection* – i.e., an evaluation of the existing set of features for the pixel data in multispectral imagery with a view to selecting the most discriminating, and discarding the rest. It is also possible to effect feature reduction by transforming the data to a new set of axes in which separability is higher in a subset of the transformed features than in any subset of the original data. This allows transformed features to be discarded. A number of image transformations could be entertained for this; however the most commonly encountered in remote sensing are the principal components or Karhunen-Loève transform and the transformation associated with so-called canonical analysis. These are treated in the following.

10.4.1 Feature Reduction Using the Principal Components Transformation

The principal components transformation (see Chap. 6) maps image data into a new, uncorrelated co-ordinated system or vector space. Moreover, in doing so, it produces a space in which the data has most variance along its first axis, the next largest variance along a second mutually orthogonal axis, and so on. The later principal components would be expected, in general, to show little variance. These could be considered therefore to contribute little to separability and could be ignored, thereby reducing the essential dimensionality of the classification space and thus improving classification speed. This is only of value however if the spectral class structure of the data is distributed substantially along the first few axes. Should this not be the case it is possible that feature reduction of the transformed data may be no more likely than with the original data. In such a case the technique of canonical analysis may be a better approach.

As an illustration of a situation of data in which principal components transformation does allow feature reduction, consider the two class two dimensional data illustrated in Fig. 10.6. Assume that the classes are not separable in either of the original data variables alone but rather both dimensions are required for separability. However, inspection indicates that the first component of a principal components transform will yield class separability. This is now demonstrated mathematically by presenting the results of hand calculations on the data.

Notwithstanding the class structure of the data the principal components transformation makes use of a global mean and global covariance. Using (6.1) and (6.2) it is shown readily that

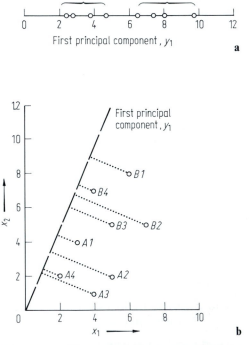

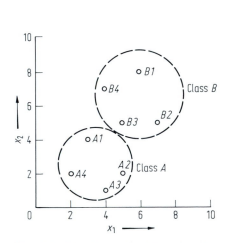

Fig. 10.6. Two dimensional, two class data in which feature reduction using principal components analysis is possible

Fig. 10.7. **a** First principal component of the image data; **b** Principal axis relative to original image components

$$m = \begin{bmatrix} 4.5 \\ 4.25 \end{bmatrix}$$

and

$$\Sigma = \begin{bmatrix} 2.57 & 1.86 \\ 1.86 & 6.21 \end{bmatrix}$$

The eigenvalues of the covariance matrix are $\lambda_1 = 6.99$ and $\lambda_2 = 1.79$ so that the first principal component will contain 79.6% of the variance. The normalised eigenvectors corresponding to these eigenvalues are

$$g_1 = \begin{bmatrix} 0.387 \\ 0.922 \end{bmatrix} \quad \text{and} \quad g_2 = \begin{bmatrix} -0.922 \\ 0.387 \end{bmatrix}$$

so that the principal components transformation matrix is

$$G = \begin{bmatrix} 0.387 & 0.922 \\ -0.922 & 0.387 \end{bmatrix}$$

Using this matrix, the first principal component of each pixel vector can be computed according to

$$y_1 = 0.387 x_1 + 0.922 x_2.$$

These are shown plotted in Fig. 10.7a wherein it is seen that the first principal component is sufficient for separation. Figure 10.7b shows the principal axes relative to the original image components.

10.4.2 Canonical Analysis as a Feature Selection Procedure

The principal components transformation is based upon the global covariance matrix of the full set of image data and thus is not sensitive explicitly to class structure in the data. The reason it often works well in remote sensing as a feature reduction tool is a result of the fact that classes are frequently distributed in the direction of maximum data scatter. This is particularly so for soils and spectrally similar cover types. Should good separation not be afforded by the principal components transformation derived from the global covariance matrix then a subset of image data could be selected that embodies the cover types of interest and this subset used to compute a covariance matrix. The resulting transformation will have its first principal axes oriented so that the cover types of interest are well discriminated. Another, more rigorous, method for generating a transformed set of feature axes, in which class separation is optimised, is based upon the procedure called canonical analysis. To illustrate this approach consider the contrived two dimensional, two class data shown in Fig. 10.8. By inspection, the classes can be seen not to be separable in either of the original feature axes on their own. Nor will they be separable in only one of the two principal

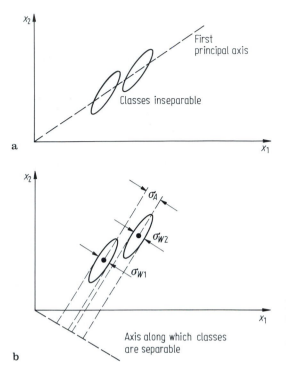

Fig. 10.8. a Hypothetical two dimensional, two class data illustrating lack of separability in either original band or in either principal component; **b** Axis along which classes can be separated

component axes because of the nature of the global data scatter compared with the scatter of data within the individual classes.

Inspection shows however that the data of Fig. 10.8a can be separated by a single feature if an axis rotation (i.e. an image transformation) such as that shown in Fig. 10.8b is adopted. A little thought reveals that the primary axis in this new transformation should be so-oriented that the classes have the largest possible separation between their means when projected onto that axis, while at the same time they should appear as small as possible in their individual spreads. If we characterise the former by a measure σ_A as illustrated in the diagram (which can be referred to as the standard deviation among the classes – it is as if the classes themselves were data points at their mean positions) and the spread of data within classes as seen on the new axis as σ_{w1}, σ_{w2} as illustrated (these are the standard deviations of the classes) then our interest is in finding a new axis for which

$$\frac{\sigma_A^2}{\sigma_w^2} = \frac{\text{among categories variance}}{\text{within categories variance}} \tag{10.10}$$

is as large as possible. Here σ_w^2 is the average of σ_{w1}^2 and σ_{w2}^2 for the example of Fig. 10.8.

10.4.2.1 Within Class and Among Class Covariance Matrices

To handle data with any number of dimensions it is necessary to define average data scatter within the classes, and the scatter of the classes themselves around the multispectral space, by covariance matrices.

The average within class covariance matrix is defined as

$$\Sigma_w = \frac{1}{M} \sum_{i=1}^{M} \Sigma_i \qquad (10.11\,a)$$

where Σ_i is the covariance matrix of the data in class i and where M is the total number of classes. The boldface sigma is printed for the summation to distinguish it from the symbol for covariance. Equation (10.11 a) applies only if the classes have equal populations. A better expression is

$$\Sigma_w = \left\{ \sum_{i=1}^{M} (n_i - 1) \Sigma_i \right\} / S_n \qquad (10.11\,b)$$

where n_i is the population of the ith class and $S_n = \sum_{i=1}^{M} n_i$.

The among class covariance matrix is given by

$$\Sigma_A = \mathscr{E} \left\{ (\boldsymbol{m}_i - \boldsymbol{m}_0)(\boldsymbol{m}_i - \boldsymbol{m}_0)^t \right\} \qquad (10.12)$$

where $\boldsymbol{m}_i$ is the mean of the ith class, $\mathscr{E}$ is the expectation operator and $\boldsymbol{m}_0$ is the global mean, given by

$$\boldsymbol{m}_0 = \frac{1}{M} \sum_{i=1}^{M} \boldsymbol{m}_i \qquad (10.13\,a)$$

where the classes have equal populations, or

$$\boldsymbol{m}_0 = \sum_{i=1}^{M} n_i \boldsymbol{m}_i / S_n \qquad (10.13\,b)$$

in general.

10.4.2.2 A Separability Measure

Let $y = D^t x$ be the required transformation that generates the new axes y in which the classes have optimal separation. The transposed form of the transformation matrix is chosen here to simplify the following expressions. By the same procedure that was used for the principal components transformation in Sect. 6.1.2 it is possible to show that the within class and among class covariance matrices in the new co-ordinate system are

$$\Sigma_w^y = D^t \Sigma_w^x D \qquad (10.14\,a)$$

$$\Sigma_A^y = D^t \Sigma_A^x D \qquad (10.14\,b)$$

where the superscripts x and y have been used to identify the matrices with their respective co-ordinates. It is significant to realise here, unlike with the case of principal components analysis, that the two new covariance matrices are not necessarily diagonal. However, as with principal components the row vectors of D^t define the axis directions in y-space. Let d^t be one particular vector (say the one that defines the first so-called canonical axis, along which the classes will be optimally separated), then the corresponding within class and among class variances will be

$$\sigma_w^2 = d^t \Sigma_w^x d$$

$$\sigma_A^2 = d^t \Sigma_A^x d.$$

What we wish to do is to find the d, (and in fact ultimately the full transformation matrix D^t) for which

$$\lambda = \sigma_A^2 / \sigma_w^2 = d^t \Sigma_A^x d / d^t \Sigma_w^x d \qquad (10.15)$$

is maximised. In the following the superscripts on the covariance matrices have been dropped for convenience.

10.4.2.3 The Generalised Eigenvalue Equation

The ratio of variances λ in (10.15) is maximised by the selection of d if

$$\frac{\partial \lambda}{\partial d} = 0.$$

Noting the identity that $\dfrac{\partial}{\partial x} \{x^t A x\} = 2 A x$ then

$$\frac{\partial \lambda}{\partial d} = \frac{\partial}{\partial d} \{(d^t \Sigma_A d)(d^t \Sigma_w d)^{-1}\}$$

$$= 2\Sigma_A d (d^t \Sigma_w d)^{-1} - 2\Sigma_w d (d^t \Sigma_A d)(d^t \Sigma_w d)^{-2}$$

$$= 0.$$

This reduces to

$$\Sigma_A d - \Sigma_w d (d^t \Sigma_A d)(d^t \Sigma_w d)^{-1} = 0.$$

Which can be written as

$$(\Sigma_A - \lambda \Sigma_w) d = 0 \qquad (10.16)$$

Equation (10.16) is called a *generalised eigenvalue equation* and has to be solved now for the unknowns λ and d. The first canonical axis will be in the direction of d and λ will give the associated ratio of among class to within class variance along that axis.

In general (10.16) can be written

$$(\Sigma_A - \Lambda \Sigma_w) D = 0 \qquad (10.17)$$

where Λ is a diagonal matrix of the full set of λ's and D is the matrix of vectors d.

The development to this stage is usually referred to as discriminant analysis. One additional step is included in the case of canonical analysis.

As with the equivalent step in the principal components transformation, solution of (10.16) amounts to finding the set of eigenvalues λ and the corresponding eigenvectors, d. While unique values for λ can be determined the components of d can only be found relative to each other. In the case of principal components we introduced the additional requirement that the vectors have unit magnitude, thereby allowing the vectors to be determined uniquely. For canonical analysis, the additional constraint used is

$$D^t \Sigma_w D = I. \tag{10.18}$$

This says that the within class covariance matrix after transformation must be the identity matrix (i.e. a unit diagonal matrix). In other words, after transformation, the classes should appear spherical.

For M classes and N bands of multispectral data, if $N > M{-}1$ there will only be $M{-}1$ non-zero roots of (10.17) and thus $M{-}1$ canonical axes (Seal, 1964). For this example, in which $N = 2$, $M = 2$ one of the eigenvalues of (10.16) will be zero and thus the corresponding eigenvector will not exist. This implies that the dimensionality of the transformed space will be less than that of the original data. Thus canonical analysis provides separability with reduced dimensionality. In general, in the first canonical axis, corresponding to the largest λ, the classes will have maximum separation. The second axis, corresponding to the next largest λ, will provide the next best degree of separation, and so on. Campbell and Atchley (1981) review canonical analysis with a particular emphasis on a geometrical interpretation.

10.4.2.4 An Example

Consider the two dimensional, two category data shown in Fig. 10.9. Both of the original features x_1 and x_2 are required to discriminate between the categories. We will now perform a canonical analysis transformation on the data to show that the categories can be discriminated in the first canonical axis.

The individual covariance matrices of the classes are

$$\Sigma_A = \begin{bmatrix} 2.25 & 2.59 \\ 2.59 & 4.25 \end{bmatrix} \qquad \Sigma_B = \begin{bmatrix} 4.25 & 3.00 \\ 3.00 & 6.67 \end{bmatrix}$$

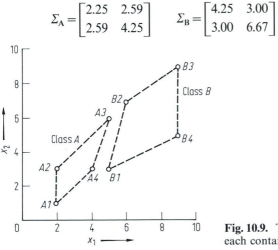

Fig. 10.9. Two classes of two dimensional data, each containing 4 data points

so that the within class covariance is

$$\Sigma_w = \frac{1}{2}\{\Sigma_A + \Sigma_B\} = \begin{bmatrix} 3.25 & 2.80 \\ 2.80 & 5.46 \end{bmatrix}.$$

The among class covariance matrix is

$$\Sigma_A = \begin{bmatrix} 8.00 & 5.50 \\ 5.50 & 3.78 \end{bmatrix}.$$

The canonical transformation matrix D^t is given by a solution to (10.17) where D is a matrix of column vectors. These vectors are the axes in the transformed space, along the first of which the ratio of among categories variance to within categories variance is greatest. Along this axis there is most chance of separating the classes. Λ is a diagonal matrix of scalar constants that are the eigenvalues of (10.17); numerically these are the ratios of variances along each of the canonical axes.

Each λ and the accompanying d can be found readily by considering the individual component equation (10.16) rather than the more general form in (10.17). For (10.16) to have a non-trivial solution it is necessary that the determinant

$$|\Sigma_A - \lambda \Sigma_w| = 0.$$

Using the values for Σ_A and Σ_w above this is

$$\begin{vmatrix} 8.00 - 3.25\,\lambda & 5.50 - 2.80\,\lambda \\ 5.50 - 2.80\,\lambda & 3.78 - 5.46\,\lambda \end{vmatrix} = 0$$

which gives $\lambda = 2.54$ or 0. Thus there is only one canonical axis defined by the vector d corresponding to $\lambda = 2.54$. This is given as the solution to

$$[\Sigma_A - 2.54\,\Sigma_w]\,d = 0$$

i.e.

$$\begin{bmatrix} -0.26 & - & 1.61 \\ -1.61 & -10.09 \end{bmatrix} \begin{bmatrix} d_1 \\ d_2 \end{bmatrix} = 0$$

whereupon $d_1 = -6.32\,d_2$.

At this stage we use (10.18), which for one vector d in D is

$$[d_1 \quad d_2] \begin{bmatrix} 3.25 & 2.80 \\ 2.80 & 5.46 \end{bmatrix} \begin{bmatrix} d_1 \\ d_2 \end{bmatrix} = 1$$

i.e. $3.25\,d_1^2 + 5.60\,d_1\,d_2 + 5.46\,d_2^2 = 1$.

Using $d_1 = -6.32\,d_2$ gives

$$d_1 = 0.632, \ d_2 = -0.100$$

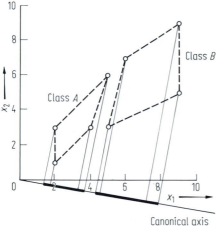

Canonical axis

Fig. 10.10. The first canonical axis for the two class data of Fig. 10.9 showing class discrimination

so that

$$d = \begin{bmatrix} 0.632 \\ -0.100 \end{bmatrix}$$

This vector is shown plotted in Fig. 10.10 wherein the projections of the patterns onto the axis defined by that vector show the classes to be separable. The brightness values of the pixels in that first axis are given by

$$y = d^t x$$
$$= [0.632 \ -0.100] \, x.$$

10.4.3 Arithmetic Transformations

Depending upon the application, feature reduction prior to classification can sometimes be carried out using simpler arithmetic operations than the transformations treated in the foregoing sections. As an illustration, taking the differences of multispectral imagery from different dates can yield difference data that can be processed for change, by comparison with the need to classify *all* the data if the preprocessing step is not adopted.

A second example is the use of the simple ratio of infrared to visible data as a vegetation index. This allows vegetation classification to be performed on the ratio data alone. More sophisticated vegetation indices exist and these can be considered as data reduction transformations. The most commonly encountered are the following in which the bands designated are those from the multispectral scanners on Landsats 1–3. The band numbers need to be redefined to refer to Landsats 4 onwards, and other sensors.

$$VI = (\text{band } 7 - \text{band } 5)/(\text{band } 7 + \text{band } 5) \quad (\text{vegetation index})$$

$$TVI = \sqrt{VI + 0.5} \quad (\text{transformed vegetation index})$$

These and others are discussed in Myers (1983).

References for Chapter 10

For more mathematical details on measures of divergence and Jeffries-Matusita distance the reader is referred to Duda and Hart (1973) and Kailath (1967). A detailed discussion on transformed divergence will be found in Swain and King (1973).

A good introductory discussion on canonical analysis in remote sensing is given in the paper by Jensen and Waltz (1979). More detailed descriptions from a remote sensing viewpoint have been given by Merembeck et al. (1977), Merembeck and Turner (1980) and Eppler (1976). These also contain results that illustrate its suitability as a feature reduction tool.

Detailed mathematical descriptions of canonical analysis, as a statistical tool, will be found in Seal (1964) and Tatsuoka (1971). Seal in particular gives the results of hand calculations on two and three dimensional data sets.

N. A. Campbell and W. R. Atchley, 1981: The Geometry of Canonical Variate Analysis. Systematic Zoology, 30, 268–280.

R. O. Duda and P. E. Hart, 1973: Pattern Classification and Scene Analysis, N. Y., Wiley.

W. Eppler, 1976: Canonical Analysis for Increased Classification Speed and Channel Selection, IEEE Trans. Geoscience Electronics, GE-14.

S. K. Jensen and F. A. Waltz, 1979: Principal Components Analysis and Canonical Analysis in Remote Sensing, Proc. American Soc. of Photogrammetry 45th Ann. Meeting.

T. Kailath, 1967: The Divergence and Bhattacharyya Distance Measures in Signal Selection, IEEE Trans. Communication Theory, COM-15, 52–60.

P. W. Mausel, W. J. Kramber and J. K. Lee, 1990: Optimum Band Selection for Supervised Classification of Multispectral Data. Photogrammetric Engineering and Remote Sensing, 56, 55–60.

B. F. Merembeck, F. Y. Borden, M. H. Podwysocki and D. N. Applegate, 1977: Application of Canonical Analysis to Multispectral Scanner Data, Proc. 14th Symp. Applications of Computer Methods in the Mineral Industry, N. Y., AIMMPE.

B. F. Merembeck and B. J. Turner, 1980: Directed Canonical Analysis and the Performance of Classifiers under its Associated Linear Transformation, IEEE Trans. Geoscience and Remote Sensing, GE-18, 190–196.

V. I. Myers, 1983: Remote Sensing Applications in Agriculture. In R. N. Colwell (Ed.) Manual of Remote Sensing, 2e, American Soc. of Photogrammetry, Falls Church.

H. Seal, 1964: Multivariate Statistical Analysis for Biologists. London, Menthuen.

P. H. Swain and S. M. Davis, (Eds.), 1978: Remote Sensing: The Quantitative Approach, N. Y., McGraw-Hill.

P. H. Swain and R. C. King, 1973: Two Effective Feature Selection Criteria for Multispectral Remote Sensing, Proc. First Int. Joint Conf. on Pattern Recognition, 536–540, November.

P. H. Swain, T. V. Robertson and A. G. Wacker, 1971: Comparison of Divergence and B-Distance in Feature Selection, Information Note 020871, Laboratory for Applications of Remote Sensing, Purdue University, West Lafayette.

M. M. Tatsuoka, 1971: Multivariate Analysis, N. Y., Wiley.

A. G. Wacker, 1971: The Minimum Distance Approach to Classification, Ph. D. Thesis, Purdue University, West Lafayette.

Problems

10.1 Kailath (1967) shows that the probability of making an error in labelling a pattern as belonging to one of two classes with equal prior probabilities is bounded according to

$$\frac{1}{16}(2 - J_{ij})^2 \le P_E \le \frac{1}{4}(2 - J_{ij})$$

where J_{ij} is the Jeffries-Matusita distance between the classes. Determine and plot the upper and lower bounds on classification accuracy for a two class problem, as a function of J_{ij}. You may wish to compare this to an empirical relationship between classification accuracy and J_{ij} found by Swain and King (1973).

10.2 Consider the training data given in problem 8.1. Suppose it is required to use only one feature to characterise each spectral class. By computing pairwise transformed divergence measures ascertain the best feature to retain if:

(a) only classes 1 and 2 are to be considered
(b) only classes 2 and 3 are to be considered
(c) all three classes are to be considered.

In each case estimate the maximum possible classification accuracy.

10.3 Using the same data as in problem 10.2, perform feature reductions if possible using principal component transformations if the covariance matrix is generated using

(a) only classes 1 and 2
(b) only classes 2 and 3
(c) all three classes.

10.4 Using the same data as in problem 10.2, compute a canonical analysis transformation for all three classes of data and see whether the classes have better discrimination in the transformed axis.

10.5 Suppose the mean vectors and covariance matrices have been determined, using training data, for a particular image of an agricultural region. Because of the nature of the land use, the region consists predominantly of fields that are large compared with the effective ground dimensions of a pixel, and within each field there is a degree of similarity among the pixels, owing to its use for a single crop type.

Suppose you delineate a field from the rest of the image (either manually or automatically) and then compute the mean vector and covariance matrix for all the pixels in that field. Describe how pairwise divergence, or Jeffries-Matusita distance could be used to classify the complete *field* of pixels into one of the training classes.

Chapter 11
Image Classification Methodologies

11.1 Introduction

In principle, classification of multispectral image data should be straightforward. However to achieve results of acceptable accuracy care is required first in choosing the analytical tools to be used and then in applying them. In the following the classical analytical procedures of supervised and unsupervised classification are examined from an operational point of view, with their strengths and weaknesses highlighted. These approaches are often acceptable; however more often a judicious combination of the two will be necessary to attain optimal results. A hybrid supervised/unsupervised strategy is therefore also presented.

Other compound classification approaches are also possible including the hierarchical decision tree methods covered in Sect. 11.8.

11.2 Supervised Classification

11.2.1 Outline

As discussed in Chap. 8 the underlying requirement of supervised classification techniques is that the analyst has available sufficient known pixels for each class of interest that representative signatures can be developed for those classes. These prototype pixels are often referred to as training data, and collections of them, identified in an image and used to generate class signatures, are called training fields. The step of determining class signatures is frequently called training.

Signatures generated from the training data will be of a different form depending on the classifier type to be used. For parallelepiped classification the class signatures will be the upper and lower bounds of brightness in each spectral band. For minimum distance classification the signatures will be the mean vectors of the training data for each class, while for maximum likelihood classification both class mean vectors and covariance matrices constitute the signatures. For neural network classifiers the collection of weights define the boundaries between classes. While they do not represent class signatures as such they are the inherent properties of the classifier, learnt form training data, that allow classes to be discriminated.

By having the labelled training data available beforehand, from which the signatures are estimated, the analyst is, in a relative sense, teaching the classification algorithm to recognise the spectral characteristics of each class, thereby leading to the term *supervised* as a qualification relating to the algorithm's learning about the data with which it has to work.

As a proportion of the full image to be analysed the amount of training data would represent less than 1% to 5% of the pixels. The learning phase therefore, in which the analyst plays an important part in the *a priori* labelling of pixels, is performed on a very small part of the image. Once trained, the classifier is then asked to attach labels to *all* the image pixels by using the class estimates provided to it.

The steps in this fundamental outline are now examined in more detail, noting the practical issues that should be considered to achieve reliable results.

11.2.2 Determination of Training Data

The major step in straightforward supervised classification is in the prior identification of training pixels. This may involve the expensive enterprise of field visits, or may require use of reference data such as topographic maps and air photographs. In the latter, a skilled photointerpreter may be required to determine the training data. Once training fields are suitably chosen however they have to be related to the pixel addresses in the satellite imagery. Sometimes training data can be chosen by photointerpretation from image products formed from the multispectral data to be classified. Generally however this is restricted to major cover types and again can require a great deal of photointerpretive skill if more than a simple segmentation of the image is required.

Some image processing systems have digitizing tables that allow map data – such as polygons of training pixels, i.e. training fields – to be taken from maps and superimposed over the image data. While this requires a registration of the map and image, using the procedures of Sect. 2.4, it represents an unbiased method for choosing the training data. It is important however, as with all training procedures based upon field or reference data, that the training data be recorded at about the same time as the multispectral data to be classified. Otherwise errors resulting from temporal variations may arise.

It is necessary to identify training data at least for all classes of interest and preferably for all apparent classes in the segment of image to be analysed. In either case, and particularly if the selection of training data is not exhaustive or representative, it is prudent to use some form of threshold or limit if the classification is of the minimum distance or maximum likelihood variety; this will ensure poorly characterised pixels are not erroneously labelled. Limits in minimum distance classification can be imposed by only allowing a pixel to be classified if it is within a prespecified number of standard deviations from the nearest mean. For maximum likelihood classification a limit may be applied by the use of thresholds on the discriminant functions. Having so limited a classification, pixels in the image which are not well represented in the training data will not be classified. This will identify weaknesses in the selection of the training sets which can then be rectified and the image re-classified. Repeated refinement of the training data and reclassification in this manner can be carried out using a represented portion of the image data.

11.2.3 Feature Selection

The cost of the classification of a full image segment is reduced if bands or features that do not aid discrimination significantly are removed. After training is complete feature selection can be carried out using the separability measures presented in Chap. 10. The recommended measures are transformed divergence, if maximum likelihood signatures have been generated, or Euclidean distance if the signatures have been prepared for minimum distance classification.

Separability measures can also be used to assess whether any pair of classes are so similar in multispectral space that significant misclassification will occur if they are both used. Should such a pair be found the analyst should give consideration to merging them to form a single class.

11.2.4 Detecting Multimodal Distributions

The most common algorithm for supervised classification is that based upon maximum likelihood estimation of class membership of an unknown pixel using multivariate normal distribution models for the classes. Its attraction lies in its ability to model class distributions that are elongated to different extents in different directions in multispectral space and its consequent theoretical guarantee that, if properly applied, it will lead to minimum average classification error. However, its major limitation in this regard is that the classes must be appropriately representable as multivariate normal distributions. Often the information classes of interest will not appear as single distributions but rather are best resolved into a set of constituent spectral classes or sub-classes. Should these spectral classes not be properly identified beforehand, the accuracy of supervised maximum likelihood classification will suffer. Multimodal classes can be identified to an extent using clustering algorithms; indeed this is the basis of the hybrid classification methodology developed in Sect. 11.4 below. A simple, yet rather more limited means, by which multimodal behaviour can be assessed is to examine scatterplots of the data in each training class. A scatterplot is a two dimensional multispectral space with user defined axes. A Landsat 1 band 7 versus band 5 scatterplot for "vegetation" prototype pixels could show for example, two distinct regions of data concentration, corresponding to sub-classes of "grassland" and "trees".

Should any of the sets of training data be found to be multimodal, steps should be taken to resolve them into the appropriate sub-classes in order to minimise classification error. Again clustering of the training sets could be used to do this, although it is frequently straightforward to identify groups of image pixels corresponding to each of the data modes in a scatterplot, thereby allowing the analyst to subdivide the corresponding training fields.

11.2.5 Presentation of Results

Two types of output are available from a classification. One is the thematic (or class) map in which pixels are given a label (represented by a colour or symbol) to identify them with a class. The other output is a table that summarises the number of pixels in the image found to belong to each class. The table can be interpreted also as a table of areas, in hectares. However that requires either that the user has resampled the image data to a map grid beforehand, so that the pixels correspond to an actual area on the ground, or that the user takes account of any systematic pixel overlap such as the 23 m overlap of Landsat MSS pixels caused by the detector sampling strategy. In that case it is important to recall that the effective MSS pixel is 56 m × 79 m and thus represents an area of 0.4424 ha for Landsats 1 to 3.

11.2.6 Effect of Resampling on Classification

The utility of remote sensing image data is improved if it is registered to a map base. As discussed in Sect. 2.4.1.3 several interpolation techniques can be used to synthesise pixel values on the map grid, the most common being nearest neighbour resampling and resampling by cubic convolution. In the former, original image pixels are simply relocated onto a geometrically correct map grid whereas in the latter new pixel brightness values are synthesised by interpolating over a group of sixteen pixels.

Usually it is desirable to have the thematic maps produced by classification registered to a map base. This can be done either by rectifying the image before classification or by rectifying the actual thematic map (in which case nearest neighbour resampling is the only option). An advantage in correcting the image beforehand is that it is often easier to relate reference data and ground truth information to the image if it is in correct geometric registration to a map. However a drawback with doing this from a data analysis/information extraction point of view is that the data is then processed before classification is attempted. That preprocessing can add uncertainty to the pixel brightness values and therefore prejudice subsequent classification accuracy. Accordingly, a good rule wherever possible is not to correct the data before classification. Should it be necessary to rectify the data then nearest neighbour interpolation should be used in the resampling stage.

The influence of resampling on classification has been addressed by Billingsley (1982), Verdin (1983) and Forster and Trinder (1984) who show examples of how cubic convolution interpolation can have a major influence across boundaries such as that between vegetation and water, leading to uncertainties in classification.

When images in a multitemporal sequence have to be classified to extract change information it is necessary to perform image to image registration (which could alternatively consist of registering all the images to a reference map). Since registration cannot be avoided in this case, nearest neighbour resampling should be used.

11.3 Unsupervised Classification

11.3.1 Outline, and Comparison with Supervised Methods

Unsupervised classification is an analytical procedure based upon clustering, using algorithms such as those described in Chap. 9. Application of clustering partitions the image data in multispectral space into a number of spectral classes, and then labels all pixels of interest as belonging to one of those spectral classes, although the labels are purely symbolic (e.g. A, B, C, ..., or class 1, class 2, ...) and are as yet unrelated to ground cover types. Hopefully the classes will be unimodal; however, if simple unsupervised classification is of interest, this is not essential.

Following segmentation of the multispectral space by clustering, the clusters or spectral classes are associated with information classes – i.e. ground cover types – by the analyst. This *a posteriori* identification may need to be performed explicitly only for classes of interest. The other classes will have been used by the algorithm to ensure good discrimination but will remain labelled only by arbitrary symbols rather than by class names.

The identification of classes of interest against reference data is often more easily carried out when the spatial distribution of spectrally similar pixels has been established in the image data. This is an advantage of unsupervised classification and the technique is therefore a convenient means by which to generate signatures for spatially elongated classes such as rivers and roads.

In contrast to the *a priori* use of analyst-provided information in supervised classification, unsupervised classification is a segmentation of the data space in the absence of any information provided by the analyst. Analyst information is used only to attach information class (or ground cover type, or map) labels to the segments established by clustering. Clearly this is an advantage of the approach. However it is a time-consuming procedure computationally by comparison to techniques for supervised classification. This can be demonstrated by comparing, for example, multiplication requirements of the iterative clustering algorithm of Sect. 9.3 with the maximum likelihood classification decision rule of Sect. 8.2.3.

Suppose a particular classification exercise involves N spectral bands and C classes. Maximum likelihood classification requires $CPN(N+1)$ multiplications where P is the number of pixels in the image segment of interest. By comparison, clustering of the data requires PCI distance measures for I iterations. Each distance calculation demands N multiplications[1], so that the total number of multiplications for clustering is $PCIN$. Thus the speed comparison of the two approaches is approximately $(N+1)/I$ for maximum likelihood classification compared with clustering. For Landsat MSS data, therefore, in a situation where all 4 spectral bands are used, clustering would have to be completed within 5 iterations to be speed competitive with maximum likelihood classification. Frequently 20 times this number of iterations is necessary to achieve an acceptable clustering. Training of the classifier would add about a 10% loading to its time demand; however a significant time loading should also be added to clustering to account for the labelling phase. Often this is done by associating pixels with the nearest (Euclidean distance) cluster. However, sometimes Mahalanobis or maximum likelihood distance labelling is used. This adds substantially to the cost of clustering.

Because of the time demand of clustering algorithms, unsupervised classification is often carried out with small image sequents. Alternatively a representative subset of data is used in the actual clustering phase in order to cluster or segment the multispectral space. That information is then used to assign all the image pixels to a cluster.

When comparing the time requirements of supervised and unsupervised classification it must be recalled that a large demand on user time is required in training a supervised procedure. This is necessary both for determining training data and then identifying training pixels by reference to that data. The corresponding step in unsupervised classification is the *a posteriori* labelling of clusters. While this still requires user effort in determining labelled prototype data, not as much may be required. As noted earlier, data is only required for those classes of interest; moreover only a handful of labelled pixels is necessary to identify a class. By comparison, sufficient training pixels per class are required in supervised training to ensure reliable estimates of class signatures are generated.

---

[1] Usually distance squared is calculated avoiding the need to evaluate the square root operation in (9.1).

A final point that must be taken into account when contemplating unsupervised classification via clustering is that there is no facility for including prior probabilities of class membership. By comparison the decision functions for maximum likelihood classification can be biased by previous knowledge or estimates of class membership.

11.3.2 Feature Selection

Most clustering procedures used for unsupervised classification in remote sensing generate the mean vector and covariance matrix for each cluster found. Accordingly separability measures can be used to assess whether feature reduction is necessary or whether some clusters are sufficiently similar spectrally that they should be merged. These are only considerations of course if the clustering is generated on a sample of data, with a second phase used to allocate all image pixels to a cluster. Feature selection would be performed between the two phases.

11.4 A Hybrid Supervised/Unsupervised Methodology

11.4.1 The Essential Steps

The strength of supervised classification based upon the maximum likelihood procedure is that it minimises classification error for classes that are distributed in a multivariate normal fashion. Moreover, it can label data relatively quickly. Its major drawback lies in the need to have delineated unimodal spectral classes beforehand. This, however, is a task that can be handled using clustering, based upon a representative subset of image data. Used for this task, unsupervised classification performs the valuable function of identifying the existence of all spectral classes, yet it is not expected to perform the entire classification. Consequently, the rather logical hybrid classification procedure outlined below can be envisaged. This is due to Fleming et al. (1975).

Step 1: Use Clustering to determine the spectral classes into which the image resolves. For reasons of economy this is performed on a representative subset of data. Spectral class statistics are also produced from this unsupervised step.

Step 2: Using available ground truth or other reference data associate the spectral classes (or clusters) with information classes (ground cover types).

Step 3: Perform a feature selection evaluation to see whether all features (bands) need to be retained for reliable classification.

Step 4: Using the maximum likelihood algorithm, classify the entire image into the set of spectral classes.

Step 5: Label each pixel in the classification by the ground cover type associated with each spectral class.

It is now instructive to consider some of these steps in detail and thereby introduce some useful practical concepts. The method depends for its accuracy (as do all classifications) upon the skills and experience of the analyst. Consequently, it is not unusual in practice to iterate over sets of steps as experience is gained with the particular problem at hand.

11.4.2 Choice of the Clustering Regions

Clustering is employed in Step 1 above to determine the spectral classes, using a subset of the image data. It is recommended that about 3 to 6 small regions, or so-called candidate clustering areas, be chosen for this purpose. These should be well spaced over the image and located such that each one contains several of the cover types (information classes) of interest and such that all cover types are represented in the collection of clustering areas. An advantage in choosing heterogeneous regions to cluster, as against apparently homogeneous training areas used in supervised classification, is that mixture pixels lying on class boundaries will be identified as legitimate spectral classes.

If an iterative clustering procedure is used, the analyst will have to prespecify the number of clusters expected in each candidate area. Experience has shown that, on the average, there are about 2 to 3 spectral classes per information class. This number should be chosen, with a view to removing or rationalising unnecessary clusters at a later stage.

It is of value to cluster each region separately as this saves computation, and produces cluster maps within those areas with more distinct class boundaries than would be the case if all regions were pooled beforehand.

11.4.3 Rationalisation of the Number of Spectral Classes

When clustering is complete the spectral classes are then associated with information classes using available reference data. It is then necessary to see whether any spectral classes or clusters can be discarded, or more importantly, whether sets of clusters can be merged, thereby reducing their number and leading ultimately to a faster classification. Decisions about merging can be made on the basis of separability measures, such as those treated in Chap. 10.

During this rationalisation procedure it is useful to be able to visualise the locations of the spectral classes. For this a bispectral plot can be constructed. The bispectral plot is not unlike a two dimensional scatter plot view of the multispectral space in which the data appears. However, rather than having the individual pixels shown, the class or cluster means are located according to their spectral components. In some exercises the most significant pair of spectral bands would be chosen in order to view the relative locations of the cluster centres. These could be bands 5 and 7 for a vegetation study involving Landsat MSS data such as that shown in Fig. 11.3. Alternatively, for MSS data it is common to use the average of bands 4 and 5 on one axis and the average of bands 6 and 7 on the other. In general, the choice of bands and combinations to use in a bi-spectral plot will depend on the sensor and application. Sometimes several plots with different bands will give a fuller appreciation of the distribution of classes in multispectral space.

11.5 Assessment of Classification Accuracy

At the completion of a classification exercise it is necessary to assess the accuracy of the results obtained. This will allow a degree of confidence to be attached to those results and will serve to indicate whether the analysis objectives have been achieved.

Accuracy is determined empirically, by selecting a sample of pixels from the thematic map and checking their labels against classes determined from reference data (desirably gathered during site visits). Often reference data is referred to as ground truth. From these checks the percentage of pixels from each class in the image labelled correctly by the classifier can be estimated, along with the proportions of pixels from each class erroneously labelled into every other class. These results are then expressed in tabular form, often referred to as a *confusion or error matrix*, of the type illustrated in Table 11.1. The values listed in the table represent the number of ground truth pixels, in each case, correctly and incorrectly labelled by the classifier. It is common to average the percentage of correct classifications and regard this the overall classification accuracy (in this case 83 %), although a better measure globally would be to weight the average according to the areas of the classes in the map.

Sometimes a distinction is made between errors of omission and errors of commission, particularly when only a small number of cover types is of interest, such as in the estimation of the area of a single crop in agricultural applications. Errors of omission correspond to those pixels belonging to the class of interest that the classifier has failed to recognise whereas errors of commission are those that correspond to pixels from other classes that the classifier has labelled as belonging to the class of interest. The former refer to columns of the confusion matrix, whereas the latter refer to rows.

When interpreting an error matrix of the type shown in Table 11.1 from the point of view of a particular class, it is important to understand that different indications of class accuracies will result according to whether the number of correct pixels for a class is divided by the total number of reference (ground truth) pixels for the class (the corresponding column sum in Table 11.1) or the total number of pixels the classifier attributes to the class (the row sum in Table 11.1). Consider class B in Table 11.1, for example. As noted, 37 of the reference data pixels have been correctly labelled. This represents $37/40 \equiv 93$ % of the ground truth pixels for the class. We interpret this measure, which Congalton (1991) refers to as the Producer's accuracy, as the probability that the classifier has labelled the image pixel as B given that the actual (ground truth) class is B. As a user of a thematic map produced by a classifier we are more interested in the probability that the actual class is B given that the pixel has been labelled B (on the thematic map) by the classifier. This is what Congalton refers to as the User accuracy, and for this example is $37/50 \equiv 74$ %. Thus only 74 % of the pixels labelled B on the thematic map are correct, even though the classifier coped with 93% of the B class reference data. This distinction is important and leads one to believe that the User accuracy is the figure that should most often be adopted.

Table 11.1. Illustration of a confusion matrix used in assessing the accuracy of a classification

| | | Ground truth classes | | | Total |
		A	B	C	
Thematic	A	35	2	2	39
map classes	B	10	37	3	50
	C	5	1	41	47
Number of ground truth pixels		50	40	46	136

Some authors prefer to use the kappa coefficient as a measure of map accuracy (Hudson and Ramm 1987, Congalton 1991). This is defined in terms of the elements of the error matrix; let these be represented by x_{ij}, and suppose the total number of test pixels (observations) represented in the error matrix is P. Also, let

$$x_{i+} = \sum_j x_{ij} \text{ (ie the sum over all columns for row } i)$$

$$x_{+j} = \sum_i x_{ij} \text{ (ie the sum over all rows for column } j)$$

then the kappa estimate is defined by

$$\kappa = \frac{P\sum_k x_{kk} - \sum_k x_{k+} x_{+k}}{P^2 - \sum_k x_{k+} x_{+k}}$$

Choice of the sample of pixels for accuracy assessment is an important consideration and still subject to investigation. Perhaps the simplest strategy for evaluating classifer performance is to choose a set of testing fields for each class, akin to the training fields used to estimate class signatures. These testing fields are also labelled using available reference data, presumably at the same time as the training areas. After classification the accuracy of the classifer is determined from its performance on the test pixels. Another approach, with perhaps more statistical significance since it avoids correlated near-neighbouring pixels, is to choose a random sample of individual pixels across the thematic map for comparison with reference data. A difficulty that can arise with random sampling in this manner is that it is area-weighted. That is, large classes tend to be represented by a larger number of sample points than the smaller classes; indeed some very small classes may not be represented at all. Assessment of the accuracy of labelling small classes will therefore be prejudiced. To avoid this it is necessary to ensure small classes are represented adequately. An approach that is widely adopted is *stratified random sampling* in which the user first of all decides upon a set of strata into which the image is divided. Random sampling is then carried out within each stratum. The strata could be any convenient area segmentation of the thematic map, such as gridcells. However the most appropriate stratification to use is the actual thematic classes themselves. Consequently, the user should choose a random sample within each thematic class to assess the classification accuracy of that class.

If one adopts random sampling, stratified by class, the question that must then be answered is how many test pixels should be chosen within each class to ensure that the results entered into the confusion matrix of Table 11.1 are an accurate reflection of the performance of the classifer, and that the percentage correct classification so-derived is a reliable estimate of the real accuracy of the thematic map. To illustrate this point, a sample of one pixel from a particular class will suggest an accuracy of 0% or 100% depending on its match to ground truth. A sample of 100 pixels will clearly give a more realistic estimate. A number of authors have addressed this problem, using binomial statistics, in the following manner.

Let the pixels from a particular category in a thematic map be represented by the random variable x that takes on the value 1 if a pixel is correctly classified and 0 otherwise. Suppose the true map accuracy for that class is θ (which is what we wish to estimate by sampling). Then the probability of x pixels being correct in a random sample of n pixels from that class is given by the binomial probability.

$$p(x; n, \theta) = {}^nC_x \theta^x (1-\theta)^{n-x} \quad x = 0, 1, \ldots n. \tag{11.1}$$

Van Genderen et al. (1978) determine the minimum sample size, by noting that if the sample is too small there is a finite chance that those pixels selected could all be labelled correctly (as for example in the extreme situation of one pixel considered above). If this occurs then a reliable estimate of the map accuracy clearly has not been obtained. Such a situation is described by $x = n$ in (11.1), giving as the probability for all n samples being correct

$$p(n; n, \theta) = \theta^n.$$

Van Genderen et al. have evaluated this expression for a range of θ and n and have noted that $p(n; n, \theta)$ is unacceptably high if it is greater than 0.05 – i.e. if more than 5% of the time there is a chance of selecting a perfect sample from a population in which the accuracy is actually described by θ. A selection of their results is given in Table 11.2. In practice, these figures should be exceeded to ensure representative outcomes are obtained. Van Genderen et al. consider an extension of the results in Table 11.2 to the case of encountering set levels of error in the sampling, from which further recommendations are made concerning desirable sample sizes.

Table 11.2. Minimum sample size necessary per category (after Van Genderen et al. 1978)

Classification accuracy	Sample size
0.95	60
0.90	30
0.85	20
0.80	15
0.60	10
0.50	5

Rosenfield et al. (1982) have also determined guidelines for selecting minimum sample sizes. Their approach is based upon determining the number of samples required to ensure that the sample mean – i.e. the number of correct classifications divided by the total number of samples per category – is within 10% of the population mean (i.e. the map accuracy for that category) at a 95% confidence level. Again this is estimated from binomial statistics, although using the cumulative binomial distribution. Table 11.3 illustrates the results obtained; while these results agree with Table 11.2 for a map accuracy of 85% the trends about this point are opposite.

This perhaps is not surprising since the two approaches commence from different viewpoints. Rosenfield et al. are interested in ensuring that the accuracy indicated from the samples (i.e. sample mean) is a reasonable (constant) approximation of the actual map accuracy. In contrast, Van Genderen et al. base their approach on ensuring that the set of samples is representative. Both have their merits and in practice one may wish to choose a compromise of between 30 and 60 samples per category.

Table 11.3. Minimum sample size necessary per category (after Rosenfield et al. 1982)

Classification accuracy	Sample size
0.85	19
0.80	30
0.60	60
0.50	60

Once accuracy has been estimated through sampling it is important to place some confidence on the actual figures derived for each category. In fact it is useful to be able to express an interval within which the true map accuracy lies (with say 95% certainty). This interval can be determined from the accuracy estimate for a class using the expression (Freund, 1992).

$$p\left\{-z_{\alpha/2} < \frac{x-n\theta}{\sqrt{n\theta(1-\theta)}} < z_{\alpha/2}\right\} = 1-\alpha \tag{11.2}$$

where x is the number of correctly labelled pixels in a sample of n; θ is the true map accuracy (which we currently are estimating in the usual way by x/n) and $1-\alpha$ is a confidence limit. If we choose $\alpha = 0.05$ then the above expression says that the probability that $(x-n\theta)/\sqrt{n\theta(1-\theta)}$ will be between $\pm z_{\alpha/2}$ is 95%; $\pm z_{\alpha/2}$ are points on the *normal* distribution between which $1-\alpha$ of the population is contained. For $\alpha = 0.05$, tables show $z_{\alpha/2} = 1.960$. Equation (11.2) is derived from properties of the normal distribution; however for a large number of samples (typically 30 or more) the binomial distribution is adequately represented by a normal model making (11.2) acceptable. Our interest in (11.2) is seeing what limits it gives on θ. It is shown readily, at the 95% level of confidence, that the extreme values of θ are given by

$$\frac{x+1.921 \pm 1.960 \left\{x(n-x)/n+0.960\right\}^{\frac{1}{2}}}{n+3.842} \tag{11.3}$$

As an illustration, suppose $x = 294$, $n = 300$ for a particular category. Then ordinarily we would use $\bar{x} = x/n = 0.98$ as an estimate of θ, the true map accuracy for the category. Equation (11.3) however shows, with 95% confidence, that our estimate of θ is bounded by

$$0.9571 < \theta < 0.9908.$$

Thus the accuracy of the category in the thematic map is somewhere between 95.7% and 99.1%.

This approach has been developed by Hord and Brunner (1976) who produced tables of the upper and lower limits on the map accuracy as a function of sample size and sample mean (or accuracy) $\bar{x} = x/n$.

11.6 Case Study 1: Irrigated Area Determination

It is the purpose of this case study to demonstrate a simple classification, carried out using the hybrid strategy of Sect. 11.4. Rather than being based upon iterative clustering and maximum likelihood classification it makes use of a single pass clustering algorithm of the type presented in Sect. 9.6 and a minimum distance classifier as described in Sect. 8.3.

The problem presented was to use classification of Landsat Multispectral Scanner image data to assess the hectarage of cotton crops being irrigated by water from the Darling River in New South Wales. This was to act as a cross check of area estimates provided by ground personnel of the New South Wales Water Resources Commission and the New South Wales Department of Agriculture. More details of the study and the presentation of some alternative classification techniques for this problem will be found in Moreton and Richards (1984), from which the following sections are adapted.

11.6.1 Background

Much of the western region of the state of New South Wales in Australia experiences arid to semi-arid climatic conditions with low average annual rainfalls accompanied by substantial evapotranspiration. Consequently, a viable crop industry depends to a large extent upon irrigation from major river systems. Cotton growing in the vicinity of the township of Bourke is a particular example. With an average annual rainfall of 360 mm, cotton growing succeeds by making use of irrigation from the nearby Darling River. This river also provides water for the city of Broken Hill further downstream and forms part of a major complex river system ultimately that provides water for the city of Adelaide, the capital of the state of South Australia. The Darling River itself receives major inflows from seasonal rains in Queensland, and in dry years can run at very low levels or stop flowing altogether, leading to increased salination of the water supplies of the cities downstream. Consequently, additional demands on the river made by irrigation must be carefully controlled. In New South Wales such control is exercised by the issue of irrigation licenses to farmers. It is then necessary to monitor their usage of water to ensure licenses are not infringed. This, of course, is the situation in many parts of the world where extensive irrigation systems are in use.

The water demand by a particular crop is very closely related to crop area, because most water taken up by a plant is used in transpiration (Keene and Conley, 1980). As a result, it is sufficient to monitor crop area under irrigation as an indication of water used. In this example, classification is used to provide crop area estimates.

11.6.2 The CSIRO-ORSER Image Analysis Software

The image analysis software used for this exercise was CSIRO-ORSER, a variant of the original ORSER package developed at Pennsylvania State University. At the time of the investigation the only output device available was a line printer. Notwithstanding the current ready availability of inexpensive workstations for image analysis and display, the original, simple outputs are still used in this case study.

11.6.3 The Study Region

A band 7 Landsat Multispectral Scanner image of the region considered in the study, consisting of 927 lines of 1102 pixels, is shown in Fig. 11.1. This is a portion of scene number 30704–23201 acquired on February 1980 (Path 99, Row 81). Irrigated cotton fields are clearly evident in the central left and bottom right regions, as is a further crop in the top right. The township of Bourke is just south of the Darling River, just right of the center of the image. The white border encloses a subset of the data, shown enlarged in Fig. 11.2. This smaller region was used for signature generation.

11.6.4 Clustering

Figure 11.2 shows the location of four regions selected for clustering using the single-pass algorithm. A fifth clustering region was chosen which partially included the triangular field in the bottom right region of Fig. 11.1. These regions consist of up to 500 pixels each and were selected so that a number of the irrigated cotton fields were included, along with a choice of most of the other major ground covers thought to be present. These include bare ground, lightly wooded regions, such as trees along the Darling River, apparently non-irrigated (and/or fallow) crop land, and a light coloured sand or soil.

Each of the regions shown in Fig. 11.2 was clustered separately. With the parameters entered into the clustering algorithm, each region generated between five and 11 spectral classes. The centres of the complete set of 34 spectral classes were then located on a bispectral plot. Generally, such a plot consists of the average of the visible

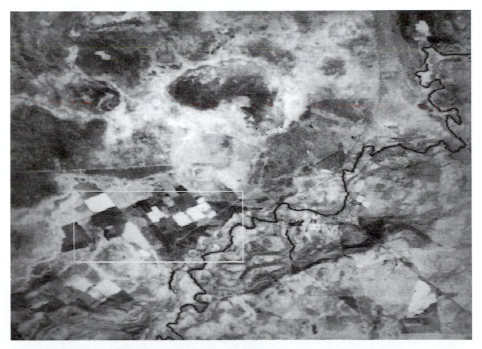

Fig. 11.1. Band 7 Landsat MSS image of the region of the investigation, showing irrigated fields (white). The area enclosed by the white border was used for signature generation. *Reproduced from Photogrammetric Engineering & Remote Sensing. Vol. 50, June 1984*

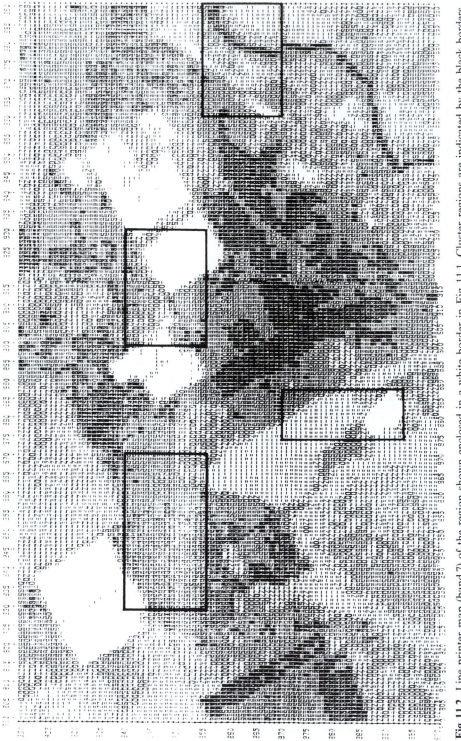

Fig. 11.2. Line printer map (band 7) of the region shown enclosed in a white border in Fig. 11.1. Cluster regions are indicated by the black borders.
Reproduced from Photogrammetric Engineering & Remote Sensing, Vol. 50, June 1984

components of the cluster means (Landsat bands 4 and 5) versus the average of the infrared components (bands 6 and 7). In this exercise, however, owing to the well-discriminated nature of the data, a band 5 versus band 7 bispectral plot was used; moreover, the subsequent classification also made use only of bands 5 and 7. This reduced the cost of the classification phase; however, the results obtained suggest that accuracy was not prejudiced. The band 5 versus band 7 bispectral plot showing the clustering results is illustrated in Fig. 11.3.

At this stage, it was necessary to rationalize the number of spectral classes and to associate spectral classes with ground cover types (so-called information classes). While a sufficient number of spectral classes must be retained to ensure classification accuracy, it is important not to have too many, because the number of class comparisons, and thus the cost of a classification, is directly related to this number. Because the classifier to be employed was known to be of the minimum distance variety, which implements linear decision surfaces between classes, spectral classes were merged into approximately circular groups (provided they were from the same broad cover type) as shown in Fig. 11.3. In this manner, the number of classes was reduced to ten. Labels were attached to each of those (as indicated in Fig. 11.3) by comparing cluster maps to black-and-white and color aerial photography, and to band 7 imagery. The

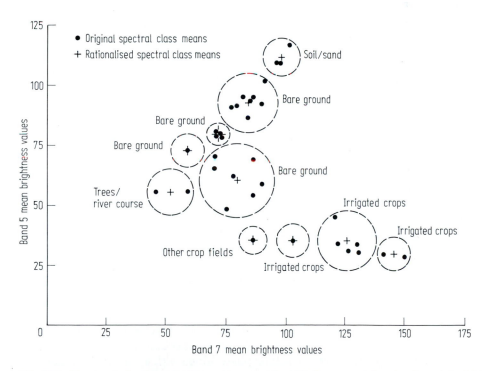

Fig. 11.3. Bispectral plot (band 5 class means versus band 7 class means) showing the original 34 cluster centers (spectral classes) generated. Also shown are the class rationalizations adopted. Original spectral classes within the dotted circles were combined to form a single class with mean positions indicated. The labels were determined from reference data and spectral response characteristics. *Reproduced from Photogrammetric Engineering & Remote Sensing, Vol. 50, June 1984*

relative band 5 and band 7 brightness values were also employed for class recognition; fields under irrigation were evident by their low band 5 values (30 on a scale of 255, indicating high chlorophyll absorption) accompanied by high band 7 reflectance (100 to 150, indicating healthy, well-watered vegetation).

11.6.5 Signature Generation

Signatures for the rationalized spectral classes were generated by averaging the means of the constituent original set of spectral classes. This was done manually, and is an acceptable procedure for the classifier used. Minimum distance classification makes use only of class means in assigning pixels and does not take any account of class covariance data. On the contrary, maximum likelihood classification incorporates both class covariance matrices and mean vectors as signatures, and merging of constituent spectral class signatures to obtain those for rationalized classes cannot readily be done by hand. Rather, a routine that combines class statistics is required. Such a procedure is available in MultiSpec (Landgrebe and Biehl, 1995). The rationalized class means are indicated in Fig. 11.3.

11.6.6 Classification and Results

With spectral class signatures determined as above, Fig. 11.1 was checked for crop fields that indicated use of irrigation. A classification map of the Fig. 11.2 (6,957 ha) region is shown in Fig. 11.4. Fields under irrigation are clearly discernible by their shape, as well as by their classification. By retaining several other ground-cover types as separate information classes (rather than giving them all a common symbol representing "non-irrigated"), other geometric features of interest are evident. For example, the Darling River is easily seen, as are some neighbouring fields that are not irrigated. This was useful for checking the results of the classification against maps and other reference data.

The results of the classification agreed remarkably well with ground-based data gathered by field officers of the New South Wales Water Resources Commission and the New South Wales Department of Agriculture. In particular, for a region of 169 651 pixels (75,000 ha) within Fig. 11.1, a measure of 803 ha given by the classifier as being under irrigation agreed to better than 1% with that given by ground data. This is well within any experimental error that could be associated with the classification and with the uncertainty regarding pixel size (in hectares), and is consistent with accuracies reported by some other investigators (Tinney et al., 1974).

11.6.7 Concluding Remarks

In general, the combined clustering/supervised classification strategy adopted works well as a means for identifying a reliable set of spectral classes upon which a classification can be based. The clustering phase, along with a construction such as a bispectral plot, is a convenient and lucid means by which to determine the structure of image data in multispectral space; this would especially apply for exercises that are as readily handled as those described here. The rationalized spectral classes used in this case correspond not so much to unimodal Gaussian classes normally associated with maximum likelihood classification, but rather are a set that match the characteristics of

Fig. 11.4 Classification map of the region of Fig. 11.2 generated using the ORSER software package. Class symbols used are: . irrigated crops; * irrigated crops; + other crop fields; × trees/river course; − soil/sand; · bare ground. *Reproduced from Photogrammetric Engineering & Remote Sensing, Vol. 50, June 1984*

the minimum distance classifier employed. This is an important general principle: the analyst should know the properties and characteristics of the classifier being used and, from a knowledge of the structure of the image, choose spectral class descriptions that match the classifier.

11.7 Case Study 2: Multitemporal Monitoring of Bush Fires

This case study demonstrates three digital image processing operations: image-to-image registration, principal components transformation and unsupervised classification. It entails the use of two Landsat multispectral scanner image segments of a region in the northern suburbs of the city of Sydney, New South Wales. The region is subject to damage by bush fires, and the images show fire events and revegetation in the region over a period of twelve months. Full details of the study can be found in Richards (1984) and Richards and Milne (1983).

11.7.1 Background

The principal components transformation developed in Chap. 6 is a redundancy reduction technique that generates a new set of variables with which to describe multispectral remote sensing data. These new variables, or principal components, are such that the first contains most of the variance in the data, the second contains the next major portion of variance and so on. Moreover, in these principal component axes the data is uncorrelated. Owing to this it has been used as a data transform to enhance regions of localised change in multitemporal multispectral image data (Byrne and Crapper 1979; Byrne et al., 1980; Ingebritsen and Lyon 1985; Fung and Le Drew 1987). This is a direct result of the high correlation that exist between image data for regions that do not change significantly and the relatively low correlation associated with regions that change substantially. Provided the major portion of the variance in a multitemporal image data set is associated with constant cover types, regions of localised change will be enhanced in the higher components of the set of images generated by a principal components transformation of the multitemporal, multispectral data. Since bushfire events will often be localised in image data of the scale of Landsat multispectral scanner imagery, the principal components transformation should therefore be of value as a pre-classification enhancement (and, as it transpires, as a feature reduction tool).

11.7.2 Simple Illustration of the Technique

Figure 11.5 shows the spectral reflectance data of healthy vegetation and vegetation damaged by fire, typical of that in the image data to be used below. As expected, the major effect is in the infrared region, corresponding to band 7 of the Landsat (1–3) MSS. To illustrate the value of principal components in highlighting changes associated with fire events suppose we consider just band 7 data from two dates. One date is prior to a fire and the other afterwards. We can construct a two date scatter diagram as shown in Fig. 11.6. Pixels that correspond to cover types that remain essentially constant between dates cluster about an elongate area as shown, representing water, vegetation and soils. Cover types that change between dates show as major departures from that general trend. For example pixels that were vegetated in the first date and burnt in the second lead to the off-diagonal cluster shown. Similarly pixels that appeared burnt in the first date and revegetated in the second appear as an off-diagonal cluster in the opposite direction.

Principal components analysis will lead to the axes shown in Fig. 11.6. As seen the band 7 variations associated with the localised changes project into both component axes. However the effect is masked in the first component by the large range of brightnesses associated with the near-constant cover types. By comparison the change effect in the second component will dominate since the constant cover types will map to a small range of brightness in the second principal component.

The same general situation occurs when all available bands of image data are used. However several of the higher order components will reflect local change information.

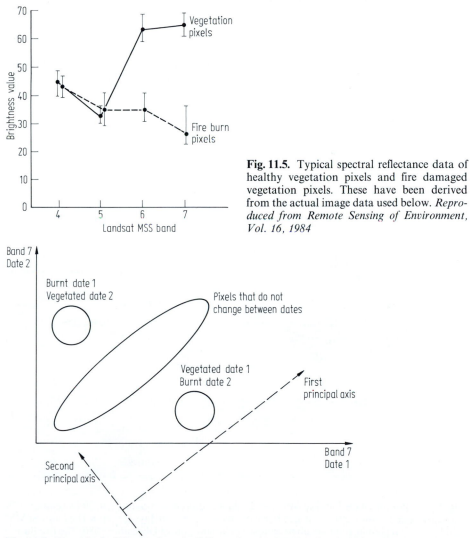

Fig. 11.5. Typical spectral reflectance data of healthy vegetation pixels and fire damaged vegetation pixels. These have been derived from the actual image data used below. *Reproduced from Remote Sensing of Environment, Vol. 16, 1984*

Fig. 11.6. Hypothetical illustration of a 2 dimensional 2 date Landsat MSS band 7 space, showing the dispersion of pixels associated with constant cover types and those that change between dates

11.7.3 The Study Area

In December 1979 the State of New South Wales in Australia experienced a number of major bushfires. While the majority of these were in mountain ranges to the west and northwest of the capital, Sydney, one particularly threatening fire occurred in Sydney's northern suburbs. Figure 11.7a shows a portion of a Landsat MSS image in this vicinity acquired on 29 December 1979. The area of bushland damaged by fire appears dark. The same region almost one year later (14 December 1980) is shown in Fig. 11.7b. The 1979 fire scar is diminished owing to partial vegetation recovery. However, two new fire burns are evident, as indicated, resulting from fires in the intervening period. The pair of images together therefore allow examination of vegetation to fire burn change and fire burn to revegetation.

11.7.4 Registration

The investigation was carried out using a Dipix Aries II image analysis system, supporting software that allows image to image registration to be performed via an interactive selection of control points in the two scenes. The task assists colocation of the control points by providing a sequential similarity detection algorithm (see Sect. 2.5.3).

Areas about two to three times larger than those shown in Fig. 11.7 were registered utilizing approximately 20 control points spaced near the scene circumference with a few scattered over the scene centre. The actual positions of some control points used are shown in Fig. 2.17a by comparison to the study area. Cubic polynomial mapping of the 1980 image to the 1979 image was performed. Resampling was based upon cubic convolution interpolation since the primary intention of the project was to examine

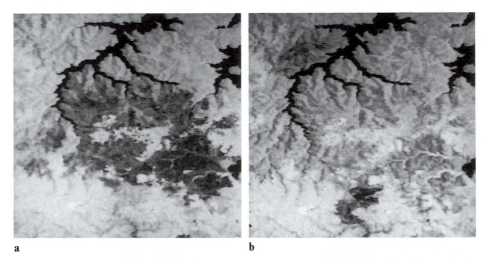

a   b

Fig. 11.7. a Portion of the Landsat MSS band 7 data acquired over Sydney on 29 December 1979. The dark region in the centre is an area burnt out by a fire several days earlier. **b** The Landsat MSS band 7 of the region in **a** but acquired almost 1 year later, on 14 December 1980. The fire burn is revegetating as evidenced by the developing light regions. Two new fire burns have occured. These show as dark regions on the north-western and southern regions of the image segment

principal components images by photointerpretation. The resulting average standard errors for the prediction of control points in the master image from those in the slave were less than $1/4$ pixel spacing in both row and column.

11.7.5 Principal Components Transformation

The registered subscenes were added to form an 8-dimensional multitemporal image data set (in the order 1979 band 4, 1979 band 5 ..., 1980 band 4, 1980 band 5 ...), from which the set of principal components was generated. Automatic polarization and scaling options were chosen in the transformation process as these gave component images with better visual dynamic range and it was felt that they would not prejudice subsequent interpretation at the level of discrimination envisaged (into major cover types and change classes but not into fine subdivision of vegetation species, etc.).

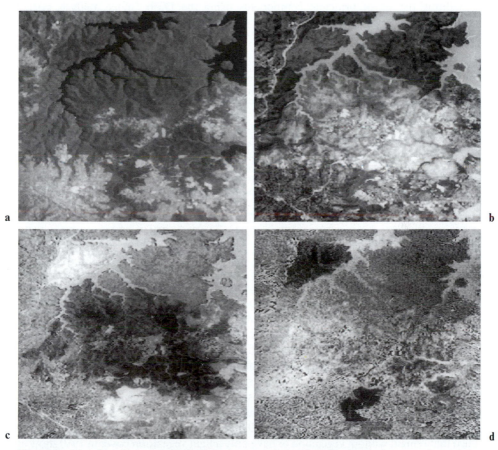

Fig. 11.8. The first four principal components of the 8-dimensional data set formed by concatenating the four Landsat MSS bands of the region of interest from each date. Components are numbered as **a** PC 1; **b** PC 2; **c** PC 3; **d** PC 4. Components 3 and 4 particularly highlight the fire-related events. *Reproduced from Remote Sensing of Environment, Vol. 16, 1984*

The first four of the eight principal component images are shown in Fig. 11.8. The remainder do not display any features of significance to the study. The first component is tantamount to a total brightness image, whereas the later components highlight changes. It is the second, third and fourth components that are most striking in relation to the fire features of interest. Pixels that have essentially the same cover type in both dates e. g., vegetation and vegetation, fire burn and fire burn, show as midgrey in the second, third and fourth components. Those that have changed, either as vegetation to fire burn or as fire burn to vegetation show as darker or brighter than midgrey, depending upon the component. These effects are easily verified by substituting typical spectral reflectance characteristics into the equations that generate the components. Each component is a linear combination of the original eight bands of data, where the weighting coefficients are the components of the corresponding eigenvector of the 8 × 8 covariance matrix. These eigenvectors along with their associated eigenvalues (which are the variances of the components) are shown in Table 11.4. In interpreting the fourth component it is necessary to account for a polarization inversion introduced by the options chosen in generating the set of principal components using the Dipix image analysis system software.

Table 11.4. Eigenvalues (variances) and eigenvectors of the 8-dimensional, original image data covariance matrix. The eigenvectors are the component weighting coefficients.

Component	Eigenvalue	Eigenvector							
1	1884	0.14	0.21	0.38	0.38	0.15	0.30	0.53	0.50
2	236	0.24	0.32	−0.21	−0.45	0.36	0.63	0.06	−0.25
3	119	0.24	0.21	0.49	0.46	0.07	0.08	−0.40	−0.53
4	19	−0.51	−0.58	−0.03	0.27	0.13	0.55	−0.04	−0.12
5	6	0.37	−0.50	0.07	−0.04	0.38	−0.30	0.49	−0.37
6	5	0.44	−0.14	−0.54	0.41	0.31	0.00	−0.37	0.32
7	4	−0.17	0.35	−0.52	0.45	−0.19	−0.05	0.42	−0.39
8	3	0.50	−0.29	−0.04	−0.02	−0.74	0.34	0.08	−0.04

The second principal component image expresses the 1979 fire burn as lighter than the average image tone, while the third principal component highlights the two fire burns. The 1979 burn region shows as darker than average whereas that for 1980 shows as slightly lighter than average. In the fourth component the 1980 fire burn shows as darker than average with the 1979 scar not evident. What can be seen, however, is revegetation in 1980 from the 1979 fire. This shows as lighter regions. A particular example is revegetation in two stream beds on the right-hand side of the image a litte over halfway down.

A colour-composite image formed by displaying the second principal component as red, the third as green, and the fourth as blue is shown in Fig. 11.9. This shows the area that was vegetated in 1979 but burned in 1980 as lime green; the regions from the 1979 burn that remain without vegetation or have only a light vegetation cover in 1980 show as bright red; revegetated regions in 1980 from the 1979 fire display as bright blue/purple whereas the vegetated, urban, and water backgrounds that remained essentially unchanged between dates show as dark green/grey.

Fig. 11.9. Colour composite multitemporal image formed by displaying the second principal component as red, the third principal component as green and the fourth principal component as blue

11.7.6 Classification of Principal Components Imagery

Because of the change enhancement offered in the principal components it should be possible to produce a change class thematic map by classification.

An initial unsupervised classification of the first four principal components produced substantial confusion between water/land and fire burn/vegetation. Owing to the nature of the first component, this is to be expected. A second test using components 2, 3, and 4 was acceptable, although some of the richly revegetated regions were unclassified. Consequently, it was decided to use just components 3 and 4 in the classification as a visual inspection indicates that they contain most of the class/change class information required.

The six major cover types were roadways, water, 1979 to 1980 fire burn, 1979 to 1980 revegetation, unchanged vegetation, and residual 1979 fire burn. Unchanged urban regions were not considered since resolution of these from other unchanged classes such as vegetation and water was not required. Maximum likelihood signatures for the six selected classes were generated. This left a significant proportion of what could be called "partial revegetation (1979 to 1980)" and "minor residual 1979 fire burn" unclassified. This situation was rectified by adding these two further classes as shown in the bispectral plot of Fig. 11.10.

The classification map of Fig. 11.11 was obtained using the eight subclasses, along with a likelihood threshold so chosen to avoid classification of regions for which signatures were not developed (such as urban). In the map only three major change classes are displayed, these being minor and major vegetation regeneration from 1979 to 1980 (the decision was made by inspection of the original standard colour composite images), the 1980 fire burn, and the residual bare effect from the 1979 fire. The latter is not strictly a change class for the pair of images considered but, nevertheless, was generated easily and is a significant class in the context of the study of fire damage and vegetation recovery. Notable in this classification is that there appears to be no confusion between burn and revegetating pixels, and water edge regions. The reason for

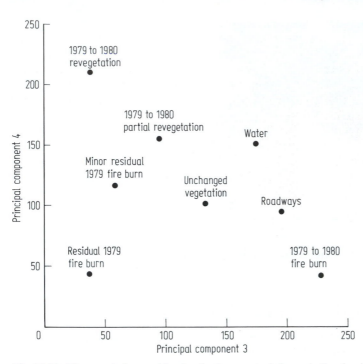

Fig. 11.10. Bispectral plot used for developing spectral classes in the classification of the third and fourth principal components into major fire-related themes. *Reproduced from Remote Sensing of Environment, Vol. 16, 1984*

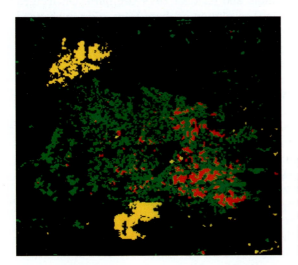

Fig. 11.11. Thematic map generated by classification of the third and fourth principal components. Only three classes are shown. These are vegetation regeneration (green), the 1980 fire burn (yellow) and the residual bare effect from the 1979 fire (red)

this is that the water edge pixels are approximately constant between dates (to the extent that tides are constant) and thus are correlated. They will map therefore to the midgrey constant background region of the higher order principal components, whereas the fire burn pixels (with which they can be confused) are vegetated in one date and burned in another and thus map to a quite different range of brightness in transformed imagery. Experience with single scene classification often shows water edge and fire burn confusion.

11.8 Hierarchical Classification

11.8.1 The Decision Tree Classifier

The classifiers treated in above have all been single stage in that only one decision is made about a pixel, as a result of which it is labelled as belonging to one of the available classes or is left unclassified. Multistage classification techniques are also possible in which a series of decisions is taken in order to determine the correct label for a pixel. Examples of such an approach are shown in Figs. 8.16 and 8.17. The more common multistage classifiers are called decision trees, examples of which are shown in Fig. 11.12. They consist of a number of connected classifiers (or decision nodes) none of which is expected to perform the complete segmentation of the image data set. Instead, each component classifier only performs part of the task, as noted in the figure. Perhaps the simplest type is the binary tree in which each component classifier, or node, is expected to perform a segmentation of the data into only one of two possible classes, or groups of classes.

The advantages of using a multistage or tree approach to classification include that different data sources, different sets of features, and even different algorithms can be used at each decision stage. Minimising the number of features to use in a decision is significant for reducing processing time and for improving the accurcy of small class training.

11.8.2 Decision Tree Design

Frequently, decision tree strategies can be designed manually, particularly when they are required to perform quite specific labelling tasks (Swain and Hauska, 1977). However, as with single stage classifier and neural network training it would be of value to have automated design procedures available.

There are three tasks in the design of a decision tree: finding the optimal structure for the tree, choosing the optimal subset of features at each node, and selecting the decision rule to use at each node. An optimal or suboptimal tree structure may aim for minimum error rate, a minimum number of nodes, or a minimum path length in deciding how to split classes at each node of the tree; consideration must be given also to means for controlling overlapping classes and for control of how many branches and layers to use.

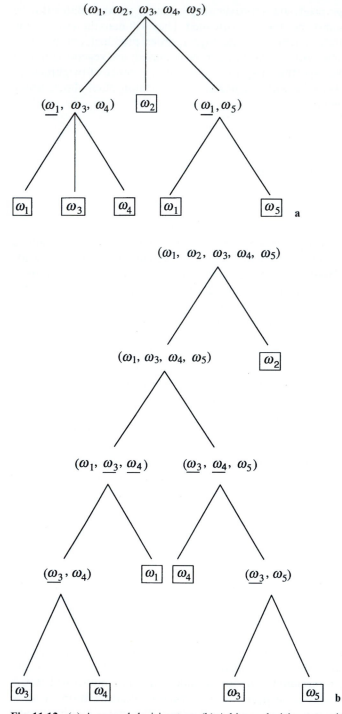

Fig. 11.12. (a) A general decision tree. (b) A binary decision tree with overlapping classes. (c) A binary tree without overlapping classes – underlines indicate class overlaps

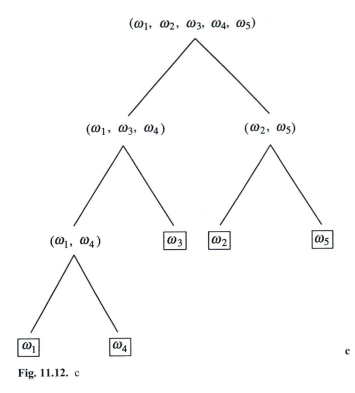

$(\omega_1, \omega_2, \omega_3, \omega_4, \omega_5)$

$(\omega_1, \omega_3, \omega_4)$ (ω_2, ω_5)

(ω_1, ω_4) $\boxed{\omega_3}$ $\boxed{\omega_2}$ $\boxed{\omega_5}$

$\boxed{\omega_1}$ $\boxed{\omega_4}$ c

Fig. 11.12. c

Since the number of possible tree structures, even for a moderately small number of classes, is astronomical, it is very difficult to design an optimal classifier (Mui and Fu, 1980). Classification accuracy and efficiency, however, rely heavily on the tree chosen. Therefore, various heuristic methods for decision tree design have been developed, details of which can be found in Safavian and Landgrebe (1991).

To make the design task easier, binary decision trees are often adopted. Discrimination ability is not necessarily weakened by choosing a binary approach, since a general decision tree can be uniquely transformed into an equivalent binary tree (Rounds, 1980).

One method for binary decision tree design is a "bottom up" approach similar to the agglomerative hierarchical clustering algorithm discussed in Sect. 9.7. If we replace the pixel data by class mean vectors, the bottom up method can be implemented by that process. Initially, the pairwise class separations are computed using a suitable distance metric, such as Euclidean distance. The two most similar classes are merged and a new mean is estimated for this combined data. This is continued until all the classes lie in a single, large class. The history of margins provides the inverse order of classes split in the decision tree.

Figure 11.13 shows the decision tree corresponding to the data given in Fig. 9.7. (The 9 pixels are treated as 9 class mean vectors.) As seen, the two most separable groups of classes are processed first, and the most subtle class pair will be discriminated at the bottom of the tree. By so doing, the cumulative error will also be minimised.

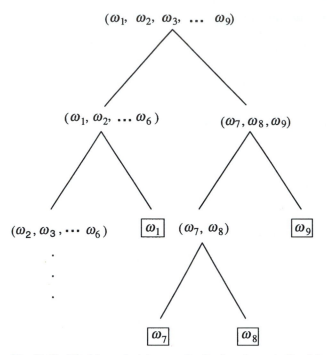

Fig. 11.13. The binary decision tree for the data shown in Fig. 9.7

This method assumes that the same set of features and the same classification algorithm are used at each decision node. A more general design philosophy is difficult to devise. However, analyst knowledge often helps in structuring a tree. For example, it is logical to separate data into water and croplands, and then croplands further into wheat, corn, etc. A user might also be able to use algorithm knowledge such as that minimum distance classification is preferred when small classes need to be identified. Moreover, some GIS data, e.g. elevation, can be segmented by a one dimensional parallelepiped algorithm.

11.8.3 Progressive Two-Class Decision Classifier

Figure 11.14 shows a progressive two-class decision classifier (Jia and Richards, 1998). Suppose there are six classes, represented by ω_a, ω_b, ω_c, ω_d, ω_e, and ω_f. The scheme focuses on one class pair at a time (at a node). The function of the first layer is to check the potential membership of an unknown pixel vector x to class ω_a and ω_b and the vector is classified temporarily as either class ω_a or class ω_b using the decision rule, D_{ab}. Class ω_b will be rejected for further consideration for those vectors labelled into the ω_a category, and class ω_a is rejected for further consideration for those vectors labelled into the ω_b category. At the second layer, there are two nodes, and two new class pairs (ω_a and ω_c for the left side node and ω_b and ω_c for the right side node) are considered, respectively. This process continues until a pure class labelling has been reached at the last layer, which is the final assignment.

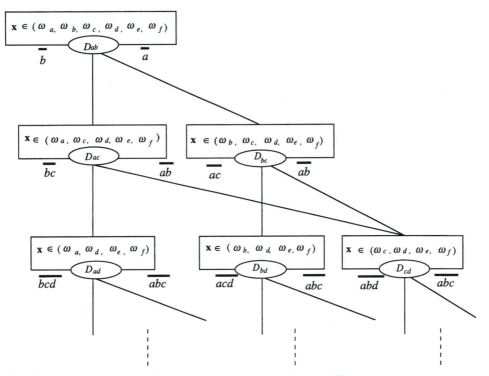

Fig. 11.14. A schematic chart for progressive two-class decision classifier

Since the class pair considered at each node is clear, one can concentrate on making decisions on which algorithm and which subset features to use for the particular class pair. An optimal environment for discriminating individual class pairs may result and thus maximum separation between them achieved.

11.8.4 Error Accumulation in a Decision Tree

A single layer classification (for example, maximum likelihood on the complete set of features to resolve the data into all the information classes in one step) can be represented by a binary decision tree. When a fixed set of features and decision rule are used at every node, the binary tree can in fact be shown to be identical to single layer classification (Mui and Fu, 1980). When optimal or suboptimal features and the most appropriate decision boundaries are used at each stage, classification performance might be improved as demonstrated by Swain and Hauska (1977), Iikura and Yasuoka (1991), Lee and Richards (1985) and Kim and Landgrebe (1991).

However, improved performance is not always achieved. Unfortunately with a decision tree there will be an accumulation of error, thereby requiring very good decisions at each node if acceptable classification errors are to be maintained at the terminal nodes. This can be seen in the following analysis for a binary tree.

Suppose the probability of error for both outcomes at every node of a binary tree is p. Although this is simplistic it serves to illustrate the problem. For a classification requiring only a single decision node then the error will be p.

However if two decision nodes are crossed in determining the labelling for a pixel then the (accumulated) error will be

$$p_E = p + (1-p)p = 2p - p^2 \qquad \text{(two decision nodes)}$$

The first term in this expression represents erroneous decisions from the previous node while the second is the probability of correct classification from the previous node $(1-p)$ multiplied by the probability of making an error on those correctly classified pixels. For a separation requiring three nodes of decision the accumulated error, proceeding in the same manner, will be

$$p_E = 2p - p^2 + (1 - 2p + p^2)p = 3p - 3p^2 + p^3 \qquad \text{(three decision nodes)}$$

Generalising, it can be shown that the error accumulated after N nodes of decision will be

$$p_E = 1 - (1 - p)^N$$

As an example, if $p = 10\%$, $N = 5$, the final error will be 59%. Thus, owing to the effect of error accumulation, it is critical to ensure very high classification accuracy at individual nodes to maintain a satisfactory accuracy at the end of the tree. Hopefully this effect will be obviated to an extent by the fact that the algorithm and features used at each node may be optimal or near optimal for the separation to be performed at the node.

A further reason as to why binary tree classifiers do not necessarily improve the correct recognition rate is that when some classes are merged into a group at a node, the decision boundaries become less specific in the discrimination between mixtures of classes (Landeweerd et al., 1983). Kim and Landgrebe (1991) point out that, if there were no Hughes phenomenon, the single-layer maximum likelihood classifier would have better performance than any decision tree classifier.

References for Chapter 11

A good discussion on choice of sampling strategies and means for determining reliable reference samples for assessing thematic map accuracy will be found in Stehman and Czaplewski (1998). Richards (1996) draws attention to the distinction between the performance of a classifier and the accuracy of the resulting thematic map, and notes conditions under which the two are the same.

Many of the international journals and conferences devoted to remote sensing technology contain case studies dealing with the use of satellite and aircraft acquired digital image data in a variety of applications. Journals that could be consulted include Remote Sensing of Environment, Photogrammetric Engineering and Remote Sensing, International Journal of Remote Sensing, ISPRS Journal of Photogrammetry and Remote Sensing, Geocarto and the IEEE Transactions on Geoscience and Remote Sensing. Relevant conferences include Purdue University's Symposia on Machine Processing of Remotely Sensed Data from the 1960's to the 1980's that place emphasis on digital processing techniques, the International Symposia on Remote Sensing of Environment run by the Environmental Research Institute of Michigan, and the IEEE International Geoscience and Remote Sensing Symposia.

The decision trees treated in this Chapter can be viewed as specific examples of the general class of layered classifiers (Nilsson, 1965). Another is a piecewise linear classifier, such as that proposed by Lee and Richards (1985), which is based upon a decision tree of the threshold logic units discussed in Sect. 8.10.1.

F.C. Billingsley, 1982: Modeling Misregistration and Related Effects on Multispectral Classification. Photogrammetric Engineering and Remote Sensing, 48, 421–430.

G.R. Byrne and P.F. Crapper, 1979: An Example of the Detection of Changes between Successive Landsat Images by Numerical Methods in an Urban Area. Proc. 1st Australasian Conference on Remote Sensing (Landsat '79), Sydney.

G.R. Byrne, P.F. Crapper and K.K. Mayo, 1980: Monitoring Land Cover Changes by Principal Components Analysis of Multitemporal Landsat Data. Remote Sensing of Environment, 10, 175–184.

R. Congalton, 1991: A Review of Assessing the Accuracy of Classifications of Remotely Sensed Data. Remote Sensing of Environment, 37, 35–46.

M.D. Fleming, J.S. Berkebile and R.M. Hoffer, 1975: Computer Aided Analysis of Landsat-1 MSS Data: A Comparison of Three Approaches including a Modified Clustering Approach. Information Note 072475, Laboratory for Applications of Remote Sensing, West Lafayette, Indiana.

J.E. Freund, 1992: Mathematical Statistics 5e, Englewood Cliffs, N.J., Prentice-Hall.

B.C. Forster and J.C. Trinder, 1984: An Examination of the Effects of Resampling on Classification Accuracy. Proc 3rd Australasian Conf. on Remote Sensing (Landsat '84), Gold Coast, Queensland, 106–115.

T. Fung and E. LeDrew, 1987: Application of Principal Components Analysis to Change Detection. Photogrammetric Engineering and Remote Sensing, 53, 1649–1658.

R.M. Hord and W. Brooner, 1976: Land-Use Map Accuracy Criteria. Photogrammetric Engineering and Remote Sensing, 42, 671–677.

W.D. Hudson and C.W. Ramm, 1987: Correct Formulation of the Kappa Coefficient of Agreement. Photogrammetric Engineering and Remote Sensing, 53, 421–422.

Y. Iikura and Y. Yasuoka, 1991: Utilisation of a Best Linear Discriminant Function for Designing the Binary Decision Tree. Int. J. Remote Sensing, 12, 55–67.

S.E. Ingebritsen and R.J.P. Lyon, 1985: Principal Components Analysis of Multitemporal Image Pairs: Int. J. Remote Sensing, 6, 687–696.

X. Jia and J.A. Richards, 1998: Progressive Two-class Decision Classifier for Optimization of Class Discriminations. Remote Sensing of Environment, 63, 289–297.

K.M. Keene and C.D. Conley, 1980: Measurement of Irrigated Acreage in Western Kansas from Landsat Images. Environmental Geology, 3, 107–116.

B. Kim and D.A. Landgrebe, 1991: Hierarchical Classifier Design in High-dimensional, Numerous Class Cases. IEEE Trans. on Geoscience and Remote Sensing, 29, 518–528.

G.H. Landeweerd, T. Timmers, E.S. Gelsema, M. Bins and M.R. Halie, 1983: Binary Tree Versus Single Level Tree Classification of White Blood Cells. Pattern Recognition, 16 571–577.

D.A. Landgrebe and L. Biehl, 1995: An Introduction to MultiSpec, West Lafayette, Purdue Research Foundation. (http://dynamo.ecn.purdue.edu/rbiehl/MultiSpec)

T. Lee and J.A. Richards, 1985: A Low Cost Classifier for Multitemporal Applications. Int. J. Remote Sensing, 6, 1405–1417.

G.E. Moreton and J.A. Richards, 1984: Irrigated Crop Inventory by Classification of Satellite Image Data. Photogrammetric Engineering and Remote Sensing, 50, 729–737.

J.K. Mui and K.S. Fu, 1980: Automated Classification of Nucleated Blood Cells Using a Binary Tree Classifier. IEEE Trans. Pattern Analysis and Machine Intelligence, PAMI-2, 429–443.

N.J. Nilsson, 1965: Learning Machines. N.Y., McGraw-Hill.

J.A. Richards and A.K. Milne, 1983: Mapping Fire Burns and Vegetation Regeneration Using Principal Components Analysis. Proc. 1983 Int. Geoscience and Remote Sensing Symposium. San Francisco.

J.A. Richards, 1984: Thematic Mapping from Multitemporal Image Data Using the Principal Components Transformation. Remote Sensing of Environment, 16, 35–46.

J.A. Richards, 1996: Classifier Performance and Map Accuracy. Remote Sensing of Environment, 57, 161–166.

G.H. Rosenfield, K. Fitzpatrick-Lins and H.S. Ling, 1982: Sampling for The Thematic Map Accuracy Testing. Photogrammetric Engineering and Remote Sensing, 48, 131–137.

E.M. Rounds, 1980: A Combined Nonparametric Approach to Feature Selection and Binary Decision Tree Design. Pattern Recognition, 12, 313–317.

S.R. Safavian and D.A. Landgrebe, 1991: A Survey of Decision Tree Classifier Methodology. IEEE Trans. on System, Man, and Cybernetics, 21, 660–674.

S.V. Stehman and R.L. Czaplewski, 1998: Design and Analysis for Thematic Map Accuracy Assessment: Fundamental Principles, Remote Sensing of Environment, 64, 331–344.

P.H. Swain and H. Hauska, 1977: The Decision Tree Classifier: Design and Potential. IEEE Trans. on Geoscience Electronics, GE-15, 142–147.

L.R. Tinney, J.E. Estes, K.H. Thaman and R.R. Thaman, 1974: Operational Use of Satellite and High Altitude Remote Sensing for the Generation of Input Data for Water Demand Models. Proc. 9th Int. Symp. on Remote Sensing of Environment, Michigan, 739–757.

J.L. Van Genderen, B.F. Lock and P.A. Vass, 1978: Remote Sensing: Statistical Testing of Thematic Map Accuracy. Remote Sensing of Environment, 7, 3–14.

J. Verdin, 1983: Corrected vs Uncorrercted Landsat 4 MSS Data. Landsat Data Users Notes, Issue No. 27, Sioux Falls, NOAA, June 4–8.

Problems

11.1 (a) What is the difference between an *information class* and a *spectral class?*

(b) Four analysts use different quantitative methods for analysing multispectral satellite data. These are summarised below. Comment on the merits and shortcomings of the four approaches and indicate which one you think is most effective.

Analyst 1

1. Chooses training data from homogeneous regions for each cover type.
2. Develops statistics for a maximum likelihood classifier.
3. Classifies image.

Analyst 2

1. Performs a clustering of the whole image and attaches labels to each cluster type afterwards.

Analyst 3

1. Chooses several regions within the image, each of which includes more than one cover type. Clusters each region.
2. Identifies the cluster types.
3. Uses statistics from the clustering process to perform a maximum likelihood classification of the whole image.

Analyst 4

1. Chooses training fields within apparent homogenous regions for each cover type. Clusters those regions to identify spectral classes.
2. Uses statistics from the clustering process to perform a maximum likelihood classification of the whole image.

(c) For the method you have identified in (b) as best, discuss how separability measures could be included to advantage.

11.2 The spectral classes used with the maximum likelihood decision rule in supervised classification are assumed to be representable by single multivariate normal probability distributions. Geometrically, this implies that they will have a hyperellipsoidal distribution in multispectral space. Do you think clustering by the iterative moving means algorithm will generate spectral classes of this nature? (See problem 9.2). You may care to extend this discussion by considering how best to generate spectral classes for maximum likelihood, minimum distance and parallelepiped classification. This concept is discussed in J.A. Richards and D.J. Kelly, 1984: On the Concept of Spectral Class, Int. J. Remote Sensing, 5, 987–991.

11.3 A maximum likelihood classifier can be developed using training data in the usual way by estimating class statistics. Describe how a threshold can be used to assist in the determination of the spectral class structure of the data.

11.4 Spaceborne microwave remote sensing depends necessarily on the use of synthetic aperture radar (SAR) techniques. SAR images of agricultural regions display a substantial "speckle" owing to the coherent nature of the radiation employed. Comment on the effect speckle would have in trying to obtain accurate automated classification of agricultural radar images.

11.5 This question relates to the effect of resampling on classification. Consider a single line of infrared image data, such as that corresponding, say, to Landsat 3 band 7 responses over a region that is vegetation to the left and water to the right. Imagine the vegetation/water boundary is sharp. Resample your single line of data onto a grid with the same centres as the original. However use both nearest neighbour and cubic convolution interpolation, the latter according to (2.11 a) with $j' = 1$. Comment on the results of classifying each of the resampled lines of data given that the classifier could have been trained on classes that have infrared brightnesses between those of vegetation and water.

11.6 Frequently texture is used as an element in the photointerpretation of airphotos or satellite images. It can be used, for example, in the discrimination of forested and grassy regions. When dealing with digital data using machine-assisted classificated methods, texture can only be used if a means for computing the texture in the neighbourhood of a pixel can be determined. A simplistic measure is the standard deviation of pixel brightnesses in a 3×3 neighbourhood about a pixel. Discuss how this texture measure can be incorporated into standard classification methods, noting any computational burdens involved.

11.7 Sometimes the spectral domain for a particular sensor and scene does consists of a set of distinct clusters of data. As an illustration, a Landsat visible red versus near infrared two dimensional space of an image of a region of just water, sand and mangrove vegetation would appear to have three groups of pixels. More often than not however, especially for images of natural vegetated and soil regions, the spectral domain will be very much a continuum, owing to the different degrees of mixing of the various cover types that can occur in nature. One is then led to question the distinctness and uniqueness, not only of spectral classes, but information classes as well for regions such as these. In view of these remarks comment on the issues involved in the classification of natural regions both in terms of the definition of the set of information classes to be used and in terms of the training procedures to be employed.

Chapter 12

Data Fusion

Frequently the need arises to analyse mixed spatial data bases, such as that depicted in Fig. 1.13. Such data sets could consist of satellite spectral, topographic and other point form data, all registered geometrically, as might be found in a geographic information system.

Labelling pixels by drawing inferences from several available sources of data is referred to as *data fusion*. As with the treatment of single image data sets, analysis of mixed data types can be carried out photointerpretatively or by using machine analysis.

Sometimes data fusion is relatively straightforward, particularly if the different data sources are substantially of the same type and thus can be handled by the same sorts of photointerpreter knowledge or machine algorithm. In many cases, though, the problem is complex, especially when the analyst wishes to apply quantitative methods to data types that are quite different from each other. Manipulating satellite multispectral data with labelled map data is an example.

It is the purpose of this chapter to present some of the more useful techniques for addressing the interpretation task quantitatively. Analysis by photointerpretation is not treated as such, since it depends largely on the analyst's skills with the range of data types present. Improving image quality for photointerpretative data fusion, however, is discussed by Gross and Schott (1997) and van der Meer (1997).

Numerically based quantitative methods are treated first, following which procedures based on evidential theory and expert systems are covered. The benefit of the latter is that the data sources do not need to be all in numerical form.

Clearly, the data to be analysed must first be registered. If the data has been retrieved from a geographic information system then that step will already have been performed. However, if spatial registration has not been carried out then the analyst will have to undertake that task using the procedures of Chapter 2. A word of caution is in order: the accuracy with which the interpretation of the mixed data can be performed will be influenced by the accuracy of the registration process as well as the effectiveness of the analytical procedures (such as classification) employed.

12.1 The Stacked Vector Approach

A straightforward way to classify mixed data is to form extended pixel vectors by stacking together the individual vectors that describe the various spectral and non-spectral data. This stacked vector will be of the form

$$X = [x_1{}',x_2{}', \ldots , x_J{}']^t \tag{12.1}$$

where J is the total number of individual data sources with corresponding data vectors $x_1 .. x_J$, and the superscript "t" denotes a vector transpose operation. The stacked vector X can, in principle, now be approached using standard classification techniques. This presents a number of difficulties if statistical methods such as maximum likelihood classification are considered. These include the incompatible statistics of the disparate data types, with some data unable to be represented by normal class models, and the quadratic cost increase with data dimensionality. Parallelepiped classification could be an appropriate algorithm to adopt since it depends only on the application of thresholds to components of the data vector X.

12.2 Statistical Methods

The single data source decision rule of (8.1) can be restated, for data describable by the extended vector of (12.1), as

$$X \in \omega_i \text{ if } p(\omega_i|X) > p(\omega_j|X) \text{ for all } j \neq i$$

As in single source methods, if the posterior probabilites $p(\omega_i|X)$ etc. are known, or can be derived from an application of Bayes' theorem, it should be possible to generate a practical decision rule. This is not straightforward, and requires assumptions of independence among the sources. These and other details of implementation will be found in Lee, Richards and Swain (1987).

Benediktsson et al (1997) also present an implementation based on statistical modelling of each of the data sources independently. A joint inference is derived from the data specific posterior probabilities using consensus theory in which weighted arithmetic or geometric (logarithmic) averages of the single data sources recommendations form an "opinion pool". The single source recommendations are weighted by a set of numbers that control the relative influences of the sources (based for example on an impression of the goodness of each data source, including its goodness in relation to specific classes being analysed).

If there are only two sources of data, such as a multispectral image data set and some pixel-specific collateral or ancillary data, several statistical approaches are possible. One involves the use of the prior probabilities in the maximum likelihood algorithm as in (8.3) and (8.7). While prior probabilities are usually assumed to express the likelihood of class membership before spectral data is utilised in the maximum likelihood rule, they can be thought of more generally as the probability of class membership as expressed from any source other than spectral. Strahler (1980) has used this rationale to carry the effect of elevation data into the classification of multispectral imagery of forested regions. Bruzzone et al (1997) have also adopted this approach.

A second method for handling multispectral and ancillary data involves a modification of the probabilistic label relaxation rule of Sect. 8.8.4. In a variation known as supervised relaxation (Richards, Landgrebe and Swain, 1982) the label probabilities at each iteration, after having been updated by reference to the spatial neighbourhood function, are also modified by reference to the likelihoods of class membership derived from the ancillary data. Iteration then develops consistency among the available spectral, spatial and ancillary information.

12.3 The Theory of Evidence

A restriction with the previous methods for handling multisource data is that all the data must be in numerical form. Yet many of the data types encountered in a spatial data base are inherently non-numerical. The mathematical Theory of Evidence is a field in which the data sources are treated separately and their contributions combined to provide a joint inference concerning the correct label for a pixel, but does not, of itself, require the original data variables to be numerical. While it involves numerical manipulation of quantitative measures of evidence, the bridge between these measures and the original data is left largely to the user.

12.3.1 The Concept of Evidential Mass

The essence of the technique involves the assignment of a so-called mass of evidence to various labelling propositions for a pixel. The total mass of evidence available for allocation over the candidate labels for the pixel is unity. To see how this is done suppose a classification exercise, involving for the moment just a single source of image data, has to label pixels as belonging to one of just three classes: ω_1, ω_2 and ω_3. It is important that the set of classes be exhaustive (i.e. cover all possibilities) so that ω_3 for example might be the class "other". Suppose some means is available by which labels can be assigned to a pixel (which could even include maximum likelihood methods if desired) which tells us that the three labels have likelihoods in the ratios $2 : 1 : 1$. However, suppose we are a little uncertain about the labelling process or even the quality of the data itself, so that we are only willing to commit ourselves to classifying the pixel with about 80 % confidence. Thus we are about 20 % uncertain about the labellings, even though we are reasonably happy about the relative likelihoods. Using the symbolism of the Theory of Evidence, the distribution of the unit mass of evidence over the three possible labels, and our uncertainty about the labelling, is expressed:

$$m(< \omega_1, \omega_2, \omega_3, \theta >) = \; < 0.4, 0.2, 0.2, 0.2 > \tag{12.2}$$

where the symbol θ is used to signify the uncertainty in the labelling[1]. Thus the mass of evidence assigned to label ω_1 as being correct for the pixel is 0.4, etc. (Note that if we were using straight maximum likelihood classification, without accounting for uncertainty, the probability that ω_1 is the correct class for the pixel would have been 0.5.) We now define two further evidential measures. First, the *support* for a labelling proposition is the sum of the mass assigned to the proposition and any of its subsets. Subsets are considered later. The *plausibility* of the proposition is one minus the total support of any contradictory propositions. Support is considered to be the minimum amount of evidence in favour of a particular labelling for a pixel whereas plausibility is the maximum possible evidence in favour of the labelling. The difference between the measures of plausibility and support is called the *evidential interval*; the true likelihood that the label under

---

[1] Strictly, in the Theory of Evidence, θ represents the set of all possible labels. The mass associated with uncertainty has to be allocated somewhere; thus it is allocated to the set as a whole.

consideration is correct for the pixel is assumed to lie somewhere in that interval. For the above example, the supports, plausibilities and evidential intervals are:

$$s(\omega_1) = 0.4 \qquad p(\omega_1) = 0.6 \qquad p(\omega_1) - s(\omega_1) = 0.2$$

$$s(\omega_2) = 0.2 \qquad p(\omega_2) = 0.4 \qquad p(\omega_2) - s(\omega_2) = 0.2$$

$$s(\omega_3) = 0.2 \qquad p(\omega_3) = 0.4 \qquad p(\omega_3) - s(\omega_3) = 0.2$$

In this simple case the evidential intervals for all labelling propositions are the same and equal to the mass allocated to the uncertainty in the process or data as discussed above, i.e. $m(\theta) = 0.2$. Consider another example involving four possible spectral classes for the pixel, one of which represents our belief that the pixel is in either of two classes. This will demonstrate that, in general, the evidential interval is different to the mass allocated to uncertainty. Suppose the mass distribution is:

$$m(<\omega_1, \omega_2, \omega_1 \vee \omega_2, \omega_3, \theta>) = \; <0.35, 0.15, 0.05, 0.3, 0.15>$$

where $\omega_1 \vee \omega_2$ represents ambiguity in the sense that, for the pixel under consideration, while we are prepared to allocate 0.35 mass to the proposition that it belongs to class ω_1 and 0.15 mass that it belongs to class ω_2, we are prepared also to allocate some additional mass to the fact that it belongs to either of those two classes and not any others.

For this example the support for ω_1 is 0.35 (being the mass attributed to it) whereas the plausibility that ω_1 is the correct class for the pixel is one minus the support for the contradictory propositions. There are two – i.e. ω_2 and ω_3. Thus the plausibility of ω_1 is 0.55, and the corresponding evidential interval is 0.2 (different now from the mass attributed to uncertainty). The support given to the mixture class $\omega_1 \vee \omega_2$ is 0.55, being the sum of masses attributed to that class and its subsets.

To see how the Theory of Evidence is able to cope with the problem of multisource data, return now to the simple example given by the mass distribution in (12.2). Suppose there is available a second data source which is also able to be labelled into the same set of spectral classes. Again, however, there will be some uncertainty in the labelling process which can be represented by a measure of uncertainty as before; also, for each pixel there will be a set of likelihoods for each possible label. For a particular pixel suppose the mass distribution after analysing the second data source is

$$\mu(<\omega_1, \omega_2, \omega_3, \theta>) = \; <0.2, 0.45, 0.3, 0.05> \tag{12.3}$$

Thus, the second analysis seems to be favouring ω_2 as the correct label for the pixel, whereas the first data source favours ω_1. The Theory of Evidence now allows the two mass distributions to be merged in order to combine the evidences and thus come up with a label which is jointly preferred and for which the overall uncertainty should be reduced. This is done through the mechanism of the orthogonal sum.

12.3.2 Combining Evidence – the Orthogonal Sum

Dempster's orthogonal sum is illustrated in Fig. 12.1. It is performed by constructing a unit square and partitioning it vertically in proportion to the mass distribution from one source and horizontally in proportion to the mass distribution from the other source. The areas of the rectangles thus formed are calculated. One rectangle is formed from the masses attributed to uncertainty (θ) in both sources; this is considered to be the remaining uncertainty in the labelling process after the evidences from both sources have been combined. Rectangles formed from the masses attributed to the same class have their resultant (area) mass assigned to that class. Rectangles formed from the product of mass assigned to a particular class in one source and mass assigned to uncertainty in another source have their resultant mass attributed to the specific class. Similarly, rectangles formed from the product of a specific label, say ω_2, and an ambiguity, say $\omega_1 \vee \omega_2$, are allocated to the specific class. Rectangles formed from different classes in the two sources are contradictory and are not used in computing merged evidence. In order that the resulting mass distribution sums to unity a normalising denominator is computed as the sum of the areas of all the rectangles that are not contradictory. For the current example this factor is 0.47. Thus, after the orthogonal sum has been computed the resulting (combined evidence) mass distribution is:

$$m(\omega_1) = (0.08 + 0.02 + 0.04)/0.47 = 0.298$$

$$m(\omega_2) = (0.09 + 0.01 + 0.09)/0.47 = 0.404$$

$$m(\omega_3) = (0.06 + 0.01 + 0.06)/0.47 = 0.277$$

$$m(\theta) = 0.01/0.47 \qquad\qquad = 0.021$$

Thus class 2 is seen to be recommended jointly. The reason for this is that source 2 favoured class 2 and had less uncertainty. While source 1 favoured class 1, its higher level of uncertainty meant that it was not as significant in influencing the final outcome.

The orthogonal sum can also be expressed in algebraic form (Lee et al 1987, Garvey et al 1981) which would be incorporated in a software implementation of evidental methods for multisource analysis. If two mass distributions are denoted m_1 and m_2 then their orthogonal sum is:

$$m_{12}(z) = \kappa \sum_{(x \cap y = z)} m_1(x).m_2(y)$$

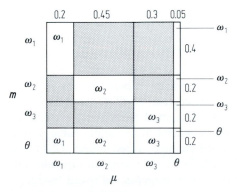

Fig. 12.1. Graphical illustration of the application of the Dempster orthogonal sum for merging the evidences from two data sources

where

$$\kappa^{-1} = \sum_{(x \cap y \neq \varphi)} m_1(x).m_2(y)$$

in which φ is the null set. In applying these formulas it is important to recognise that

$$(x \vee y) \cap y = y$$

$$\theta \cap y = y$$

For more than two sources, the orthogonal sum can be applied repetitively since the expression is both commutative (the order in which the sources are considered is not important) and associative (can be applied to any pair of sources and then a third source, or equivalently can be applied to a different pair and a third source).

12.3.3 Decision Rule

After the orthogonal sum has been applied the user can then compute the support for and plausibility of each possible ground cover class for a pixel. Two following steps are then possible. First a decision rule might be applied in order to generate a single thematic map in the same manner as is done with statistical classification methods.

A number of candidate decision rules are possible including a comparison of the supports for the various candidate classes and a comparison of plausibilities as discussed in Lee et al (1987). Generally a maximum support decision rule would be used, although if the plausibility of the second most favoured class is higher than the support for the preferred label, the decision must be regarded as having a degree of risk.

Rather than produce a single thematic map, it is possible to produce a map for each category showing the distribution of supports (or plausibilities). This might be particularly appropriate in a situation where the ground cover classes are not well resolved (such as in a geological classification, for an illustration of which see Moon (1990)).

12.4 Knowledge-Based Image Analysis

Techniques for the analysis of mixed data types, such as the multisource statistical classification and evidential methods treated above, have their limitations. Apart from their complexities, most are restricted to data that is inherently in numerical form, such as that from multispectral and radar imaging devices, along with quantifiable terrain data like digital elevation maps. Yet, in the image data base of a Geographic Information System (GIS), for example, there are many spatial data sets that are non-numerical but which would enhance considerably the results expected from an analysis of a given geographical region if they could be readily incorporated into the decision process. These include geology and soil maps, planning maps and even maps showing power, water and road networks. It is clear therefore that quite a different approach for handling non-numerical data is required, particularly when a user wishes to exploit the richness of information imbedded in the multisource, multisensor data environment of a GIS. The Theory of Evidence treated in Sect. 12.3 is one possibility, but it still requires the analysis task to be expressed in a quantifiable form so that numerical manipulation of evidence is possible. To avoid having to establish this bridge, a method for *qualitative* reasoning would be a particular value.

The adoption of expert systems or knowledge-based methods offers promise in this regard. It is the role of this section to outline some of the fundamental aspects of such processes and to demonstrate their potential. The field is very diverse and, as will become clear in reading the following, the use of one particular approach may be guided by individual preferences and available software rather than a perception of what is the most appropriate algorithm for a given purpose. What will become clear however is that the use of (often qualitative) interpreter knowledge greatly aids analysis; moreover, quite simple knowledge-based methods can yield surprisingly good results.

12.4.1 Knowledge Processing: Emulating Photointerpretation

To develop the theme of a knowledge-based approach it is of value to return to the comparison of the attributes of photointerpretation and quantitative analysis developed in Table 3.1. However, rather than making the comparison solely on the basis of a single source of multispectral data, as was the case in Chapter 3, consider now that the data to be analysed consists of three parts: a Landsat multispectral image, a radar image of the same region and a soil map of that region. From what has been said above, standard methods of quantitative analysis cannot cope with trying to draw inferences about the cover types in the region since they will not function well on two numerical sources of quite different characteristics (multispectral and radar data) and also since they cannot handle non-numerical data at all.

In contrast, consider how a skilled photointerpreter might approach the problem of analysing this multiple source of spatial data. Certainly he or she would not wish to work at the individual pixel level, as discussed in Sect. 3.1, but would more likely concentrate on regions. Suppose a particular region was observed to have a predominantly pink tone on a standard false colour composite print of the multispectral data, leading the photointerpreter to infer initially that the region is vegetated; whether it is a grassland, crop or forest region may not yet be clear. However the photointerpreter could then refer to the radar imagery. If its tone is dark, then the region would be thought to be almost smooth at the radar wavelength being used. Combining this evidence with that from the multispectral source, the photointerpreter is then led to consider the region as being either grassland or a small crop. He or she might then resolve this conflict by referring to the soil map of the region. Noting that the soil type is not that normally associated with agriculture, the photointerpreter would then conclude that the region is some form of natural grassland.

In practice the process of course may not be so straightforward, and the photointerpreter may need to refer backwards and forwards over the data sets in order to finalise an interpretation, especially if the multispectral and radar tones were not uniform for the region. For example, some spots on the radar imagery may be bright. The photointerpreter would probably regard these as indicating shrubs or trees, consistent with the overall region being labelled as natural grassland. The photointerpreter will also account for differences in data quality, placing most reliance on data that is seen to be most accurate or most relevant to a particular exercise, and weighting down unreliable or marginally relevant data.

The question we need to ask at this stage is how the photointerpreter is able to make these inferences so easily. Even apart from spatial processing, as discussed in Table 3.1 (where the photointerpreter would also use spatial clues such as shape and texture), the key to the photointerpreter's success lies in his or her *knowledge* – knowledge about spectral reflectance characteristics, knowledge of radar response and also of how to combine the information from two or more sources (for example, pink multispectral appearance *and* dark radar tone indicates a low level vegetation type). We are led therefore to consider whether the knowledge possessed by an expert such as a skilled photointerpreter can be given to and used by a machine and so devise a method for analysis that is able to handle the varieties of spatial data type available in GIS-like systems. In other words can we emulate the photointerpreter's approach? If we can then we will have available an analytical procedure capable of handling mixed data types, and also able to work repetitively, at the pixel level if necessary. With respect to the latter point, it is important to recognise that photointerpreters generally work at a regional rather than a pixel level; knowledge-based image analysis is able to follow such an approach if segments in image data have previously been identified using region growing techniques such as that used in ECHO (Sect. 8.8.2).

12.4.2 Fundamentals of a Knowledge-Based Image Analysis System

12.4.2.1 Structure

If we were to visualise the structure of a traditional supervised classification approach to the analysis of image data we might come up with the block diagram shown in Fig. 12.2 a. The data to be analysed is fed to a processor (computer) which is also supplied with the algorithms (maximum likelihood rule, minimum distance rule etc.) appropriate to the task. The algorithms are applied pixel by pixel to produce a label for

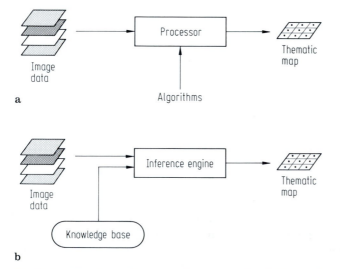

Fig. 12.2. a Traditional image analysis computing. **b** Knowledge based image analysis system.

each pixel, dependent solely on the class signatures and the characteristics of the data. It can be argued that some expert knowledge has been supplied to the process by the user in relation to the selection of algorithm to use and, more particularly, in selecting the reference data with which to train the classifier. The user, however, need not possess any detailed knowledge of spectral reflectance characteristics or other properties in order that the analysis proceed and results be produced. As we have seen in previous chapters quite good results can be achieved provided only that training regions are chosen carefully and any multimoding is removed when using maximum likelihood classification.

In contrast, Fig. 12.2b shows the structure of a knowledge-based approach to the analysis. Again, the spatial data to be analysed is fed to the processor, but so is a knowledge base. The knowledge stored in this knowledge base has been obtained from experts in the field of the analysis and stored in such a manner (see Sect. 12.4.2.2) that it can be used to analyse the data. The knowledge is applied to the data in the processor by what is called an inference mechanism, or sometimes an inference engine. This is rather simpler in principle than the full software packages one might expect to find in a remote sensing image analysis system; its role is to interpret the knowledge base, apply the knowledge to the data, and make, and keep track of, decisions about the class memberships of pixels.

12.4.2.2 Representation of Knowledge: Rules

There are several ways in which expert knowledge can be captured and recorded for use by a knowledge-based analysis system (Sell, 1985; Frost, 1986). The simplest, and perhaps most common, is to use rules (sometimes called production rules). These are of the form:

> **if** condition **then** inference.

'Condition' in the rule is a logical expression which can be either true or false. If it is true then the inference is justified otherwise no information is provided by that rule. 'Condition' can be a *simple* logical expression or can be a *compound* logical statement in which several components are linked through the logical **or** and **and** operations. These operations are defined as:

> The composite condition (condition 1 **and** condition 2) is true only if condition 1 and condition 2 are **both** true.

> The composite condition (condition 1 **or** condition 2) is true if **either** condition 1 **or** condition 2 is true.

Note also that rule-based knowledge systems can also make use of the logical **not** operation, defined by:

> **not** (condition) is false if condition is true, and vice versa.

Each single rule can be thought of as encapsulating one item of knowledge. For example, a rule which could be applied to a Landsat MSS pixel to check whether it is likely to be vegetated might be:

> **if** band 7 > band 5 **then** vegetation.

Although this is a weak rule, we know it is correct from our knowledge of the spectral reflectance characteristics of vegetation. Similarly, a rule that would reveal a region to be a smooth (specular or near specular) surface could be:

> **if** radar tone is dark **then** smooth surface.

Note that these rules need not be conclusive, but rather they should simply provide a degree of evidence in favour of pixels having the labels specified. Sometimes the knowledge contained in several rules might be necessary to enable a pixel to be identified with any degree of certainty.

A knowledge base in such an analysis system might contain many hundreds of rules of these types, obtained from experts in particular fields. When image data is presented to the inference engine for analysis, the engine goes through the rule base checking the support for or against various labelling propositions. Some rules will offer strong support while others will be weak, as illustrated above. Also, several candidate classes for a particular pixel might find support among the rules; procedures are then required for resolving among them. Possible means for doing this are described in the following sections.

As an example of a simple rule representation of knowledge, suppose a particular Landsat MSS image has to be segmented into just vegetation, water and other (unspecified) cover types. The following set of rules should be able to accomplish this task:

> **if** band 7/band 5 > threshold **then** vegetation
> **if** band 7/band 4 < 1 **then** water
> **if not** (water) **and not** (vegetation) **then** other

Notice that the third rule supposes for this particular exercise that anything that is not water or vegetation must be other. Also note that this rule has two conditions (sometimes called antecedents) that are logically 'and-ed'. Both must be true in order that the total antecedent is true and thus the inference (sometimes called the consequent) is justified. In the first rule a parameter is used – i. e. 'threshold'. This requires a numerical value to be available, which will almost certainly be scene dependent. The value could be provided to the system before the analysis starts by the user entering it manually or, alternatively, a small training region of vegetation could be used from which the value could be learnt. Many of the rules encountered in remote sensing image analysis will require parameters such as thresholds.

The rules illustrated here, and indeed most of those to be encountered in this treatment of knowledge-based methods, rely on spectral or similar pixel-specific knowledge. In many expert systems devised for the analysis of remote sensing and GIS data, spatial constraints are also used as a source of knowledge and appropriate rules are developed (Ton et al., 1991). Even spectrally derived rules may not rely on simple expressions and comparisons of bands. Spectral contrasts, such as the brightness in a given band compared with total image brightness, can also be used (Wharton, 1987).

12.4.2.3 The Inference Mechanism

The inference engine or mechanism can be quite simple if the knowledge-based system is very specific to a particular application, or can be more complex and powerful if a general expert system is required. In the simple example of the previous section all the inference mechanism has to do is to check which of the rules gives a positive response for each pixel in the image and then label the pixel accordingly. More generally, however, when large rule sets are used, the inference mechanism needs to keep track of all the rules that infer a particular cover type, along with those that infer that the pixel is not of that cover type and, similarly, the rules that suggest the pixel is or is not from other candidate classes; finally it has to make a decision about the correct class by weighing all the evidence from the rules. It may also have to account for redundant reasoning and circular arguments, and has to be able to assess whether long reasoning chains carry as much weight in the decision process as inferences that might involve only a single decision in coming up with the label for a pixel. In addition, an effective inference process will also allow uncertainties in data quality, missing data and missing rules to be accommodated. That degree of complexity is beyond this introductory treatment; a full discussion of all of these issues will be found in Srinivasan (1990) and Srinivasan and Richards (1993). However, it is of value to consider briefly something of the complexity that can be built into the inference mechanism in order to emulate more closely the reasoning process that might be adopted by a typical photointerpreter. To do this, it is instructive first to consider the approach used by Wharton (1987).

Wharton uses eight bands of Thematic Mapper Simulator data, centered on:

band 1	$0.485\,\mu m$	band 2	$0.560\,\mu m$
band 3	$0.660\,\mu m$	band 4	$0.880\,\mu m$
band 5	$1.150\,\mu m$	band 6	$1.650\,\mu m$
band 7	$2.215\,\mu m$	band 8	$11.400\,\mu m$

He then establishes spectral rule sets for each of his classes of interest. These rules are in three groups, one of which compares band combinations. For example, a rule for determining whether a pixel might belong to the green vegetation class is:

> **if** average of bands 4 and 5 > sum of bands 2 and 3
> **then** class is green vegetation [5, 20]

The figures in square brackets are measures of evidence in favour of and against the labelling proposition. Thus, when testing for the green vegetation category, if the test is positive then the evidence in support of this being the correct class for the pixel is incremented by 5. If the test fails then the evidence against the class for the pixel being green vegetation is incremented by 20. It is the function of the inference process to keep track of all the evidence in favour of and against each class for a given pixel and then, when all the rules have been used, to test the accumulated evidence and decide the most appropriate class for the pixel from the perspective of spectral data.

12.4.3 Handling Multisource and Multisensor Data

There are two approaches that might be adopted when considering the development of a knowledge-based approach for the analysis of data that comes from more than one source or sensor. If the analysis is strongly focussed on a particular application it might be appropriate to consider a single knowledge base which contains all the rules, including those rules necessary to process two or more sources together. As a simple illustration, the following rule would be used to determine if a particular region might be urban if both radar and Landsat MSS data were available:

> **if** band 5 is high **and** radar tone is high **then** urban

Of course, use of this rule requires a specification of what 'high' means for both the multispectral and radar data. However, given that those thresholds are available, rules such as this can be used to process the data sources jointly.

Possibly a more practical approach is, first, to decompose the multisource, multisensor problem into a set of individual analyses and then combine their results in a separate expert system that is able to perform the joint analysis as depicted in Fig. 12.3. Each individual analysis module and the combination module will have its own rule base and inference mechanism. The advantages of this approach are that the rule sets are each focussed on a particular sensor and that results can be updated at a later time if and when new data sources become available. This is a particularly important consideration in the context of a GIS.

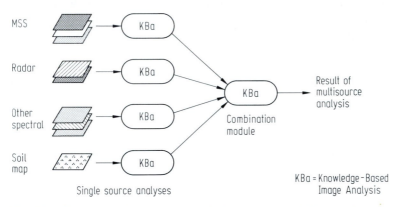

Fig. 12.3. Decomposed multisource analysis using a knowledge based approach

Separate knowledge bases that might be used for a simple segmentation could be:
For the MSS source:

> **if** band 7/band 5 > 3.0 **then** vegetation
> **if** band 7/band 4 < 0.9 **then** water
> **if** (band 4 + band 5)/(band 6 + band 7) > 0.6 **then** soil

For the radar source (see Fig. 1.3):

> **if** radar tone < threshold 1 **then** specular surface
> **if** radar tone > threshold 2 **then** corner reflector effect
> **if** threshold 1 < radar tone < threshold 2 **then** diffuse surface or volume scattering

On the basis of these rules it is presumed we are not able to discriminate diffuse surface scattering and volume scattering. The combined inference 'diffuse surface or volume scattering' is thus the best that can be done, where appropriate, in the following.

No separate rules are to be used at this stage for the soil map in Fig. 12.3, since it already consists of a set of labels for each pixel.

If the rules sets above are applied, as appropriate, to the MSS and radar data we will produce labels for each pixel from the individual analyses. For example the MSS data might specify a pixel as vegetation and the radar data classify it as a specular surface, while the soil map indicates a loam soil type. What the combination knowledge-base has to do is to process these labels to come up with a specific land cover category for the pixel, of the type needed by the user. A set of rules that might be found in the combination module therefore could be:

> **if** soil **and** specular surface **then** bare ground
> **if** soil **and** corner reflector effect **then** urban
> **if** vegetation **and** specular surface **then** low level vegetation
> **if** vegetation **and** diffuse surface or volume **then** trees or shrubs
> **if** low level vegetation **and** loam **then** crops
> **if** low level vegetation **and** sand **then** grassland
> **if** low level vegetation **and** clay **then** grassland
> **if** water **and** specular surface **then** lake
> **if** water **and** diffuse surface or volume **then** open water

Whereas all previous examples of rules have had numerical conditions to test, these combination rules have conditions defined in terms of labels. This decomposition strategy is illustrated in the following example.

12.4.4 An Example

Figure 12.4 shows Landsat MSS bands 5 and 7 and an L band SIR-B synthetic aperture radar image for a small urban area in Sydney's north-western suburbs. The Landsat data is unable to distinguish between urban areas and areas cleared for development. The radar data on the other hand, provides structural information, but no information on the actual cover type. The knowledge-based analysis system developed by Srinivasan and Richards (1993) is able to analyse the images jointly and thus develop a cover type map that resolves classes that are confused in either the Landsat or radar data alone. Full details of this and other applications of this approach will be found in Srinivasan (1990). In the following sections a summary of the expert system used is provided. It is based upon a decomposition philosophy of the style shown in Fig. 12.3 but, in its full version, also has a final module that allows spatial knowledge to be applied to the output of the combination module. The latter is an important component of photointerpretation and can be handled in knowledge-based analysis by region growing beforehand or by applying neighbourhood relations during or after analysis. For simplicity in this example, only a pixel-based approach is discussed.

12.4.4.1 Rules as Justifiers for a Labelling Proposition

In this method, production rules of the form outlined above are referred to as *justifiers* since they provide a degree of justification or evidence in favour of a particular labelling proposition. Expressing a rule in its generic form:

> **if** condition **then** inference

the approach specifies four types of rule:

> Conclusive   If the condition is true then the justification for the inference is conclusive (i. e. absolute).
>
> For example:
>
> **if** radar tone is black **then** radar shadow.

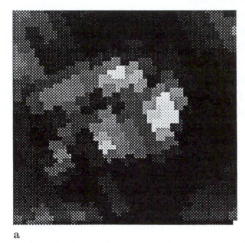

a

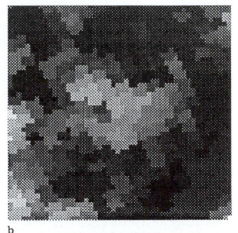

b

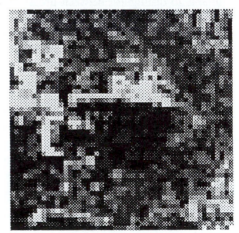

Fig. 12.4. Landsat MSS band 5 *a*, band 7 *b* and SIR-B radar data *c* of an urban region in the city of Sydney

c

Prima Facie   If the condition is true then there is reason to believe that the inference is true. If the condition is false it cannot be concluded in general that the inference is false.

For example:

if MSS band 7/MSS band 5 > 2 **then** vegetation.

Criterion   This is a special prima facie justifier for which a false condition provides prima facie justification to disbelieve the inference.

For example:

if MSS band 7 < MSS band 4 **then** water

(noting that if MSS band 7 > MSS band 4 then definitely not water).

Contingent   If the condition is true then support is provided for other, prima facie, reasons to believe the inference. These types of rule are not sufficient in themselves to justify the inference.

For example:

if MSS band 7 > MSS band 5 **then** vegetation.

This structuring of justifications is not unlike the strengths of reasoning used by photointerpreters. In some cases the evidence would suggest to a photointerpreter that the cover type simply *must* be of a particular type. In other cases the evidence might be so slight as simply to suggest what the cover type might be – indeed the photointerpreter might even withhold making a decision in such a situation until some further evidence is available.

This has been a simple review of the concept of justifiers in qualitative reasoning systems. A fuller treatment, which considers justifiers as inferences in a so-called defeasible logic, can be found in Nute (1988), Pollock (1974) and Srinivasan (1990).

12.4.4.2 Endorsement of a Labelling Proposition

The justifiers of the previous section play a major role in reasoning. At any given stage in the reasoning process an inference may have valid reasons for and against it. It is then necessary to resolve among these to determine the most supported label. This is the role of the **endorsement.**

The endorsement of a label is the final level of justification for an inference. Given a set of justifiers for and against an inference, the implementation used in this example of a scene interpretation system employs the following endorsements:

The inference **is Definitely True**   if there is at least one conclusive justifier in support.

The inference **is Likely To be True**	if there is some net prima facie evidence in support.
The inference **is Indicated**	if, in the absence of prima facie justification, there are some net contingent justifiers in its favour.
A proposition **is Null**	if all justifiers for the belief are balanced by those for opposing beliefs.
A proposition **is Contradicted**	if it has conclusive justifiers balanced for and against it.

A labelling proposition is said to be **Unknown** if nothing is known about it.

Complements of these endorsements also exist.

After all the rules in the knowledge base have been applied to a pixel under examination, each of the possible labels will have some level of endorsement. That with the strongest endorsement is chosen as the label most appropriate for the pixel. Endorsements for other labels, although weaker, may still have value: for example, the two endorsements for a pixel that 'grassland is likely to be true' and 'soil is indicated' are fully consistent – the cover type may in fact be a sparse grassland, which the analyst would infer from the pair of endorsements.

If an endorsement falls in the last three categories above the pixel would be left unclassified.

12.4.4.3 Knowledge Base and Results

The knowledge base for this exercise consisted of the following rules (Srinivasan, 1990).

For the Landsat MSS data source:

if band 7/band 5 is approximately 1	**then** contingent support for urban **and** contingent support for soil
if band 7/band 5 is moderate	**then** contingent support for urban **and** contingent support for vegetation
if band 7/band 5 is high	**then** prima facie support for vegetation

These rules have to be trained in order to establish what is meant by moderate and high.

For the SIR-B data source:

| **if** radar response is low | **then** prima facie support for specular behaviour |

if radar response is moderate

then prima facie support for volume scattering

if radar response is high

then prima facie support for corner reflector

Similarly the thresholds between low, moderate and high are established using small training areas.
The combination rules used by Srinivasan are:

if vegetation is likely to be true **and** corner reflector is likely to be true
then prima facie support for woody vegetation

if vegetation is likely to be true **and** volume scattering is likely to be true
then prima facie support for vegetation

if vegetation is likely to be true **and** specular behaviour is likely to be true
then prima facie support for grassland

if soil is likely to be true **and** specular behaviour is likely to be true
then prima facie support for cleared land

if vegetation is indicated **and** corner reflector is likely to be true
then prima facie support for residential

if vegetation is indicated **and** volume scattering is likely to be true
then prima facie support for residential

if urban is likely to be true **and** corner reflector is likely to be true
then prima facie support for buildings

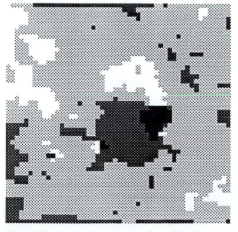

Fig. 12.5. Thematic map produced by knowledge based analysis of the data in Fig. 12.3. Classes are: black = soil, dark grey = grassland, light grey = woody vegetation, white = urban (cleared land, buildings, residential)

if vegetation is indicated **and** vegetation is not likely to be true
and specular behaviour
then contingent support for grassland

Note that the conditions tested in these rules are endorsements from the single source knowledge-base analyses.

Applying these rules yields the thematic map of Fig. 12.5, while Table 12.1 summarises the results quantitatively, using a careful photointerpretation of the data sets, and local knowledge, to provide the necessary ground truth data. Figure 12.5 demonstrates that the classifier is able to distinguish between grasslands and woody vegetation, owing to the structural information present in the radar image. Note also that not all bright regions in the radar image are classified as urban. Some actually correspond to rows of trees; the confusion has been resolved using the land-cover information present in the Landsat image.

Table 12.1. Results of Combined Multispectral and Radar Analysis
Overall accuracy = 77.3% (area weighted = 81.5%)

Ground Classes	Classes Identified by Rule-based Classification						
	Cleared land	Grassland (likely)	Grassland (indicated)	Woody vegetation	Residential	Buildings	Soil
Cleared land	82.5	2.5	5.0	2.5	0	0	7.5
Grassland	2.5	57.2	20.8	16.1	2.1	0.7	0
Woody vegetation	0	6.6	0	88.1	5.2	0	0
Urban	0	0.8	3.5	13.5	70.6	10.7	0.9

References for Chapter 12

Schistad Solberg et al (1994) develop the multisource statistical method of Section 12.2 further and provide means by which joint decisions can be made from multiple sources of data, while incorporating the effect of reliability of the data sources. Bruzzone et al (1997) provide a comparative study of maximum likelihood methods (modified to bring in ancillary information through the prior probalities) and neural networks for multisource classification.

Full details of the Theory of Evidence can be found in Shafer (1976): It has been applied to the problem of image analysis by Lee et al (1987) and to integration of geological and geophysical spatial data sets by Moon (1990) while Garvey (1987) has used the method for describing geographical areas. Gong (1996) considers both the Theory of Evidence and the application of feedword neural networks as methodologies for data fusion involving integrated spatial data types. Peddle (1995a) has incorporated evidential reasoning into a software scheme called MERCURY ⊕ as a means for multisource classification. Peddle (1995b) has also discussed how measures of evidence can be generated from histograms of class training data.

The application of knowledge-based techniques to remote sensing was demonstrated by Nagao and Matsuyama (1980). Carlotto et al (1984) describe a knowledge-based classification system for a single source of data, as does Mulder et al (1988). A spectral rule-based approach for urban land cover discrimination using Landsat TM has been demonstrated by Wharton (1987), while Ton et al. (1991) demonstate the use of both spectral and spatial knowledge for segmentation of Landsat imagery. Nicolin and Gabler (1987) describe a system for automatic interpretation of suburban scenes while Goldberg et al (1985) describe a multi-level expert system for updating forestry maps with Landsat data, that has led to the development of a general purpose shell (Goodenough et al, 1987). Schowengerdt (1989) describes a system which enables inexperienced users perform rule-based image processing tasks. Kartiken et al (1995) demonstrate the application of rule based expert systems to land cover analysis.

Knowledge of the radar response of terrain at different angles of incidence is used by Dobson et al (1996) to develop a knowledge based approach to (structural) land cover classification from two radar sensors (ERS-1 and JERS-1). Solaiman et al. (1998) show how fusion of thematic map and edge information, both obtained from the same image data, can be used to improve a final map product.

J. A. Benediktsson, J. R. Sveinsson and P. H. Swain, 1997: Hybrid Consensus Theoretic Classification. IEEE Trans. Geoscience and Remote Sensing. 35, 833–843.

L. Bruzzone, C. Conese, F. Maselli and F. Roli, 1997: Multisource Classification of Complex Rural Areas by Statistical and Neural-Network Approaches. Photogrammetric Engineering and Remote Sensing. 63, 523–533.

M. J. Carlotto, V. T. Tom, P. W. Baim and R. A. Upton, 1984: Knowledge-Based Multispectral Image Classification. SPIE Vol 504, Applications of Digital Image Processing VII, 45–53.

M. C. Dobson, L. E. Pierce and F. T. Ulaby, 1996: Knowledge-Based Land-Cover Classification Using ERS-1/JERS-1 SAR Composites. IEEE Trans Geoscience and Remote Sensing. 34, 83–99.

R. Frost, 1986: Introduction to Knowledge Base Systems, McGraw-Hill, New York.

T. D. Garvey, 1987: Evidential Reasoning for Geographic Evaluation for Helicopter Route Planning. IEEE Trans Geoscience and Remote Sensing, GE-25, 294–304.

T. D. Garvey, J. D. Lowrance and M. A. Fisher, 1981: An Inference Technique for Integrating Knowledge from Disparate-Sources. Proc 7th Int. Conf. Artificial Intelligence, Vancouver, 319–325.

M. Goldberg, D. G. Goodenough, M. Alvo and G. Karam, 1985: A Hierarchical Expert System for Updating Forestry Maps with Landsat Data. Proceedings of the IEEE, 73, 1054–1063.

P. Gong, 1996: Integrated Analysis of Spatial Data from Multiple Sources: Using Evidential Reasoning and Artificial Neural Network Techniques for Geologic Mapping. Photogrammetric Engineering and Remote Sensing, 62, 513–523.

D. G. Goodenough, M. Goldberg, G. Plunkett and J. Zelek, 1987: An Expert System for Remote Sensing. IEEE Trans Geoscience and Remote Sensing. GE-25, 349–359.

H. N. Gross and J. R. Schott, 1998: Application of Spectral Mixture Analysis and Image Fusion Techniques for Image Sharpening. Remote Sensing of Environment. 63, 85–94.

B. Kartikeyan, K. L. Majumder and A. R. Dasgupta, 1995: An Expert System for Land Cover Classification. IEEE Trans Geoscience and Remote Sensing. 33, 58–66.

T. Lee, J. A. Richards and P. H. Swain, 1987: Probabilistic and Evidential Approaches for Multisource Data Analysis. IEEE Trans Geoscience and Remote Sensing. GE-25, 283–293.

W. L. Moon, 1990: Integration of Geophysical and Geological Data Using Evidential Belief Function. IEEE Trans Geoscience and Remote Sensing, 28, 711–720.

N. J. Mulder, H. Middlekoop and J. Miltenberg, 1988: Progress in Knowledge Engineering for Image Classification. 16th Congress of the International Society for Photogrammetry and Remote Sensing, 27 (III), 395–405.

M. Nagao and T. Matsuyama, 1980: A Structural Analysis of Complex Aerial Photographs. Plenum, New York.

B. Nicholin and R. Gabler, 1987: A Knowledge-Based System for the Analysis of Aerial Images. IEEE Trans Geoscience and Remote Sensing, GE-25, 317–328

D. Nute, 1988: Defeasible Reasoning: A Philosophical Analysis in Prolog, in Aspects of Artificial Intelligence. J. H. Fetzwer (Ed) Dordrecht, Kluwer Academic Publishers.

D. R. Peddle, 1995 a: MERCURY ⊕: An Evidential Reasoning Image Classifier. Computers and Geosciences, 21, 1163–1176.

D. R. Peddle, 1995 b: Knowledge Formulation for Supervised Evidential Classification. Photogrammetric Engineering and Remote Sensing, 61, 409–417.

J. L. Pollock, 1974: Knowledge and Justification. N. J. Princeton University Press.

J. A. Richards, D. A. Landgrebe and P. H. Swain, 1982: A means of utilizing ancillary information in multispectral classification. Remote Sensing of Environment, 12, 463–477

A. H. Schistad Solberg, A. K. Jain and T. Taxt, 1994: Multisource Classification of Remotely Sensed Data: Fusion of Landsat TM and SAR Images. IEEE Trans Geoscience and Remote Sensing. 32, 768–778.

R.A. Schowengerdt, 1989: A General Purpose Expert System for Image Processing. Photogrammetric Engineering and Remote Sensing, 55, 1277–1284.

P.S. Sell, 1985: Expert Systems – a Practical Introduction. Maxmillan, Southampton.

G. Shafer, 1976: A Mathematical Theory of Evidence. NJ, Princeton UP.

B. Solaiman, R.K. Koffi, M-C Mouchot and A. Hillion, 1998: An Information Fusion Method for
Multispectral Image Classification Postprocessing. IEEE Trans Geoscience and Remote Sensing.
36, 395–406.

A. Srinivasan, 1990: An Artifical Intelligence Approach to the Analysis of Multiple Information
Sources in Remote Sensing. PhD Thesis, The University of New South Wales, Kensington.

A. Srinivasan and J.A. Richards, 1993: Analysis of GIS Spatial Data Using Knowledge-Based
Methods. Int. J. Geographic Information Systems. 7, 479–500.

A.H. Strahler, 1980: The Use of Prior Probabilities in Maximum Likelihood Classification of
Remotely Sensed Data. Remote Sensing of Environment, 10, 35–163.

J. Ton, J. Sticklen and A.K. Jain, 1991: Knowledge-Based Segmentation of Landsat Images. IEEE
Trans Geoscience and Remote Sensing, 29, 222–232

F. Van Der Meer, 1997: What Does Multisensor Image Fusion Add in Terms of Information Content
for Visual Interpretation? Int. J. Remote Sensing. 18, 445–452.

S.W. Wharton, 1987: A Spectral Knowledge Based Approach for Urban Land Cover Discrimination.
IEEE Trans Geoscience and Remote Sensing, 25, 272–282.

Problems

12.1 Compare the attributes of a knowledge-based approach to image interpretation with the
more usual approach which uses standard statistical algorithms, such as the maximum likelihood
rule. You should comment on both the training/knowledge acquisition phase and the labelling
phase.

12.2 Write a set of production rules that might be used to smooth a thematic map. The rules are to
be applied to the central labelled pixel in a 3 × 3 window. Assume the map has 5 possible classes
and that map segments with as few as 4 pixels are acceptable to the ultimate user.

12.3 Develop a set of production rules that might be applied to Landsat TM imagery to create a
thematic map with five classes: vegetation, deep clear water, shallow or muddy water, dry soil
and wet soil. To do this you may need to refer to a source of information on the spectral reflectance behaviour of these cover types in the ranges of the TM bands.

12.4 A rule-based analysis system is a very effective way of handling multi-resolution image data.
For example, rules could be applied first to the pixels of the low resolution data to see whether
there is a strong endorsement for any of the available labels. If so then the high spatial resolution
data source need not be consulted, and data processing time is saved. If, however, the rule-based
system can only give weak support to any of the available labels on the basis of the low resolution
data, then it could consult the high resolution source to see whether the smaller pixels can be
labelled at that level with certainty. This could be the case in an urban region where some low
resolution pixels (at say MSS resolution) may be difficult to classify because they are a mixture of
vegetation and concrete. The resolution of SPOT HRV may be able to resolve those classes. In
some other urban areas, which might be large vegetated regions such as golf courses, the MSS
data is quite adequate. Using the strategy of Sect. 12.5, based on justifiers and endorsements,
develop a set of rules for such a multi-resolution problem. Your approach should not go beyond
the MSS level of resolution if a pixel has a definite or likely endorsement.

This application has been developed fully by Srinivasan (1990).

12.5 Consider how the perceived quality of data might be taken into account in a qualitative
reasoning system. In the justification and endorsement approach, endorsements made on the
basis of poor quality data, for example, may lead to the down-grading of an endorsement or justification.

Chapter 13
Interpretation of Hyperspectral Image Data

13.1 Data Characteristics

The data produced by the imaging spectrometers of Section 1.4.6 is different from that of multispectral instruments owing to the enormous number of wavebands recorded – leading to the term *hyper*spectral. For a given geographical area imaged, the data produced can be viewed as a cube, as shown in Fig. 13.1, having two dimensions that represent spatial position and one that represents wavelength.

When displaying multispectral data, such as that from Landsat, both spatial dimensions are generally used, with three of the spectral bands written to the red, green and blue colour elements of the display device, as described in Fig. 3.1. Sometimes, careful band selection is required in this process to ensure the most informative display, while on other occasions multispectral transformations, such as principal components, are used to enhance the richness of the displayed data.

With hyperspectral data there are both challenges and opportunities presented in creating data displays. First, choosing the most appropriate three channels to use is not straightforward and, in any case, would invariably lead to substantial loss of the spectral benefits offered by this form of data gathering. Nevertheless, unless spectral transformations are employed, a set of three bands comparable to those used with multispectral imagery are often adopted (near IR, red, green) for simple display of the data. Secondly, because of the large number of bands available, a two dimensional display using one geographical dimension and the spectral dimension can be created as shown in Fig. 13.2. Such a representation allows changes in spectral

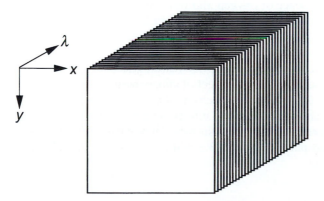

Fig. 13.1. Hyperspectral "cube" of image data such as recorded by an imaging spectrometer.

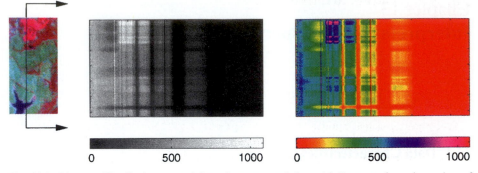

Fig. 13.2. Line profile display created from hyperspectral data. (a) Transect through portion of a hyperspectral image. (b) Greyscale display of spectral band (horizontally) versus position in the image (vertically). (c) Coloured version of (b).

profiles with position (either along track or across track) to be observed. Usually the greyscale is mapped to colour to enhance the interpretability of the displayed data.

To understand much of what is to follow it is useful to envisage how recorded hyperspectral data is affected by the presence of the atmosphere and the nature of the solar spectrum.

Imaging the region being imaged has a uniform 100% spectral response – in other words it will reflect all of the incident sunlight over all wavelengths, as depicted in Fig. 13.3a; also assume that there is no atmosphere above the surface. A detector capable to taking many spectral samples (say 200 or so) will then essentially record the solar spectrum as shown. If the spectral resolution of the detector were sufficiently fine then the recorded solar spectrum would include the Fraunhofer absorption lines, resulting from the gases in the solar atmosphere (Slater, 1980).

Now suppose there is a normal terrestrial atmosphere in the path between the sun, the surface and the detector. The spectrum recorded will be modified by the extent to which the atmosphere selectively absorbs the radiation. There are well known absorption features caused by the presence of oxygen and water vapour in the atmosphere and these appear in the recorded data as depicted in Fig. 13.3b. Also, the atmosphere scatters the solar radiation leading to the sky irradiance and path radiance terms of Fig. 2.1. So far a start, if we wished to determine the (uniform) spectrum of the ideally reflecting surface, the atmospheric absorption features need to be removed, as does the shape of the solar spectrum and the effect of atmospheric scattering.

Figure 13.3c suggests how the reflectance spectrum of a *real* surface might appear before compensation for solar and atmospheric effects. The spectrum recorded is a combination of the actual spectrum of the real surface, modulated by the effects of the solar curve and the atmosphere. Section 13.3 addresses a range of techniques used for removing those effects.

13.2 The Challenge to Interpretation

Recall from Chapter 3 that there are essentially two classes of analytical technique used with normal multispectral data – photointerpretation and machine analysis (classification). The former depends upon the use of image enhancement procedures for improving the visual interpretability of image data whereas the latter is based usually on statistical or other forms of numerical algorithms for labelling individual pixels.

When the data has hundreds of spectral bands traditional image processing and data handling techniques face difficulties. On the other hand, enough information is readily available in the data to allow analysis based on a knowledge of spectroscopic principles, as discussed in Section 13.4.1 following.

It is important to understand the limitations placed on the more traditional analytical approaches since those methods still find application with hyperspectral data, not the least reason for which is the substantial investment in image processing software. In the following Section the features which distinguish hyperspectral from multispectral data are highlighted as a precursor to a discussion on the methods of analysis that can be used with hyperspectral data, either modified or in original form.

13.2.1 Data Volume

Although data volume strictly does not pose any major data processing challenges with contemporary computing systems it is nevertheless useful to examine the relative magnitudes of data for say Landsat Thematic Mapper multispectral imagery and AVIRIS hyperspectral data.

Clearly, the major differences to note between the two is the number of wavebands (7 versus 224) and the radiometric quantisations used (8 versus 10 bits per pixel per band). Ignoring differenences in spatial resolution, the relative data volumes, per pixel, are $7 \times 8 : 224 \times 10$ – ie. $56 : 2240$. Per pixel there are 40 times as many bits therefore for AVIRIS as for TM data. Consequently, storage and transmission of hyperspectral data are issues for consideration; suitable data compression techniques are discussed in Section 13.7.

13.2.2 Redundancy

With 40 times as much data per pixel one is led to question whether 40 times as much information can be obtained about the ground cover types being imaged. Generally, of course, that is not the case – much of the additional data does not add to the inherent information content for a particular application even though it often helps in discovering that information. In other words it contains redundancies.

Much of the data we deal with in everyday life is highly redundant. Take the English language as an example. If we remove certain letters from a word we can often still understand what word is intended. For example *rmte sesng* would be recognised by most people who read this book as *remote sensing* because there are sufficient redundant letters that losing some is not critical to understanding. The same is true with remote sensing data, especially that recorded by hyperspectral sensors –

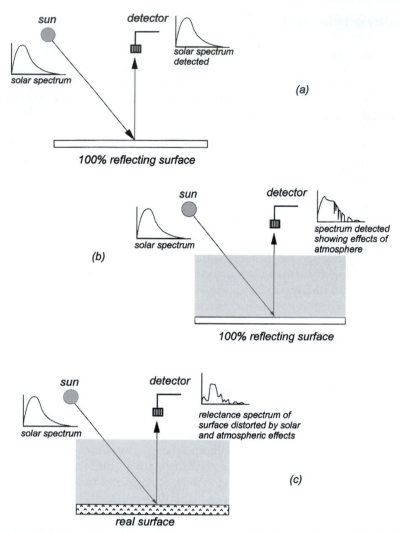

Fig. 13.3. Formation of the reflectance spectrum of a given surface, and the biasing effects of the solar spectral irradiance, atmospheric absorption and scattering.

there is often substantial overlap of information content over the bands of data re-corded for a given pixel. In such cases not all of the data is needed to characterise a pixel properly, although redundant data may be different for different applications.

In remote sensing data redundancy can take two forms: spatial and spectral. Exploiting spatial redundancy is behind the spatial context methods of Section 8.8. Spectral redundancy means that the information content of one band can be fully or partly predicted from the other bands in the data. An example of this is seen in Fig. 6.2b.

An interesting way to view spectral redundancy is to form the correlation matrix for an image (or portion of an image of interest); the correlation matrix can be deriv-

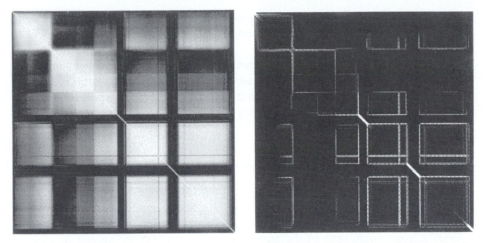

Fig. 13.4. (a) The correlation matrix for 196 wavebands[1] covering 400 nm to 2400 nm for the AVIRIS Jasper Ridge image. (white represents correlations of 1 or –1, while black indicates a correlation of 0). (b) The result of edge detecting the correlation matrix.

ed from the covariance matrix using (6.3). High correlations between band pairs indicate high degrees of redundancy. Because there are so many bands with hyperspectral data it is not practical to list all the correlations numerically, such as is done in Section 6.1.1. Instead, it is better to display the inherent correlations (redundancies) pictorially as showhn in Fig. 13.4a, where a grey scale is used to represent levels of correlation. This representation is often used with hyperspectral data and is a useful tool for identifying correlations among bands when applying traditional processing tools as will be seen later. An interesting by-product of representing the correlation (or covariance) matrix in this form is that image processing procedures can be applied to it. For example its block structure can be emphasised by using a simple edge detection filter to give the result shown in Fig. 13.4b.

Means for removing inherent redundancy are often not readily apparent, although techniques such as the principal components transformation assist in the task since decorrelation followed by a discarding of low variance components amounts to redundancy-reduction.

13.2.3 The Need for Calibration

The high spectral resolution of hyperspectral data sets means that fine atmospheric absorption features will be detected and displayed as discussed in Section 13.1. In order that they not be confused with absorption features of the ground cover type being imaged it is important to account for them and "remove" them from the data.

---

[1] Overlapping bands result from the use of four individual spectrometers in the AVIRIS instrument; these and the significant water absorption bands and the bands which have very small means (<2) have been deleted from the original 224 bands, leaving 196 bands for image processing.

Moreover, because the high spectral resolution suggests that recorded spectra can be interpreted scientifically it is important also to remove the modulating effect of the solar spectrum.

Neither of those effects has been particularly important in the processing and analysis of multispectral data because of the absence of well defined absorption features and the use of average solar irradiance over each of the recorded wavebands as suggested in (2.1). With multispectral data only the effects of atmospheric scattering and transmittance are corrected.

13.2.4 The Problem of Dimensionality: the Hughes Phenomenon

While recognised since the earliest attempts at machine processing of remotely sensed image data (Swain and Davis, 1978), the Hughes phenomenon has not been of major concern until the advent of hyperspectral data.

Briefly, a minimum ratio of the number of training pixels to number of spectral bands is needed to ensure reliable estimates of class statistics are obtained when training supervised classifiers; as the dimensionality of the data set increases the minimum number of training pixels per class must be increased to preserve the accuracy of the statistical estimates. Thus, adding more spectral bands, as in the case of AVIRIS, is not helpful unless more training pixels per class are available. This turns out to be one of the major limitations in attempting to apply traditional image classification procedures to hyperspectral data. A simple example, based upon determining a reliable linear separating surface, can be used to illustrate the problem. Figure 13.5 shows three different training sets of data for the same two dimensional (band) data set. The first (Fig. 13.5a) has only one pixel per class. As seen, while a separating surface can be found it may not be accurate. Having two training pixels per class as in the case of Fig. 13.5b provides a better estimate of the separating surfaces, but it is not until we have many pixels per class, when compared to the num-

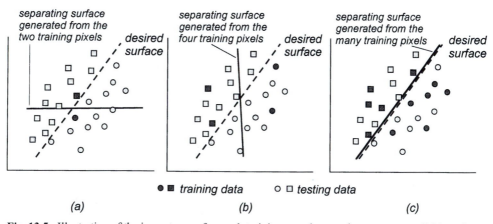

Fig. 13.5. Illustration of the importance of enough training samples per class to ensure reliable estimation of a separating surface. When too few pixels are used (a) good separation of the training data is possible but the classifier performs poorly on the testing data. Large numbers of (randomly positioned) training pixels generate a surface that also performs well for testing data (c).

ber of channels in the data, that we will obtain good estimates of the parameters of the supervised classifier (Fig. 13.5 c).

This is simply another way of looking at the material of Section 8.2.6. However, rather than increase the number of training pixels for a given number of bands, consider now the case of increasing the number of bands for a set number of training pixels; the same problem is observed as illustrated in Fig. 13.6. We note that the performance of the classifier is compromised by the poor estimates of the training statistics beyond about ten features.

13.3 Data Calibration Techniques

13.3.1 Detailed Radiometric Correction

As discussed in Sections 2.1.1 and 13.2.3, the upwelling radiance measured by a sensor results from incident solar energy scattered and reflected from the atmosphere

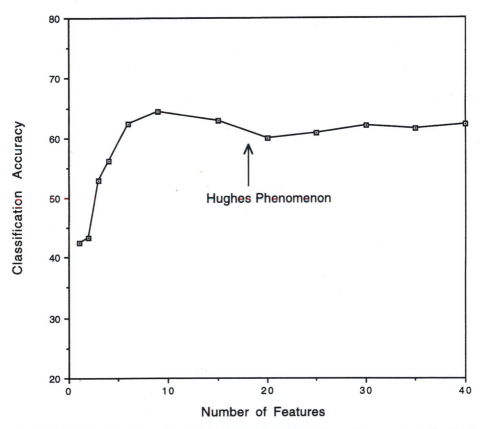

Fig. 13.6. The Hughes phenomenon, demonstrating loss of classifier performance (on testing data) with increasing data dimensionality. This graph is the result of a four category classification; the features indicated are the best sets of those sizes, selected using the Bhattacharyya separability measure.

and earth surface. Detailed radiometric correction to obtain surface reflectance for hyperspectral data follows similar procedures as for the examples given in Section 2.2.1. However, since hyperspectral data covers the whole spectral range from 0.4 to 2.4 μm, including water absorption features, and has high spectral resolution, a more systematic process is generally required, consisting of three possible steps:

- Compensation for the shape of the solar spectrum. The measured radiances are divided by solar irradiances above the atmosphere to obtain the *apparent* reflectances of the surface.
- Compensation for atmospheric gaseous transmittances and molecular and aerosol scattering. Simulating these atmospheric effects allows the *apparent* reflectances to be converted to *scaled* surface reflectances.
- Scaled surface reflectances are converted to *real* surface reflectances after consideration of any topographic effects. If topographic data is not available, real reflectance is assumed to be identical to scaled reflectance under the assumption that the surfaces of interest are Lambertian.

Procedures for solar curve and atmospheric modelling are incorporated in a number of models (Gao et al, 1993), including Lowtran 7 (Low Resolution Atmospheric Radiance and Transmittance), 5S Code (Simulation of the Satellite Signal in the Solar Spectrum) and Modtran 3 (The Moderate Resolution Atmospheric Radiance and Transmittance Model – see Anderson et al, 1995).

ATREM (Atmosphere REMoval Program, Gao et al, 1992), which is built upon 5S code, overcomes a difficulty with the other approaches in removing water vapour absorption features in AVIRIS data; water vapour effects vary from pixel to pixel and from time to time. In ATREM the amount of water vapour on a pixel-by-pixel basis is derived from AVIRIS data itself, particularly from the 0.94 μm and 1.14 μm water vapour features. A technique referred to as three-channel ratioing is developed for this purpose (Gao et al, 1993). Figure 13.7 shows an example of a corrected spectrum against the original measurements.

13.3.2 Data Normalisation

When detailed radiometric correction is not feasible (for example, because the necessary ancillary information is unavailable) normalisation is an alternative which makes the corrected data independent of multiplicative noise, such as topographic and solar spectrum effects. This can be performed using *Log Residuals* (Green and Craig, 1985), based on the relationship between radiance (raw data) and reflectance:

$$x_{i,n} = T_i R_{i,n} I_n, \quad i = 1, \dots K; n = 1, \dots N$$

where $x_{i,n}$ is radiance for pixel i in waveband n. T_i is the topographic effect, which is assumed constant for all wavelengths. $R_{i,n}$ is the real reflectance for pixel i in waveband n. I_n is the (unknown) illumination factor, which is assumed independent of pixel. K and N are the total number of the pixels in the image and the total number of bands, respectively.

There are two steps which remove the topographic and illumination effects respectively. $x_{i,n}$ can be made independent of T_i and I_n by dividing $x_{i,n}$ by its geometric

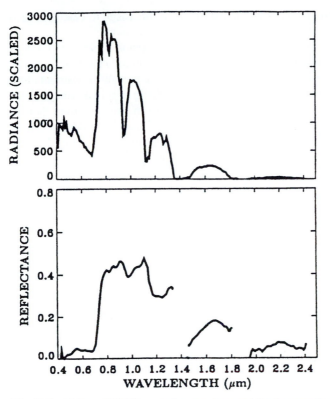

Fig. 13.7. (a) Raw AVIRIS vegetation spectrum and (b) its correction based on ATREM. (Reprinted from Gao et al., 1993 with permission from Elsevier Science permission.)

mean over all bands and then its geometric mean over all pixels. The result is not identical to reflectance but is independent of the multiplicative illumination and topographic effects present in the raw data. The procedure is carried out logarithmically so that the geometric means are replaced by arithmetic means and the final result obtained for the normalised data is

$$\log z_{i,n} = \log x_{i,n} - \log m_n - \log m_i = \log x_{i,n} - \frac{1}{N}\sum_{n=1}^{N}\log x_{i,n} - \frac{1}{K}\sum_{i=1}^{K}\log x_{i,n}$$

13.3.3 Approximate Radiometric Correction

As with multispectral data approximate correction is acceptable for some applications. One approach is the Empirical Line procedure (Roberts et al, 1985). Two spectrally uniform targets in the site of interest, one dark and one bright, are selected; their actual reflectances are then determined by field or laboratory measurements. The radiance spectra for each target are extracted from the image and then mapped to the actual reflectances using linear regression techniques. The gain and offset so-derived for each band are then applied to *all* pixels in the image to calculate their reflectances.

While the computational load is manageable with this method, field or laboratory data may not be available. The Flat Field method (Roberts et al, 1986), an approximate correction technique that relies purely on the image data itself, is then an alternative. This depends upon locating a large, spectrally uniform area in an image (such as sand or clouds) and finding its average radiance spectrum. It is assumed that the shape and the absorption features presented in this spectrum are caused by solar and atmospheric effects. The reflectance of each pixel is then obtained by dividing the average radiance spectrum into the image spectrum of the pixel.

13.4 Interpretation by Spectral Analysis

13.4.1 Spectroscopic Analysis

Having a well-defined spectrum means that a scientific approach to interpretation can be carried out, much as a sample is identified using spectroscopy in the laboratory through a knowledge of spectral features.

Absorption features (seen as localised dips) are often observed in the reflectance spectra of specific minerals provided sufficient spectral resolution is available. Characterisation and thus automatic detection of such absorption features, when they occur, is of particular interest in hyperspectral image recognition.

Absorption features can be characterised by their locations (bands), relative depths and widths (full width at half the maximum depth), and used in pixel identification in the following manner. A complete spectrum is divided into several regions and absorption features are detected within each region. A feature is defined as an absorption feature if it is a local minimum, and below the pixel spectral mean by a user-specified threshold. An unknown spectrum is labelled as belonging to a given class if its absorption features match those of the spectrum for that class held in a reference library.

13.4.2 Spectral Angle Mapping

In N dimensional multi-(hyper-)spectral space a pixel vector x has both magnitude (length) and an angle measured with respect to the axes that define the coordinate system of the space (see Appendix C). In the spectral angle mapper (SAM) technique for identifying pixel spectra only the angular information is used. Figure 13.8a shows a two dimensional (ie. two band) example where spectra are characterised entirely by their angles from the horizontal axis. The spectra can be distinguished from each other provided the angles are sufficiently different. Using this concept, angular decision boundaries can be set up (from library information or training data) that segment the space as shown in Fig. 13.8b. Spectra are then labelled according to the sector within which they fall.

Clearly the SAM technique will fail if the vector magnitude is important in providing discriminating information, which it will in many instances. However, if the pixel spectra from the different classes are well distributed in the space there is a high

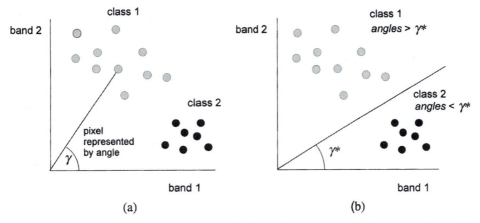

Fig. 13.8. (a) Representing pixels by their angles from the band axes. (b) Segmenting the multi-spectral space by angle.

likelihood that angular information alone will provide good separation. The technique functions well in the face of scaling noise. Details for implementing SAM will be found in Kruse et al (1993).

13.4.3 Library Searching Techniques

Because the pixel spectrum is so well-specified and can be corrected for atmospheric and solar distortions, spectral comparison is possible – either with previously recorded data or with laboratory spectra – for pixel identification. The reference spectra are usually stored in a spectral library.

It is clear that the searching and matching processes must be efficient in such a procedure. Full spectral matching using original radiometric data is not practical. However, given the degree of redundancy spectrally and radiometrically that one would anticipate with the data recorded by an imaging spectrometer, coding techniques can be employed to represent a pixel spectrum in a simple and effective manner so that fast library searching and matching can be achieved.

13.4.3.1 Binary Spectral Codes

A simple binary code for a reflectance spectrum can be formed according to

$$h(n) = 0 \quad \text{if } x(n) \leq T$$
$$ 1 \quad \text{otherwise} \quad n = 1, \dots N$$

where $x(n)$ is the brightness value of a pixel in the nth channel, T is a user specified threshold for forming the binary code, and $h(n)$ is the resulting binary code symbol for the pixel in the nth spectral band. Usually T is chosen as the average brightness value of the spectrum. Figure 13.9 demonstrates a typical spectrum encoded in this

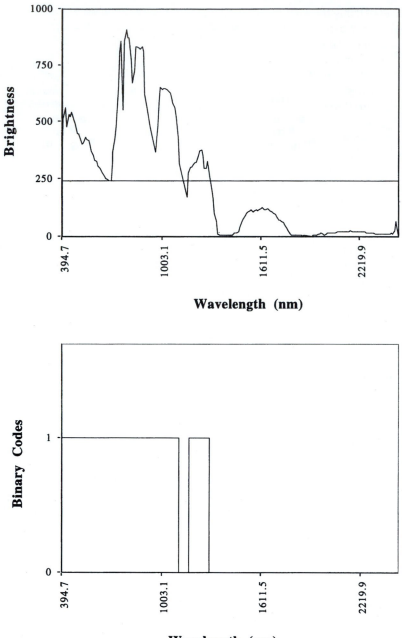

Fig. 13.9. Formation of a simple binary code for an AVIRIS spectrum.

manner. Instead of using the average brightness of the complete spectrum as a threshold, the local average over the adjacent channels could be employed.

Such a simple binary code will not always provide reasonable separability between the spectra in a library, nor will it guarantee that a measured spectrum will match either with only one or a small number of library spectra. Consequently, more sophisticated codes may need to be adopted. For example, more than one threshold could be used. With three thresholds a two binary digit code for the brightness of a pixel will be created:

$$h(n) = 00 \quad \text{if } x(n) \leq T1$$
$$01 \quad \text{if } T1 < x(n) \leq T2$$
$$11^2 \quad \text{if } T2 < x(n) \leq T3$$
$$10 \quad \text{if } T3 < x(n).$$

The mean brightness over the spectrum can be one threshold; the other two are chosen above and below this value.

Spectral slope can also be used as part of a code. One binary representation of the local slope, $s(n)$, at each waveband is:

$$s(n) = 0 \quad \text{if } (x(n+1) - x(n-1)) \leq 0$$
$$1 \quad \text{otherwise} \qquad n = 1, \ldots N$$

Another variation is to develop separate codewords for different regions of the spectrum. The resulting codes accommodate better spectral coarse structure. Uniformly spaced regions could be used or perhaps those regions of the spectrum suspected as being most significant in differentiating cover types could be adopted. The latter is based upon the knowledge that in different wavelength ranges the reflectance spectrum is dominated by different physical characteristics of the surface being imaged. Figure 13.10 shows an example of coding on selected bands with 3 thresholds.

13.4.3.2 Matching Algorithms

Comparison of binary coded spectra can be made by measuring the Hamming distance between them, defined as

$$D_H(h_i, h_j) = \sum_{l=1}^{L} (h_i(l) \oplus h_j(l))$$

where h_i and h_j are two spectral codewords of length (i.e. number of bits) L. For simple amplitude coding, $L = N$, the number of bands. $L = 2N$ if slope coding is also used or three thresholds are employed. $\oplus$ denotes the exclusive OR operator. It is applied on a bit-by-bit basis for a pair of binary code words and records a difference as '1' and no difference as '0'. For example, the exclusive OR of two spectral codewords 01110011 and 00101011 becomes 01011000. Hamming distance is then cal-

---

[2] The third level code word is chosen as 11 rather than 10 so that there is only one binary digit difference between levels.

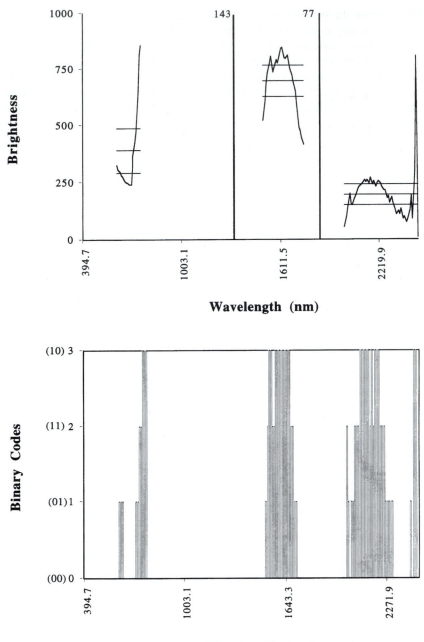

Fig. 13.10. Formation of a binary code using three thresholds, chosen differently in different regions of the spectrum.

culated by summing the number of times the binary digits are different. In this example, the Hamming distance is 3. If the distance is within a user-specified threshold, the two pixels are identified as belonging to the same class. When one, say $h_i(n)$, is a class signature code, the comparison leads to labelling for pixel j.

13.5 Hyperspectral Interpretation by Statistical Methods

13.5.1 Limitations of Traditional Thematic Mapping Procedures

Traditional supervised and unsupervised classification techniques will require very long processing times for hyperspectral data because of the dependence on the number of wavebands (see Sections 8.5 and 9.3). A more serious problem, however, is the need to estimate class signatures – i.e. the mean vector and covariance matrix – when using algorithms, such as maximum likelihood, based on second order statistics. The difficulty lies in the small number of available training pixels per class compared with the number of wavebands used, and is related directly to the Hughes phenomenon of Section 13.2.4. If too few training samples are used then the class model may be very accurate for the training data and classification accuracy on training data can be very high. However, classification accuracy on testing data will be poor. In this case, the classifier is overtrained and the statistics estimated are unreliable. This difficulty is analogous to that of curve fitting illustrated in Section 2.4.1.4. To avoid the problem of unreliable class statistics and thus poor classifier performance the number of training pixels per class should be at least ten times the dimensionality of the data, with desirably 100 times as discussed in Section 8.2.6.

The following sections treat some techniques developed for dealing with the small training set problem.

13.5.2 Block-based Maximum Likelihood Classification

In general, correlations between neighbouring bands in hyperspectral data sets are higher than for bands further apart and highly correlated bands appear in groups. As a result, the correlation matrix is roughly block diagonal in form as shown in Fig. 13.4, in which a greyscale is used to represent degree of correlation. Figure 13.11 shows the data of Fig. 13.4 a but, for purposes of illustration, with the correlations averaged within identifiable blocks demonstrating the strongly block diagonal form of the correlation and thus the covariance matrix. Those blocks can be identified visually or with the assistance of edge detection on the correlation maxtrix as shown in Fig. 13.4 b.

Now assume that the low off-diagonal correlations are zero. The matrix is then fully block diagonal as depicted in general terms in Fig. 13.12. By assuming that the subgroups of bands within each block are independent of those in other subgroups, maximum likelihood classification can then be applied to each subgroup independently.

Noting that the block diagonal form of the correlation matrix leads to a covariance matrix of the same structure, the discriminant function becomes the sum of the logarithmic discriminant values of the individual groups of wavebands (blocks):

$$g_i(\boldsymbol{x}) = -\sum_{k=1}^{K} \{\ln |\Sigma_{ik}| + (\boldsymbol{x}_k - \boldsymbol{m}_{ik})^t \, \Sigma_{ik}^{-1} \, (\boldsymbol{x}_k - \boldsymbol{m}_{ik})\} \quad i = 1, \dots M; k = 1, \dots K \quad (13.1)$$

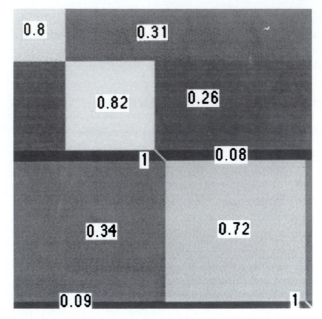

Fig. 13.11. Average correlations within diagonal blocks and within selected off-diagonal segments of Fig. 13.4 illustrating the pseudo block diagonal nature of the matrix.

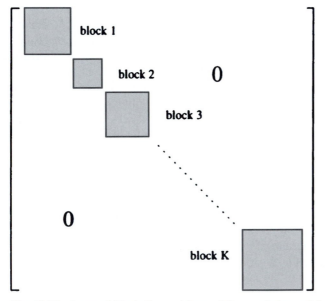

Fig. 13.12. Assumed block diagonal form of the correlation and thus covariance matrix.

In (13.1) the dimensions of x, m_i, and Σ_i are reduced to $n_k(n_k < N)$, the size of the kth subgroup of bands, so that advantage can be taken of the corresponding quadratic reduction in classification time (see Section 8.5). Also, the number of training pixels required per class for reliable statistics, determined by the size of *the biggest* subgroup, is much smaller than when all bands are used.

The sizes of subgroups to use are generally guided by observation of the boundaries of the high correlation blocks along the principal diagonal of a correlation matrix which will be different for different images.

If training data is limited some relatively high correlations may have to be ignored. However, this approach will still be better than, say, minimum distance classification (often used when training pixels are limited – see Section 8.3.1) since at least some correlations are included.

With some data sets highly correlated blocks of bands will occur away from the diagonal. They can be moved onto the diagonal simply by reordering the bands before the correlation matrix is computed. Such an operation makes no difference to the information contained in the matrix or to subsequent image analysis operations. However, it does mean that a reconstructed pixel spectrum will have some bands out of order in the sequence of wavelengths.

13.6 Feature Reduction

Given that hyperspectral data is often highly redundant, feature reduction will be an important pre-processing step to image analysis. However, feature reduction itself for hyperspectral data is a time consuming process and feature extraction via linear transform relies, as with classification, on good estimates of class statistics. To solve this problem the block-based technique presented in Section 13.5.2 can be extended to deal with hyperspectral feature reduction.

13.6.1 Feature Selection

Separability measures, such as the JM distance of (10.5) and (10.6), provide metrics of the average distance between two class density functions, and are thus used to find the best subsets of features.

When the complete set of bands is treated as K independent blocks as discussed in 13.5.2, the JM distance or other separability measures can be simplified; (10.6) for example becomes

$$B = \sum_{k=1}^{K} \left\{ \frac{1}{8} (m_{ik} - m_{jk})^t \left[\frac{\Sigma_{ik} + \Sigma_{jk}}{2} \right]^{-1} (m_{ik} - m_{jk}) + \frac{1}{2} \ln \left\{ \frac{|(\Sigma_{ik} + \Sigma_{jk})/2|}{|\Sigma_{ik}|^{1/2} |\Sigma_{jk}|^{1/2}} \right\} \right\}$$

Thus the Bhattacharyya distance between a class pair is the sum of the distances computed for each block (group of bands).

13.6.2 Spectral Transformations

The principal components transformation, which uses global statistics to determine the transformation operation, is sometimes used in multispectral data analysis as a tool for feature reduction. The main concern is employing it with hyperspectral data is its high computational load.

Implementing the transformation consists of two tasks: eigenanalysis to generate the transformation matrix G in (6.4), and pixel by pixel linear transformation. The former requires an insignificant amount of work. However, the latter is a time consuming process which requires $N \times N$ multiplications and $N \times (N-1)$ additions per pixel. Moreover, the process can be biased by high variance bands. For example, the data recorded by AVIRIS is affected in shape by the solar spectrum as shown in Fig. 13.3 (c). This indicates that a spectral weighting is imposed. As a result, the variances of the spectral bands in the short wavelength region are much higher than the remaining bands if the data is not calibrated. A conventional principal components transform will be dominated, therefore, by the visible and near infrared bands.

When the original bands are highly correlated, the principal components transform works effectively, while for poorly correlated data there may be little change after application of the transform. Recall, for hyperspectral data, high correlations generally occur in blocks. If the conventional principal components transform is modified so that the low correlations between the highly correlated blocks are avoided, the efficiency of the transformation will be improved while the results should be little affected. This leads to the formation of a segmented principal components transformation.

Figure 13.13 shows the process schematically. The complete data set is first partitioned into K subgroups of highly correlated bands. Denote by $n_1, n_2, \ldots, n_K$ the number of bands in subgroups 1, 2 ..., K, respectively. The principal components transformation is now conducted separately on each subgroup of data. Feature selection on each of the transformed data sets is carried out by either making use of variance information in each component as is common in multispectral data processing, or by pursuing single band separabilities (see Section 13.6.3). The features selected can be regrouped and transformed again to compress the data further. Generally, the steps can be repeated until the required data reduction ratio is achieved for classification or storage purposes. For colour composite display the most informative three features will be used.

Segmenting the principal components transform in this manner requires $n_k \times n_k$ multiplications for each subgroup and thus a total of $\sum_{k=1}^{K} n_k^2$ multiplications for each pixel vector in contrast to $N \times N$ multiplications for each pixel vector if transformation over the full set of bands is performed. As an example, 2/3 of the total time is saved when three subgroups of uniform size are used (i.e., $K = 3$, and $n_1 = n_2 = n_3 = N/3$).

So long as all the new transformed components are kept, there is no variance (information) loss by transforming sub-vectors separately. When the new components obtained from each segmented transform are gathered and transformed again, the resulting data variance and covariance are identical to those for the conventional principal components transform.

The segmentation idea can be extended to canonical analysis. The complete set of bands is segmented into K groups. Then conventional canonical analysis is applied to each group, with up to $M - 1$ best features selected from each transformed set, where M is the number of classes. By so doing, class statistics involving the complete set of bands are no longer needed (which otherwise presents the difficulties under limited training pixels discussed in Section 13.5.1).

13.6.3 Feature Selection from Principal Components Transformed Data

For original, untransformed data, feature selection is based on pairwise separability measures such as the Bhattacharyya distance (10.6). If the covariance matrices, Σ_i and Σ_j, are diagonal (following transformation) then (10.6) becomes

$$B = \sum_{n=1}^{N} \left[\frac{(m_i(n) - m_j(n))^2}{4(\sigma_i^2(n) + \sigma_j^2(n))} + 0,5 \ln \frac{(\sigma_i^2(n)/2 + \sigma_j^2(n)/2)}{\sqrt{\sigma_i^2(n)\sigma_j^2(n)}} \right]$$

where $m_i(n)$, $\sigma_i^2(n)$ represent, respectively, the mean and variance of the nth band for class i. This suggests that when the data has low correlation (close to zero), following transformation, class separability is determined largely by individual feature separabilities and can be estimated by summing those single feature separabilities. Therefore, single band separability can be used as an approximate measure for feature selection from features that are poorly correlated.

Generally, high data variance is usually needed for separating different classes in an image and, thus, higher order principal components with small variances provide little significant information. Therefore, it is possible simply to select the first few high variance features and ignore the higher ordered principal components. However, it is important to recognise that some features selected in this way may be misleading. For example, original noisy bands will lead to some principal components with high variance but low separability.

13.7 Compression of Hyperspectral Data

Owing to the large data volumes involved, storage and transmission of data from imaging spectrometers benefit from the application of procedures that will reduce data volume without substantially affecting the information content. Those procedures are generally in the form of codes that represent the spectra in reduced form. The binary codes of Section 13.4 are typical of codes that could be used, although with such reductions in the spectra significant information loss (allowing the spectra to be used over a large number of applications) could be expected.

More sophisticated codes minimise information loss while compressing the data. The principal components transformation is an example. The higher order components with low variance can be discarded without significant information loss and yet with a reduction in storage requirement in proportion to the number of bands discarded. Also, the original spectral or image data can be reconstructed from the reduced representation (using an inverse principal components transform) although with loss of information. Sometimes the information loss is referred to as distortion since the reconstructed data will differ, depending on the level of loss of detail, from the original.

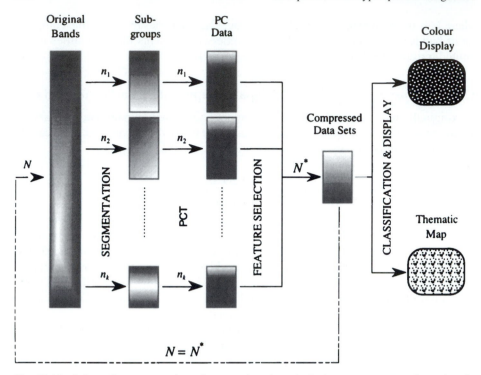

Fig. 13.13. Schematic representation of segmenting the principal components transformation for feature reduction.

An alternative transformation widely used in the television and video industry is the Discrete Cosine Transform (Rao and Yip, 1990). The DCT is similar in principle to the Discrete Fourier Transform of Section 7.7, but with cosine expansion functions instead of complex exponentials as seen in (7.16).

If the user can tolerate substantial amounts of distortion then significant compression of remote sensing imagery is possible; figures as high as 100 times reduction in volume have been reported, but one is then led to question the integrity of the compressed data. Generally, those compression schemes that allow the original image to be reconstructed without error (so-called lossless compression algorithms) will give compression ratios of about 2 to 3.

A compression scheme well matched to the needs of remote sensing is referred to as vector quantisation, based upon the use of a so-called code book. That book contains a number of representative pixel vectors (for example class means) that could be obtained from training data, or possibly could even be prototypical reference spectra. Each code book vector is given a label (such as a number or even a class symbol).

Now imagine an image has to be transmitted over a telecommunications channel. If the spectrum matches exactly one of the stored spectra then only the label need be transmitted. The receiver also has a copy of the code book and can retrieve the spectrum in question through matching the label. If the spectrum does not match a code book entry exactly then transmitting the label of the nearest match will incur an error. Whether that error is acceptable, or whether a correction needs to be transmitted with the label of closest match, will depend on the application. The efficacy of the scheme depends upon how well the code book represents the range of pixel vectors in the image. A good code book will give rise to small differences (errors) between code book entries and pixel vectors to be transmitted. Such small differences can be encoded using a small number of bits (substantially smaller than the number of bits in the original pixel vector), so that good data compression is achieved.

A simple illustration is given in Table 13.1 in which 10 SPOT multispectral vectors are to be sent over a channel. Ordinarily, with each band represented by 8 bits, the ten pixels require $10 \times 3 \times 8 = 240$ bits to be transmitted. However, recognising there are two clusters in the data and using the cluster means as code book vectors, it is possible to represent each of the pixels to be transmitted by their difference (error) from the nearest mean. There are 8 distinct differences (between 0 and 7); they can all be accommodated (i.e. distinguished from each other) by allowing a 3 bit word for coding them. Thus the number of bits then to be transmitted is $10 \times 3 \times 3 = 90$ bits, plus one bit per pixel to indicate the code book vector label (one bit is enough to represent just two labels – i.e. 0 or 1) and $2 \times 3 \times 8 = 48$ bits to transmit the code book beforehand. Thus the vector quantised scheme requires $90 + 10 + 48 = 148$ bits for the 10 pixels. Thus the "compression ratio" is $240/148 = 1.6$ with the ability to reconstruct the original pixel vectors without loss (distortion).

Table 13.1. Simple Illustration of Vector Quantisation

Cluster 1 (e.g. vegetation)					Cluster 2 (e.g. soil)				
Original pixel vectors:									
1	*2*	*3*	*4*	*5*	*6*	*7*	*8*	*9*	*10*
50	55	60	58	48	48	49	55	53	51
10	11	12	9	9	70	69	73	71	68
150	152	148	154	160	171	163	165	167	160

Code book entries (cluster means):

54	51
10	70
153	165

Differences between pixel vectors and nearest code book entry:

1	*2*	*3*	*4*	*5*	*6*	*7*	*8*	*9*	*10*
4	1	6	4	6	3	2	4	2	0
0	1	2	1	1	0	1	3	1	2
3	1	5	2	7	6	2	0	2	5

Further compression of the data is possible by using a more efficient coding process on the errors. Rather than simply allocating (in this example) 3 bits per difference (based on the observation that there are 8 different errors to transmit) shorter code words (in terms of numbers of bits) can be ascribed to the most commonly encountered errors (in this example 1 and 2). Details of this refinement, vector quantisation in general and the overall issue of compression in remotely sensed data can be found in Ryan and Arnold (1997a, b).

13.8 Spectral Unmixing: End Member Analysis

A challenge that has faced interpreters throughout the history of remote sensing has been the need to handle mixed pixels – i.e. those pixels that represent a mixture of cover types or information classes. Several early studies attempted to resolve the proportions of pure cover types within mixed pixels by assuming that the measured radiance is a linear combination of the radiances of the "pure" constituents in each of the imaging wavebands used.

With low resolution (multispectral) data the approach generally did not meet with a great deal of success because most cover types are not well differentiated in the small number of wavebands used. However, with hyperspectral data, the prospect of uniquely characterising a vast number of earth cover types, and thus differentiating them from each other spectroscopically, suggests that the mixing approach should be re-visited as a means for establishing mixture proportions of pure cover types in pixels. This has particular relevance in minerals mapping where abundance maps for minerals of interest can then be produced based upon the proportions determined for all pixels in a given image.

The process can be developed mathematically in the following manner. Assume there are M pure cover types in the image of interest. In the nomenclature of mixing models there are referred to as *endmembers*. Let the proportions of the various endmembers in a pixel be represented by f_m, $m = 1, \ldots M$. These are the unknowns in the process which we wish to find, based on observation of the hyperspectral reflectance of the pixel.

Let R_n, $n = 1, \ldots N$ be the observed reflectance of the pixel in the nth spectral band of the sensor and $a_{n, m}$ be the spectral reflectance in the nth band of the mth endmember. Then we assume

$$R_n = \sum_{m=1}^{M} f_m a_{n, m} + \xi_n \qquad n = 1, \ldots N$$

where ξ_n is an error in band n. The equation says that the observed reflectance in each band is the linear sum of the reflectances of the endmembers; the extent to which that does not work exactly in a given situation is encapsulated in the error term.

An assumption that allows us to assume linear mixing in this form is that the incident energy is scattered only once to the sensor from the landscape and does not undergo multiple scatterings among, for example, foliage components.

The above mixing equation can be expressed in matrix form as

$$\boldsymbol{R} = A\boldsymbol{f} + \boldsymbol{\xi}$$

where f is a column vector of size M, R and ξ are column vectors of size N and A is an $N \times M$ matrix of endmember spectral signatures (by column).

Spectral unmixing, as the process is called, involves finding a set of endmember proportions that will minimise the error vector ξ. On the assumption that the correct set of endmembers has been chosen the problem then becomes one of solving the simpler equation

$$R = Af.$$

Normally there are more equations than unknowns so that simple inversion of the last equation to find the vector of mixing proportions is not possible. Instead, a least squares solution is found by using the pseudo inverse

$$f = (A^t A)^{-1} A^t R.$$

It should be mentioned that there are two constraints that the mixing proportions of the endmembers are expected to satisfy. The first is that the proportions should sum to unity and the second is that they should all be non-negative:

$$R_n = \sum_{m=1}^{M} f_m = 1$$

$$0 \le f_m \le 1 \text{ for all } m$$

As discussed by Gross and Schott (1998) these constraints are sometimes violated if the endmembers are derived from average cover type spectra or the endmember selection is poor.

References for Chapter 13

Information on radiometric correction of imaging spectrometer data using atmospheric and solar curve models can be found in Gao et al (1993). Roberts et al (1985, 1986) can be consulted for details of the empirical line and flat field methods of correction, while Green and Craig (1985) develop the Log Residuals technique in detail.

Piech and Piech (1987) have proposed a general method to detect all the absorption features in a pixel spectrum automatically and order them by their depth. The original spectrum is convolved with Gaussian masks over a range of standard deviations (called scale parameters). In so doing, absorption features disappear with increasing scale parameter and, as a result, a set of progressively smoothed spectra are obtained. Then, the derivative of the smoothed spectrum is calculated and the zero crossings indicate the local minima.

Binary coding of hyperspectral imagery is treated in Mazer et al (1988), Vane and Goetz (1993), Jia (1996) and Jia and Richards (1993), while Jia (1996) and Jia and Richards (1994) cover the block based methods for treating data of such high dimensionality.

The problem of small training sets is also treated by Shahshahani and Landgrebe (1994) who demonstrate that unlabelled pixels can be used to supplement labelled training data in an endeavour to develop more reliable training statistics for Bayesian classification.

Anderson, G.P., Wang, J. and Chetwynd, J.H., 1995: MODTRAN3: An Update and Recent Validations Against Airborne High Resolution Interferometer Measurements, 5th Annual JPL Airborne Sciences Workshop, Vol 1 (AVIRIS Workshop), 5–8.

Gao, B.C., Heidebrecht, K.B. and Goetz, A.F.H., 1992: Atmospheric Removal Program (ATREM) Users Guide, Centre for the Study of Earth from Space, University of Colorado, Boulder.

Gao, B.C., Heidebrecht, K.B. and Goetz, A.F.H., 1993: Derivation of Scaled Surface Reflectance from AVIRIS Data, Remote Sensing of Environment, 44, 165–178.

Green, A.A. and Craig, M.D., 1985: Analysis of Aircraft Spectrometer Data with Logarithmic Residuals, Proc. AIS workshop, JPL Publication 85–41, Jet Propulsion Laboratory, Pasadena, California, April 8–10, 111–119.

Gross, H.N. and Schott, J.R., 1998. Application of Spectral Mixture Analysis and Image Fusion Techniques for Image Sharpening. Remote Sensing of Environment, 63, 85–94.

Jia, X., 1996: Classification Techniques for Hyperspectral Remote Sensing Image Data, PhD Thesis, School of Electrical Engineering, University College, ADFA, The University of New South Wales.

Jia, X. and J.A. Richards, 1993: Binary Coding of Imaging Spectrometer Data for Fast Spectral Matching and Classification, Remote Sensing of Environment, 43, 47–53.

Jia, X. and J.A. Richards, 1994: Efficient Maximum Likelihood Classification for Imaging Spectrometer Data Sets, IEEE Trans on Geoscience and Remote Sensing, 32, 274–281.

Kruse, F.A., Lefkoff, A.B., Boardman, J.W., Heidebrecht, K.B., Shapiro, A.T., Barloon, P.J. and Goetz, A.F.H., 1993: The Spectral Image Processing System (SIPS) – interactive visualisation and analysis of imaging spectrometer data, Remote Sensing Environment, 44, 145–163.

Mazer, A.S., Martin, M., Lee, M. and Dolomon, J.E., 1988: Image Processing Software for Imaging Spectrometry Data Analysis, Remote Sensing of Environment, 24, 201–211.

Piech, M.A. and Piech, K.R., 1987: Symbolic Representation of Hyperspectral Data, Applied Optics, 26, 4018–4026.

Rao, K.R. and Yip, P., 1990: Discrete Cosine Transform. New York: Academic.

Roberts, D.A., Yamaguchi Y. and Lyon R.J.P., 1985: Calibration of Airborne Imaging Spectrometer Data to percent Reflectance Using Field Spectral Measurements, 19th International Symposium on Remote Sensing of Environment, Ann Arbor, Michigan, October 21–25.

Roberts, D.A., Yamaguchi, Y. and Lyon, R.J.P., 1986: Comparison of Various Techniques for Calibration of AIS Data, in Proceedings, 2nd AIS workshop, JPL Publication 86–35, Jet Propulsion Laboratory, Pasadena, California, 21–30.

Ryan, M.J. and Arnold, J.F., 1997a: The Lossless Compression of AVIRIS Images by Vector Quantization. IEEE Trans Geoscience and Remote Sensing, 35, 546–550.

Ryan, M.J. and Arnold, J.F., 1997b: Lossy Compression of Hyperspectral Data Using Vector Quantization. Remote Sensing of Environment, 61, 419–436.

Shahshahani, B.M. and Landgrebe, D.A., 1994: The Effect of Unlabeled Samples in Reducing the Small Sample Size Problem and Mitigating the Hughes Phenomenon, IEEE Trans on Geoscience and Remote Sensing, 32, 1087–1095.

Slater, P.N., 1980: Remote Sensing: Optics and Optical Systems. Reading Mass: Addison-Wesley.

Swain, P.H. and Davis, S.M. (eds), 1978: Remote Sensing: The Quantitative Approach, New York: McGraw-Hill.

Vane, G. and Goetz, A.F.H., 1993: Terrestrial Imaging Spectrometry: Current Status, Future Trends, Remote Sensing of Environment, 44, 117–126.

Problems

13.1 The block based maximum likelihood classification scheme of Sect. 13.5.2 requires decisions to be taken about what blocks to use. From your knowledge of the spectral responses of the three common ground cover types of vegetation, soil and water, recommend an acceptable set of block boundaries that might always be used with AVIRIS data. You may wish also to take note of the major water absorption features in AVIRIS spectra as seen in Fig. 1.10.

13.2 Using the results of question 1 or otherwise discuss how the canonical analysis transformation might take advantage of partitioning the covariance matrices into blocks.

13.3 Does partitioning the covariance matrix into blocks assist minimum distance classification?

13.4 (a) Consider the block based approach to principal components analysis as developed in Sect. 13.6.2. Suppose several stages of transformation without feature reduction are used as depicted in Fig. 13.13. Prove that the overall data variance after the final transformation is the same as that generated had a single stage principal components transform been carried out.

(b) (This requires significant matrix analysis skills) As in part (a), but, a principal components transform without segmentation is finally performed on the data which are obtained after several stages of segmented principal components transform (without feature reduction) are used. Prove that the data variance of each feature is the same as that generated had conventional PCT been carried out for the original data.

13.5 Consider the simple binary coding scheme (with one threshold) developed in Sect. 13.4.3.1 and illustrated in Fig. 13.9. How many distinct codes are possible for AVIRIS, TM and SPOT HRV data sets? Why would binary codes not be a sufficient representative form for SPOT and MSS data?

13.6 Suppose a particular image contains just two cover types – vegetation and soil. A pixel identification exercise is carried out to attempt to attach either a soil or vegetation label to each pixel and thereby come up with an estimate of the proportion of vegetation in the region being imaged. From homogeneous regions of the image it is possible easily to label pure soil and pure vegetation pixels. Clearly the image also contains a number of mixed pixels and so end member analysis is considered as a matter of resolving their soil/vegetation proportions. Is the additional work justified if the approximate proportion of vegetation to soil is $1:100$, $50:50$ or $100:1$?

13.7 Explain the concept of transmittance and name the main gases which cause markedly low atmospheric transmittance at wavelength(s) between $400-2400$ nm.

13.8 Describe briefly spectral library searching techniques, stating why they are feasible to use with imaging spectrometer data and noting their advantages over statistical classification methods.

13.9 The principal components transform and the Bhattacharyya distance can both be used for band reduction. Comment on the main differences between the two methods for this application.

13.10 Two-threshold coding is normally not recommended. Explain why.

13.11 When three-threshold coding is used, make suggestions on how to define the three thresholds, particularly the upper and lower thresholds if the mean brightness value over the spectrum is used as the middle threshold.

13.12 In the spectral angle mapper technique, the angle of a pixel vector in the spectral space needs to be determined (Fig. 13.8). Write down the formula for calculating the angle for the general case with number of bands of N.

Appendix A
Satellite Altitudes and Periods

Civilian remote sensing satellites are generally launched into circular orbits. By equating centripetal acceleration in a circular orbit with the acceleration of gravity it can be shown (Duck & King, 1983) that the orbital period corresponding to an orbital radius r is given by

$$T = 2\pi \sqrt{r^3/\mu} \tag{A-1}$$

where μ is the earth gravitational constant, with a value of $3.986 \times 10^{14}\,\mathrm{m^3 s^{-2}}$. The corresponding orbital angular velocity is

$$\dot{\theta} = \sqrt{\mu/r^3}\ \mathrm{rad \cdot s^{-1}} \tag{A-2}$$

The orbital radius r can be written as the sum of the earth radius r_e and the altitude of a satellite above the earth, h:

$$r = r_e + h \tag{A-3}$$

where $r_e = 6.378$ Mm. Thus the effective velocity of a satellite over the ground (at its sub-nadir point) *ignoring earth rotation*, is given by

$$v = r_e \dot{\theta} = r_e \sqrt{\mu/(r_e + h)^3} \tag{A-4}$$

The actual velocity over the earth's surface taking into account the earth's rotation depends upon the orbit's inclination (measured as an angle i anticlockwise from the equator on an ascending pass – i.e. at the so-called ascending node) and the latitude at which the velocity is of interest.

Let the earth rotational velocity at the equator be v_e (to the east). Then at latitude ϕ, the surface velocity of a point on the earth will be $v_e \cos \phi \cos i$. Therefore the actual ground track velocity of a satellite at altitude h and orbital inclination i is given by

$$v_s = r_e \sqrt{\mu/(r_e + h)^3} \pm v_e \cos \phi \cos i \tag{A-5}$$

where the $+$ sign applies when the component of earth rotation opposes the satellite motion (i.e. on descending nodes for inclinations less than 90° or for ascending nodes with inclinations greater than 90°). Otherwise the negative sign is used.

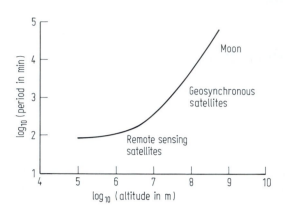

Fig. A-1. Satellite periods versus altitude above the earth's surface, for circular orbits

Equations (A-1) to (A-5) can be used to derive some numbers of significance for remote sensing missions, although it is to be stressed here that the equations apply only for circular orbits and a spherical earth.

Figure A-1 shows a plot of orbital period (in minutes) as a function of satellite altitude plotted on logarithmic coordinates. This has been derived from (A-1) and (A-3). Some significant altitudes to note are (i) $h = 907$ km, at which $T = 103$ min; being the approximate period of the first three Landsat satellites, (ii) $h = 35,800$ km at which $T = 24$ hours, being the so-called geosynchronous orbit – if this is established over the equator then the satellite appears stationary to a point on the ground; this is the orbit used by many communication satellites, (iii) $h = 380$ Mm at which $T = 28$ days – this is the orbit of the moon.

Consider now a calculation of the time taken for Landsat 1 to acquire a 185 km frame of MSS data. This can be found by determining the local velocity. For the Landsat satellite the orbital inclination is $100°$; at Sydney Australia the latitude is $34°$S. From (A-5) this gives

$$v_s = 6.392 \text{ km s}^{-1}.$$

Therefore 185 km requires 28.9 s to record.

References for Appendix A

K. I. Duck and J. C. King, 1983: Orbital Mechanics for Remote Sensing. In R. N. Colwell (Ed.). Manual of Remote Sensing, 2e, American Society of Photogrammetry, Falls Church.

Appendix B

Binary Representation of Decimal Numbers

In digital data handling we frequently refer to numbers in binary form; this is because computers and their associated storage media represent data in this format. In the binary system the numbers are arranged in columns that represent powers of 2 while in the decimal system numbers are arranged in columns that are powers of 10. Thus whereas we can count up to 9 in each column in the decimal system we can only count up to one in each binary column. From the right, the columns represent $2^0, 2^1, 2^2$ etc., so that the decimal numbers between 0 and 7 have the binary versions:

Decimal	Binary 2^2	2^1	2^0	
0	0	0	0	
1	0	0	1	
2	0	1	0	
3	0	1	1	(i.e. 2+1)
4	1	0	0	
5	1	0	1	(i.e. 4+1)
6	1	1	0	(i.e. 4+2)
7	1	1	1	(i.e. 4+2+1)

The digits in the binary system are referred to as bits. In the above example it can be seen that by using just 3 binary digits it is not possible to represent decimal numbers beyond 7 – i.e. a total of 8 decimal numbers altogether, including 0. To represents 16 decimal numbers, which could be 16 levels of brightness in remote sensing image data between 0 and 15, it is necessary to have a binary "word" with 4 bits. In that case the word 1111 is equivalent to decimal 15. In this way it is readily shown that the numbers of decimal values that can be represented by various numbers of binary digits are:

Number of bits	Number of decimal levels	
1	2	(i.e. 0, 1)
2	4	(0, 1, 2, 3)
3	8	(0, ..., 7)
4	16	(0, ..., 15)
5	32	etc.
6	64	
7	128	
8	256	
9	512	
10	1024	
11	2048	
12	4096	

An eight bit word, which can represent 256 decimal numbers between 0 and 255, is referred to as a *byte* and is a fundamental data unit used in computers.

Appendix C
Essential Results from Vector and Matrix Algebra

C.1 Definition of a Vector and a Matrix

The pixels in an image can be plotted in a rectangular co-ordinate system according to their brightness values in each band. For example, for Landsat MSS bands 5 and 7 a vegetation pixel would appear somewhat as shown in Fig. C-1. The pixel can be described by its co-ordinates (10, 40); this will be a set of four numbers if all 4 MSS bands are considered, in which case the co-ordinate system also is four dimensional. For the rest of this discussion only two dimensions will be used for illustration but the results apply to any number. For example a 7 dimensional space would be required for Landsat TM data.

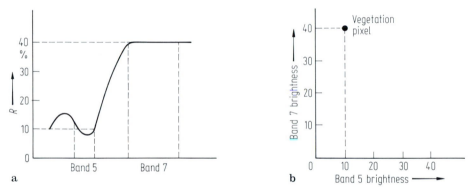

Fig. C-1. a Spectral reflectance characteristic of vegetation; **b** Typical vegetation pixel plotted in a rectangular co-ordinate system

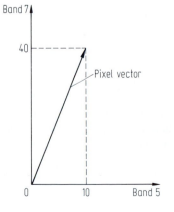

Fig. C-2. Representation of a pixel point in multispectral space by a vector drawn from the origin

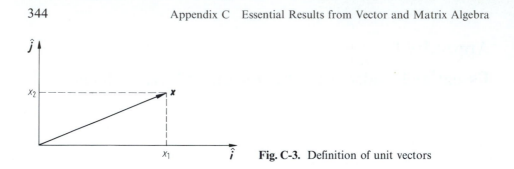

Fig. C-3. Definition of unit vectors

An alternative but *equivalent* means by which the pixel point can be represented is as a *vector* drawn from the origin, as illustrated in Fig. C-2. In this context the vector is simply an arrow that points to the pixel. While we never actually draw the vector as such it is useful to remember that it is implied in much of what follows.

Mathematically a vector from the origin is described in the following way. First we define so-called *unit vectors* along the co-ordinate directions. These are simply direction indicators, which for the two dimensional case are as shown in Fig. C-3. With these, the vector is written as

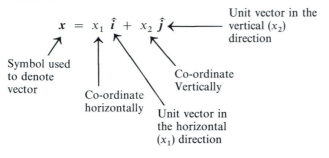

In "shorthand" form we represent the vector as

$$x = \begin{bmatrix} x_1 \\ x_2 \end{bmatrix}$$

which is properly referred to as a column vector, owing to its vertical arrangement. Note that the unit vectors, and thus the corresponding co-ordinate directions are implied by the ordering in the column. Sometimes a row version is used. This is called the *transpose* of *x* and is written as

$$x' = [x_1 \quad x_2].$$

Recall that for Landsat MSS data *x* will have 4 column entries representing the four response values for the pixel that *x* describes.

Sometimes we might wish to create another vector *y* from an existing vector *x*. For illustration, if we take both to be just two dimensional then the components of *y* can be obtained most generally according to the pair of equations

$$y_1 = m_{11} x_1 + m_{12} x_2$$

$$y_2 = m_{21} x_1 + m_{22} x_2$$

i.e. the components of y are just (linear) combinations of those of x. In shorthand this *transformation* of the vector is expressed as

$$\begin{bmatrix} y_1 \\ y_2 \end{bmatrix} = \begin{bmatrix} m_{11} & m_{12} \\ m_{21} & m_{22} \end{bmatrix} \begin{bmatrix} x_1 \\ x_2 \end{bmatrix}$$

or

$$y \quad = \quad M \quad x$$

where M is referred to as a *matrix* of coefficients. By comparing the previous expressions, note how a multiplication of a matrix by a vector is carried out.

C.2 Properties of Matrices

The *inverse* of M is called M^{-1} and is defined by

$$MM^{-1} = I$$

where I is the *identity matrix*

$$\begin{bmatrix} 1 & 0 & \\ 0 & 1 & \\ & & \cdot \cdot \cdot 1 \end{bmatrix}$$

which, if used to transform the vector x, will leave it unchanged. This can be seen if it is used in place of M in the equations above. The inverse of a matrix is not always easily computed. It should be noted however that it can be expressed as

$$M^{-1} = M^*/|M|$$

where M^* is called the *adjoint* of M and $|M|$ is called its *determinant*. The adjoint, in theory, is a *transposed matrix of cofactors*. This is not important in general for remote sensing since all the calculations are usually performed with software that includes a procedure for inverting a matrix. However it is useful for illustration purposes to know that the adjoint for a 2×2 matrix is:

$$M^* = \begin{bmatrix} m_{22} & -m_{12} \\ -m_{21} & m_{11} \end{bmatrix} \quad \text{when} \quad M = \begin{bmatrix} m_{11} & m_{12} \\ m_{21} & m_{22} \end{bmatrix}$$

Similarly large order determinant calculations are carried out by computer. It is only necessary to note, for illustration, that

$$\begin{vmatrix} m_{11} & m_{12} \\ m_{21} & m_{22} \end{vmatrix} = m_{11}m_{22} - m_{21}m_{12} = \text{a scalar constant}.$$

C.3 Multiplication, Addition and Subtraction of Matrices

If M and N are two matrices, chosen as 2×2 for illustration and defined as

$$M = \begin{bmatrix} m_{11} & m_{12} \\ m_{21} & m_{22} \end{bmatrix} \qquad N = \begin{bmatrix} n_{11} & n_{12} \\ n_{21} & n_{22} \end{bmatrix}$$

then

$$M \pm N = \begin{bmatrix} m_{11} \pm n_{11} & m_{12} \pm n_{12} \\ m_{21} \pm n_{21} & m_{22} \pm n_{22} \end{bmatrix}$$

and

$$MN = \begin{bmatrix} m_{11} & m_{12} \\ m_{21} & m_{22} \end{bmatrix} \times \begin{bmatrix} n_{11} & n_{12} \\ n_{21} & n_{22} \end{bmatrix}$$

$$= \begin{bmatrix} m_{11}n_{11} + m_{12}n_{21} & m_{11}n_{12} + m_{12}n_{22} \\ m_{21}n_{11} + m_{22}n_{21} & m_{21}n_{12} + m_{22}n_{22} \end{bmatrix}$$

Note that the last expression is obtained by multiplying, term by term, the rows of the first matrix by the columns of the second. Within each multiplication the terms are summed. This pattern holds for larger matrices.

Division is not defined as a matrix operation. Rather its place is taken by the definition of a matrix inverse, as in the above section.

C.4 The Eigenvalues and Eigenvectors of a Matrix

We have discussed the matrix M above as a matrix that transforms one vector to another, (alternatively it can be used to transform the co-ordinate system in which a point or vector is described). It is relevant at this stage to ask if there is a vector that can be multiplied by a simple (but in general complex) number and thus be transformed in exactly the same manner as it would be had it been multiplied by the matrix M. In other words can we find a vector x in our co-ordinate space and a (complex) number λ such that

$$M x = \lambda x \quad \text{(i.e. } y = \lambda x \text{ is equivalent to } y = M x \text{).}$$

This implies

$$M x - \lambda x = 0$$

or

$$(M - \lambda I) x = 0 \tag{C-1}$$

The theory of simultaneous equations tells us that for this equation to be true it is necessary to have either $x = 0$ or

$$|M - \lambda I| = 0 \tag{C-2}$$

This expression is a polynominal equation in λ. When evaluated it yields values for λ. When these are substituted into (C-1) the vectors x corresponding to those λ will be found. Those λ's are called the *eigenvalues* of M and the associated x's are called the *eigenvectors*.

C.5 Some Important Matrix, Vector Operations

If x is a column vector, say $x = \begin{bmatrix} x_1 \\ x_2 \end{bmatrix}$

then $\quad x x^t = \begin{bmatrix} x_1 \\ x_2 \end{bmatrix} \begin{bmatrix} x_1 & x_2 \end{bmatrix}$

$$= \begin{bmatrix} x_1^2 & x_1 x_2 \\ x_1 x_2 & x_2^2 \end{bmatrix}$$

i.e. a matrix

and

$$x^t x = \begin{bmatrix} x_1 & x_2 \end{bmatrix} \begin{bmatrix} x_1 \\ x_2 \end{bmatrix}$$

$$= x_1^2 + x_2^2$$

i.e. a constant (this is often referred to as the dot product, scalar product or inner product).

C.6 An Orthogonal Matrix – the Concept of Matrix Transpose

The inverse of an *orthogonal* matrix is identical to its transpose. Thus, if M is orthogonal then

$$M^{-1} = M^t \quad \text{and} \quad M^t M = I,$$

where M^t is the transpose of the matrix, given by rotating all the elements about the principal diagonal (that which runs through the elements $m_{11}, m_{22}, m_{33} \ldots$).

C.7 Diagonalisation of a Matrix

Consider a transformation matrix M such that

$$y = M x.$$

As before the eigenvalues λ_i of M and their associated eigenvectors x_i are defined by the expression

$$\lambda_i x_i = M\, x_i, \quad i = 1, \dots n$$

where n is the number of distinct eigenvalues.
These n different equations can be expressed in the compact manner

$$X \Lambda = M X$$

where Λ is the diagonal matrix

$$\begin{bmatrix} \lambda_1 & & 0 \\ & \ddots & \\ 0 & & \lambda_n \end{bmatrix}$$

and X is the matrix of eigenvectors $(x_1, x_2, \dots x_n)$.
Consequently $\Lambda = X^{-1} M X$

This is referred   This matrix
to as the diagonal   diagonalises M.
form of M.

References for Appendix C

Pipes, L. A.: Applied Mathematics for Engineers and Physicists, 2nd. ed., McGraw-Hill, N.Y., 1968.
Faddeeva, V. N.: Computational Methods of Linear Algebra, Dover, N.Y., 1959 (see particularly pp. 37–38, 44–45, 46–49).

Appendix D

Some Fundamental Material from Probability and Statistics

D.1 Conditional Probability

It is the purpose of this Appendix to outline some of the fundamental statistical concepts commonly used in remote sensing theoretical developments. Remote sensing terminology is used throughout and an emphasis is placed on understanding rather than theoretical rigour.

The expression $p(x)$ is interpreted as the probability that the event x occurs. In the case of remote sensing, if x is a pixel vector, $p(x)$ is the probability that a pixel can be found at position x in multispectral space.

Often we wish to know the probability of an event occuring conditional upon some other event or circumstance. This is written as $p(x|y)$ which is expressed as the probability that x occurs given that y is specified. As an illustration $p(x|\omega_i)$ is the probability of finding a pixel at position x in multispectral space, given that we are interested in class ω_i – i.e. it is the probability that a pixel from class ω_i exists at position x. These $p(x|y)$ are referred to as *conditional probabilities*; the available y generally form a complete set. In the case of remote sensing the set of ω_i, $i = 1, \dots M$ are the complete set of spectral classes used to describe the image data for a particular exercise. If we know the complete set of $p(x|\omega_i)$ – which are often referred to as the *class conditional probabilities* – then we can determine $p(x)$ in the following manner. Consider the product $p(x|\omega) \, p(\omega_i)$ where $p(\omega_i)$ is the probability that class ω_i occurs in the image (or that a pixel selected at random will come from class ω_i). The product is the probability that a pixel at position x in multispectral space *is* an ω_i pixel. The probability that a pixel from any class can be found at position x clearly is the sum of the probabilities that pixels will be found there from all the available classes. In other words

$$p(x) = \sum_{i=1}^{M} p(x|\omega_i) \, p(\omega_i) \tag{D-1}$$

The product $p(x|\omega_i) \, p(\omega_i)$ is called the *joint probability* of the "events" x and ω_i. It is interpreted strictly as the probability that a pixel occurs at position x and that the class is ω_i (this is different from the probability that a pixel occurs at position x given that we are interested in class ω_i). The joint probability is written

$$p(x, \omega_i) = p(x|\omega_i) \, p(\omega_i) \tag{D-2a}$$

We can also write

$$p(\omega_i, x) = p(\omega_i | x)\, p(x) \tag{D-2b}$$

where $p(\omega_i | x)$ is the conditional probability that expresses the likelihood that the class is ω_i given that we are examining a pixel at position x in multispectral space. Often this is called the *posterior* probability of class ω_i. Again $p(\omega_i, x)$ is the probability that ω_i and x exist together, which is the same as $p(x, \omega_i)$. As a consequence, from (D-2a) and (D-2b)

$$p(\omega_i | x) = p(x | \omega_i)\, p(\omega_i) / p(x) \tag{D-3}$$

which is known as Bayes' theorem (Freund, 1992).

D.2 The Normal Probability Distribution

D.2.1 The Univariate Case

The class conditional probabilities $p(x|\omega_i)$ in remote sensing are frequently assumed to belong to a normal probability distribution. In the case of a one dimensional spectral space this is described by

$$p(x|\omega_i) = (2\pi)^{-1/2}\, \sigma_i^{-1}\, \exp\left\{ -\frac{1}{2}(x - m_i)^2 / \sigma_i^2 \right\} \tag{D-4}$$

in which x is the single spectral variable, m_i is the mean value of x and σ_i is its standard deviation; the square of the standard deviation, σ_i^2, is called the variance of the distribution. The mean is referred to also as the expected value of x since, on the average, it is the value of x that will be observed on many trials. It is computed as the mean value of a large number of samples of x. The variance of the normal distribution is found as the expected value of the difference squared of x from its mean. A simple average of this squared difference gives a biased estimate. An unbiased estimate is obtained from (Freund, 1992)

$$\sigma_i^2 = \frac{1}{q_i - 1} \sum_{j=1}^{q_i} (x_j - m_i)^2 \tag{D-5}$$

where q_i is the number of pixels in class ω_i, and x_j is the jth sample.

D.2.2 The Multivariate Case

The one dimensional case just outlined is seldom encountered in remote sensing, but it serves as a basis for inducing the nature of the multivariate normal probability distribution, without the need for theoretical development. Several texts treat the bivariate case – i.e. that where x is two dimensional – and these could be consulted should a simple multivariate case be of interest without vector and matrix notation (Nilsson, 1965; Swain and Davis, 1978; Freund, 1992).

Consider (D-4) and see how it can be modified to accommodate a multidimensional x. First, and logically, x must be replaced by $\boldsymbol{x}$. Likewise the univariate mean m_i must be replaced by its multivariate counterpart $\boldsymbol{m}_i$. The variance σ_i^2 in (D-4) must be modified, not only to take account of multidimensionality but also to include the effect of correlation between spectral bands. This role is filled by the covariance matrix Σ_i defined by

$$\Sigma_i = \mathscr{E}\{(\boldsymbol{x}-\boldsymbol{m}_i)(\boldsymbol{x}-\boldsymbol{m}_i)^t\} \tag{D-6a}$$

where $\mathscr{E}$ is the expectation operator and the superscript "t" is the vector transpose operation. An unbiased estimate for Σ_i is given by

$$\Sigma_i = \frac{1}{q_i-1} \sum_{j=1}^{q_i} \{(\boldsymbol{x}_j-\boldsymbol{m}_i)(\boldsymbol{x}_j-\boldsymbol{m}_i)^t\} \tag{D-6b}$$

Inside the exponent in (D-4) the variance σ_i^2 appears in the denominator. In its multivariate extension the covariance matrix is inverted and inserted into the numerator of the exponent. Moreover the squared difference between $\boldsymbol{x}$ and $\boldsymbol{m}_i$ is expressed using the vector transpose expression $(\boldsymbol{x}-\boldsymbol{m}_i)^t(\boldsymbol{x}-\boldsymbol{m}_i)$. Together these allow the exponent to be recast as $-\frac{1}{2}(\boldsymbol{x}-\boldsymbol{m}_i)^t\Sigma_i^{-1}(\boldsymbol{x}-\boldsymbol{m}_i)$. We now turn our attention to the pre-exponential term. First we need to obtain a multivariate form for the reciprocal of the standard deviation. This is achieved first by using the determinant of the covariance matrix as a measure of its size (see Appendix C) – giving a single number measure of variance – and then taking its square root. Finally the term $(2\pi)^{-1/2}$ needs to be replaced by $(2\pi)^{-N/2}$, leading to the complete form of the multivariate normal distribution for N spectral dimensions

$$p(\boldsymbol{x}|\omega_i) = (2\pi)^{-N/2}|\Sigma_i|^{-1/2} \exp\left\{-\frac{1}{2}(\boldsymbol{x}-\boldsymbol{m}_i)^t\Sigma_i^{-1}(\boldsymbol{x}-\boldsymbol{m}_i)\right\}. \tag{D-7}$$

References for Appendix D

Along with vector and matrix analysis and calculus, a sound understanding of probability and statistics is important in developing a high degree of skill in quantitative remote sensing. This is necessary not only to appreciate algorithm development but also because of the role of statistical sampling techniques and the like when dealing with sampled data. The depth of treatment in this appendix and in the body of the book has been sufficient for a first level appreciation of quantitative methods. The reader wishing to develop further understanding, particularly of important concepts in probability and statistics, would be well advised to consult well-known standard treatments such as Freund (1992) and Feller (1967).

W. Feller, 1967: An Introduction to Probability Theory and its Applications, 2e, N.Y., Wiley.
J.E. Freund, 1992: Mathematical Statistics, 5e, Englewood Cliffs, N.J., Prentice Hall.
N.J. Nilsson, 1965: Learning Machines, N.Y., McGraw-Hill.
P.H. Swain and S.M. Davis (Eds), 1978: Remote Sensing: The Quantitative Approach, N.Y., McGraw-Hill.

Appendix E
Penalty Function Derivation
of the Maximum Likelihood Decision Rule

E.1 Loss Functions and Conditional Average Loss

The derivation of maximum likelihood classification in Sect. 8.2 is generally acceptable for remote sensing applications and is used widely. However it is based implicity on the understanding that misclassifying any particular pixel is no more significant than misclassifying any other pixel in an image. The more general approach presented in the following allows the user to specify the importance of making certain labelling errors compared with others. For example, for crop classification involving two sub-classes of wheat it would probably be less of a problem if a particular wheat pixel was erroneously classified into the other sub-class than it would if it were classified as water.

To develop the general method we introduce the penalty function, or loss function

$$\lambda(i|k) \quad i, k = 1, \ldots M \tag{E-1}$$

This is a measure of the loss or penalty incurred when an algorithm erroneously labels a pixel as belonging to class ω_i when in reality the pixel is from class ω_k. It is reasonable to expect that $\lambda(i|i) = 0$ for all i: this implies there is no penalty for a correct classification. In principle, there are M^2 distinct values of $\lambda(i|k)$ where M is the number of classes.

The penalty incurred by erroneously labelling a pixel at position x in multispectral space into class ω_i is

$$\lambda(i|k)\, p(\omega_k|x)$$

where the pixel comes correctly from class ω_k and $p(\omega_k|x)$ is the posterior probability that ω_k is the correct class for pixels at x. Averaging this over all possible ω_k we have the average loss, correctly referred to as the *conditional average loss*, associated with labelling a pixel as belonging to class ω_i. This is given by

$$L_x(\omega_i) = \sum_{k=1}^{M} \lambda(i|k)\, p(\omega_k|x) \tag{E-2}$$

and is a measure of the accumulated penalty incurred given the pixel could have belonged to any of the available classes and that we have available the penalty functions relating all the classes to class ω_i. Clearly, a useful decision rule for assigning a label to a pixel is to choose that class for which the conditional average loss is the smallest, viz

$$x \in \omega_i \quad \text{if} \quad L_x(\omega_i) < L_x(\omega_j) \quad \text{for all} \quad j \neq i. \tag{E-3}$$

An algorithm that implements (E-3) is often referred to as a Bayes' optimal algorithm.

Even if the $\lambda(i|k)$ were known, the $p(\omega_k|x)$ usually are not. Therefore, as in Sect. (8.2.2) we adopt Bayes' theorem which allows the posterior probabilities to be expressed in terms of the class probability distribution functions $p(x|\omega_k)$; viz

$$p(\omega_k|x) = p(x|\omega_k) \, p(\omega_k)/p(x)$$

where $p(\omega_k)$ is the class prior probability. Using this in (E-2) gives

$$L_x(\omega_i) = \frac{1}{p(x)} \, l_x(\omega_i)$$

with

$$l_x(\omega_i) = \sum_{k=1}^{M} \lambda(i|k) \, p(x|\omega_k) \, p(\omega_k). \tag{E-4}$$

Since $p(x)$ is common to all classes it is sufficient to decide class membership on the basis of the $l_x(\omega_i)$.

E.2 A Particular Loss Function

Suppose $\lambda(i|k) = 1 - \Phi_{ik}$ with $\Phi_{ii} = 1$ and $\Phi_{ik}(k \neq i)$ to be defined. Then (E-4) can be expressed

$$l_x(\omega_i) = \sum_{k=1}^{M} p(x|\omega_k) \, p(\omega_k) - \sum_{k=1}^{M} \Phi_{ik} p(x|\omega_k) \, p(\omega_k)$$
$$= p(x) - g_i(x)$$

with

$$g_i(x) = \sum_{k=1}^{M} \Phi_{ik} p(x|\omega_k) \, p(\omega_k) \tag{E-5}$$

Again since $p(x)$ is common to all classes it does not aid discrimination and thus can be removed from the conditional average loss expression, leaving just $l_x(\omega_i) = -g_i(x)$. Because of the minus sign in this expression we can then decide the "least cost" labelling of a pattern at position x in multispectral space according to maximisation of the *discriminant function* $g_i(x)$, viz

$$x \in \omega_i \quad \text{if} \quad g_i(x) > g_j(x) \quad \text{for all} \quad j \neq i \tag{E-6}$$

It is of interest at this stage to put

$$\Phi_{ik} = \delta_{ik}, \text{ the Kroneker delta function,}$$

defined by

$$\delta_{ik} = 1 \quad \text{for} \quad i = k$$
$$= 0 \quad \text{for} \quad i \neq k.$$

Equation (E-5) then becomes

$$g_i(x) = p(x|\omega_i)\, p(\omega_i)$$

so that the decision rule in (E-6) is

$$x \in \omega_i \quad \text{if} \quad p(x|\omega_i)\, p(\omega_i) > p(x|\omega_j)\, p(\omega_j) \quad \text{for all} \quad j \neq i$$

which is the classification rule adopted in (8.3) in Chap. 8. Frequently this is referred to as the unconditional maximum likelihood decision rule.

References for Appendix E

Nilsson (1965) gives an excellent account of the derivation of the maximum likelihood decision rule based upon penalty functions in the manner just derived. Other treatments that could be consulted include Duda and Hart (1973) and Andrews (1972).

H. C. Andrews, 1972: Introduction to Mathematical Techniques in Pattern Recognition, N.Y., Wiley.
R. O. Duda and P. E. Hart, 1973: Pattern Classification and Scene Analysis N.Y., Wiley.
N. J. Nilsson, 1965: Learning Machines, N.Y., McGraw-Hill.

Subject Index

Springer
and the
environment

At Springer we firmly believe that an international science publisher has a special obligation to the environment, and our corporate policies consistently reflect this conviction.

We also expect our business partners – paper mills, printers, packaging manufacturers, etc. – to commit themselves to using materials and production processes that do not harm the environment. The paper in this book is made from low- or no-chlorine pulp and is acid free, in conformance with international standards for paper permanency.

 Springer

Printing: Saladruck, Berlin
Binding: Buchbinderei Lüderitz & Bauer, Berlin